International Symposium on Interfacial Joining and Surface Technology

AA002699

(IJST 2013)

IOP Conference Series: Materials Science and Engineering Volume 61

Osaka, Japan
27 – 29 November 2013

ISBN: 978-1-63439-872-5
ISSN: 1757-8981

Printed from e-media with permission by:

Curran Associates, Inc.
57 Morehouse Lane
Red Hook, NY 12571

Some format issues inherent in the e-media version may also appear in this print version.

Copyright© by the Institute of Physics
All rights reserved.

Printed by Curran Associates, Inc. (2015)

For permission requests, please contact the Institute of Physics
at the address below.

Institute of Physics
Dirac House, Temple Back
Bristol BS1 6BE UK

Phone: 44 1 17 929 7481
Fax: 44 1 17 920 0979

techtracking@iop.org

Additional copies of this publication are available from:

Curran Associates, Inc.
57 Morehouse Lane
Red Hook, NY 12571 USA
Phone: 845-758-0400
Fax: 845-758-2634
Email: curran@proceedings.com
Web: www.proceedings.com

International Symposium on Interfacial Joining and Surface Technology

(IJST 2013)

IOP Conference Series: Materials Science
and Engineering Volume 61

Osaka, Japan
27 – 29 November 2013

ISBN: 978-1-63439-872-5
ISSN: 1757-8981

IOP Institute of Physics

TABLE OF CONTENTS

IOP Conf. Series: Materials Science and Engineering **61** (2014) 011001 doi:10.1088/1757-899X/61/1/011001

International Symposium on Interfacial Joining and Surface Technology --- Solid State Joining, Brazing and Soldering, Surface Treatment, Advanced Packaging ---

IJST 2013

Preface

Interfacial joining (bonding) is a widely accepted welding process and one of the environmentally benign technologies used in industrial production. As the bonding temperature is lower than the melting point of the parent materials, melting of the latter is kept to a minimum. The process can be based on diffusion bonding, pressure welding, friction welding, ultrasonic bonding, or brazing–soldering, all of which offer many advantages over fusion welding. In addition, surface technologies such as surface modification, spraying, coating, plating, and thin-film formation are necessary for advanced manufacturing, fabrication, and electronics packaging. Together, interfacial joining and surface technology (IJST) will continue to be used in various industrial fields because IJST is a very significant form of environmentally conscious materials processing.

The international symposium of IJST 2013 was held at Icho Kaikan, Osaka University, Japan from 27 November to 29 November, 2013. A total of 138 participants came from around the world to attend 56 oral presentations and 36 posters presented at the symposium, and to discuss the latest research and developments on interfacial joining and surface technologies. This symposium was also held to commemorate the 30th anniversary of the Technical Commission on Interfacial Joining of the Japan Welding Society.

On behalf of the chair of the symposium, it is my great pleasure to present this volume of IOP Conference Series: Materials Science and Engineering (MSE). Among the presentations, 43 papers are published here, and I believe all of the papers have provided the welding community with much useful information. I would like to thank the authors for their enthusiastic and excellent contributions.

Finally, I would like to thank all members of the committees, secretariats, participants, and everyone who contributed to this symposium through their support and invaluable effort for the success of IJST 2013.

Y. Takahashi

Yasuo Takahashi
Chair of IJST 2013

November 27-29, 2013 Venue: Icho Kaikan, Osaka University, Japan

Content from this work may be used under the terms of the Creative Commons Attribution 3.0 licence. Any further distribution of this work must maintain attribution to the author(s) and the title of the work, journal citation and DOI.
Published under licence by IOP Publishing Ltd

IOP Conf. Series: Materials Science and Engineering **61** (2014) 011001 doi:10.1088/1757-899X/61/1/011001

IJST2013 Organization

Conference Chairman
Yasuo Takahashi (Osaka Univ.)

Organized by
Technical Commission on Interfacial Joining, Japan Welding Society (JWS)

Co-organized by
Joining and Welding Research Institute (JWRI), Osaka University
Center for Environmental Innovation Design for Sustainability (CEIDS), Osaka University

Co-operated by
The Japan Welding Engineering Society (JWES)
Smart Processing Society for Materials, Environment & Energy

Organizing Committee:
Yasuo Takahashi (Chair, Osaka Univ.)
Yasuyuki Miyazawa (Vice Chair, Tokai Univ.)
Takehiko Watanabe (Niigata Univ.)
Masahiro Fukumoto (Toyohashi Univ. of Technol.)

Executive Committee:
T. Fukai (neis Co., Ltd.)
M. Fukumoto (Toyohashi Univ. of Technol.)
T. Ikeshoji (Tokyo Institute of Technol.)
M. Maeda (Osaka Univ.)
K. Matsu (Tokyo Braze Co., Ltd.)
Y. Miyazawa (Vice Chair, Tokai Univ.)
S. Nambu (The Univ. of Tokyo)
K. Nezaki (IHI Corporation)
S. Ozawa (Chiba Institute of Technol.)
T. Sasaki (Niigata Univ.)
K. Takahashi (Tokyo Institute of Technol.)
M. Takahashi (Toshiba Corporation)
Y. Takahashi (Chair, Osaka Univ.)
T. Yamazaki (Tokyo Institute of Technol.)
T. Yasui (Toyohashi Univ. of Technol.)

National Advisory Board:
K. Aota (Hitachi, Ltd.)
A. Fukushima (Mitsubishi Heavy Industries, Ltd.)
S. Katayama (Osaka Univ.)
S. Mori (Mitsubishi Materials Corporation)
K. Nakata (Osaka Univ.)
K. Ogawa (Nippon Steel & Sumitomo Metal Corporation)
O. Ohashi (WELLBOND)
K. Saida (Osaka Univ.)
A. Suzumura (Tokyo Institute of Technol.)
S. Tanaka (Tohoku Univ.)
T. Tomimura (Kumamoto Univ.)
T. Tsukamoto (Hitachi, Ltd.)
M. Ueda (Metal Technol. Co., Ltd.)

International Advisory Board:
T. Gang (Harbin Institute of Technol., China)
J. P. Jung (The Univ. of Seoul, Korea)
S. B. Jung (Sungkyunkwan Univ., Korea)
C. Y. Kang (Pusan National Univ., Korea)
J. H. Kim (Korea Institute of Industrial Technol., Korea)
Y. C. Lee (Univ. of Colorado Boulder, USA)
C. Li (Xi'an Jiao Tong Univ., China)
M. Li (Harbin Institute of Technol., China)
J. C. Schuster (Univ. of Vienna, Austria)
W. Tillmann (TU Dortmund, Germany)
D. Zhang (Shanghai Jiao Tong Univ., China)
K. Zhang (Henan Univ. of Science and Technol., China)

Program Committee:
T. Ikeshoji (Tokyo Institute of Technol.)
M. Maeda (Vice Chair, Osaka Univ.)
K. Matsu (Tokyo Braze Co., Ltd.)
Y. Miyazawa (Tokai Univ.)
Y. Takahashi (Chair, Osaka Univ.)
T. Yamazaki (Tokyo Institute of Technol.)

Editorial Board:
M. Fukumoto (Vice Chair, Toyohashi Univ. of Technol.)
T. Haga (Osaka Institute of Technol.)
Y. Hiraoka (Okayama Univ. of Science)
T. Ikeshoji (Tokyo Institute of Technol.)
K. Ito (Osaka Univ.)
M. Kato (Hiroshima Univ.)
M. Kimura (Univ. of Hyogo)
M. Maeda (Osaka Univ.)
K. Matsu (Tokyo Braze Co., Ltd.)
Y. Miyazawa (Vice Chair, Tokai Univ.)
K. Nishikawa (Kagoshima Univ.)
N. Suematsu (Tohoku Univ.)
K. Takahashi (Tokyo Institute of Technol.)
Y. Takahashi (Chair, Osaka Univ.)
W. Takahara (Osaka Univ.)
T. Tomimura (Kumamoto Univ.)
T. Yamazaki (Tokyo Institute of Technol.)
K. Yasuda (Osaka Univ.)

Conference Secretariats:
Y. Nakagawa (Osaka Univ.)
A. Ou (Osaka Univ.)

Peer review statement

All papers published in this volume of *IOP Conference Series: Materials Science and Engineering* have been peer reviewed through processes administered by the proceedings Editors. Reviews were conducted by expert referees to the professional and scientific standards expected of a proceedings journal published by IOP Publishing.

Content from this work may be used under the terms of the Creative Commons Attribution 3.0 licence. Any further distribution of this work must maintain attribution to the author(s) and the title of the work, journal citation and DOI.
Published under licence by IOP Publishing Ltd

Influence of surface asperities on interfacial extension during solid state pressure welding

Yasuo Takahashi[1], Terumi Nakamura[2], Yoshihiro Asakura[3] and Masakatsu Maeda[1]

[1] Joining and Welding Research Institute, Osaka University, Japan
[2] National Institute for Materials Science, Ibaraki, Japan
[3] Dept. Mechanical Eng. Kobe City College of Technology

E-mail: takasy@jwri.osaka-u.ac.jp

Abstract. The interfacial contact and extension of two joining surfaces are numerically studied using a finite element method. The joining surface is assumed to have two-dimensional triangular asperities. Surface-asperity-induced voids are formed at the initial contact. The interfacial extension of the initial contact area is very different from that of void surface area. The initial void surface area shrank once before full contact was attained. The large asperity angle facilitates interfacial extension and void-surface shrinkage, resulting in the formation of numerous metallic bonds. The pressure-welding-induced interfacial extension and metallic bond formation are largely influenced by the asperity angle, whose effects are experimentally confirmed.

1. Introduction

The solid-state bonding of metals has many advantages, which cannot be obtained using fusion welding [1-3]. Because the bonding temperature, T, is lower than the melting points, the best use of the material properties can be obtained through solid-state bonding. From the view of energy saving and environmental conscious processing, it is necessary to decrease the bonding temperature. As the bonding temperature decreases, the surface oxide film becomes to prevent a good bondaliability [4]. Plastic-deformation-induced interfacial extension is very important for strongly bonding new clean metallic surfaces. The probability of the metallic contact area, M_a, after full void-free contact is given by

$$M_a = \left(\frac{s - s_o}{s}\right)^2 = Y^2, \qquad (1)$$

where s_o represents the length of the initial segment (*i.e.* local length) at the faying surface at time $t = 0$, s represents the length of the pressure-welded segment, and $Y = (s\text{-}s_o)/s$ represents the surface exposure [5], which is expressed by $Y = \varepsilon_i/(\varepsilon_i+1)$, where $\varepsilon_i = (s\text{-}s_o)/s$ represents the interfacial extension (*i.e.* strain). The power of 2 in equation (1) means the probability that intimate contact is produced between clean surfaces of similar metals because oxide films on two faying surfaces are not necessarily broken at the same places [5], and $(1\text{-}M_a)$ expresses the probability of the area where the oxide film exists. If the ductility, G_d, of the surface oxide film or contamination (*i.e.* cover) layer on the faying surfaces is taken into account [5], equation (1) is modified to $M_a' = \{(1\text{-}G_d)Y\}^2$, and G_d ranges from 0.1 to 0.32 [5]. If the oxide film is scrubbed by being stirred around the bond interface during friction welding or

Content from this work may be used under the terms of the Creative Commons Attribution 3.0 licence. Any further distribution of this work must maintain attribution to the author(s) and the title of the work, journal citation and DOI.
Published under licence by IOP Publishing Ltd

ultrasonic bonding [6] then G_d is apparently negative. The parameters M_a (or M_a`), Y, and ε_i are useful for discussing the oxide film behaviour during frictionless solid-state pressure welding.

Surface asperities largely influence interfacial contacting [7-10]. The surface asperity angle α_o (figure 1), therefore, largely affects the metallic bond ratio. Because the extension behaviour of the void surface has not previously been reported, it must be investigated by increasing the bulk strain, ε_B. Hence, we numerically analysed the effects of ε_B and α_o on interfacial extension. The breaking behaviour of the surface oxide film was also experimentally surveyed to better understand the distribution of superficial- and interfacial-extension-induced broken oxide films.

2. Calculation procedure

Figure 1 illustrates the cross-sections of two faying surfaces with long triangular ridges. Figure 1 (a) is the pre-joining profile. The initial contact is determined by the instantaneous plastic flow, as illustrated in figure 1 (b) [7]. Half the initial contact width, X_o, is given by

$$X_o = \frac{L_{oo} \cdot P}{(1+\varphi)\cdot \sigma_y(T)}, \tag{2}$$

from the slip line analysis, as shown in figure 1 (c), where L_{oo} represents half the interval of the surface asperity; P, the bonding pressure; φ, the angle defined in figure 1 (c); and $\sigma_Y(T)$, the yield stress for the instantaneous plastic deformation at T (measured in Kelvin). L_o is supposed to be equal to L_{oo} at time t = +0 (*i.e.* immediately after the initial contact).

The post-initial contact between faying surfaces is assumed to be dominated by viscoplastic deformation under plane strain. Non-linear 8-node isoparametric finite elements were used for the numerical simulation. The calculation procedure is detailed elsewhere in the literature [7].

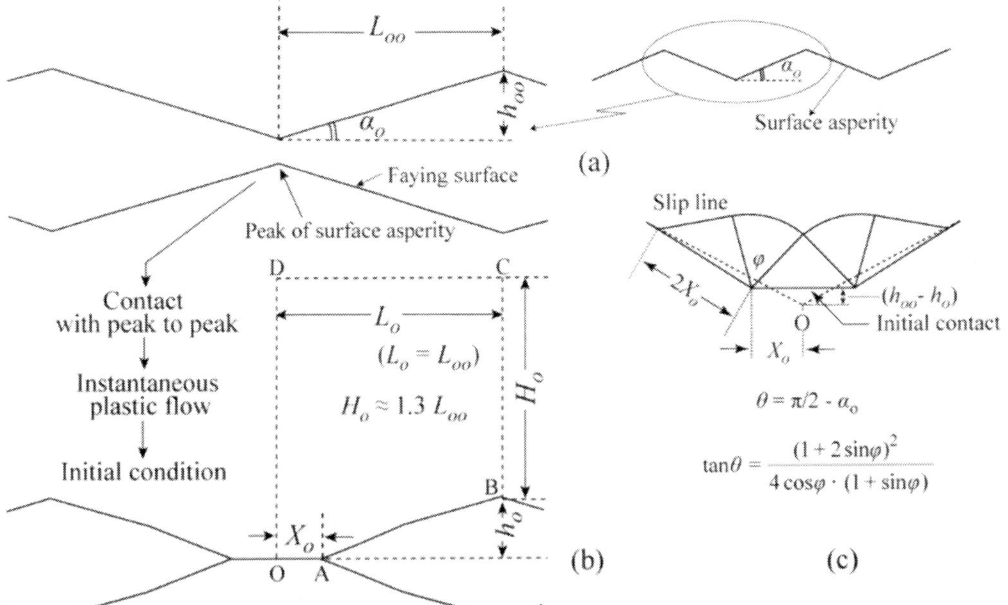

Figure 1. Model of contact between two faying surfaces during solid-state bonding: (a) surface asperity model, (b) instantaneous-plastic-flow-induced post-initial-contact bond interface profile, and (c) slip line field method used to calculate half initial contact width, X_o.

The equivalent strain rate, $\dot{\bar{\varepsilon}}$, is given by

$$\dot{\bar{\varepsilon}} = A_o\left(\frac{D_o bG}{kT}\right)\left(\frac{\bar{\sigma}}{G}\right)^n \exp\left(\frac{Q_c}{RT}\right), \tag{3}$$

where A_o is a dimensionless constant; D_o, the frequency factor; k, Boltzmann's constant; $\bar{\sigma}$, the equivalent stress; b, a Burgers vector; G, the shear modulus; n, the stress exponent; Q_c, the activation energy; and R, the gas constant. Table 1 shows the material constants for pure copper, which can be used for $T/T_m = 0.35$–0.5, where T_m represents the melting point (measured in Kelvin) [7, 11].

Figure 2 illustrates the mesh pattern for area OABCD in figure 1 (b) at $t = +0$ for $\alpha_o = 15°$. L represents half the post-initial-contact asperity width. The height of calculation region H_o is given by $H_o = 1.3\ L_o$. The stress in the y-direction on the DC plane, σ_y, is given by the bulk stress σ_x ($= -L_o P/L$), where P represents a nominal bonding pressure at $t = 0$. The BC plane is assumed to have been displaced at $\dot{u}_x$, as shown in figure 2. The bulk strain rate in the x-direction, $\dot{\varepsilon}_\infty = (3^{1/2}/2)\dot{\bar{\varepsilon}}$, where $\dot{\bar{\varepsilon}}$ can be obtained by substituting $\bar{\sigma} = (3^{1/2}/2)\ \sigma_x$ into equation (3). Two horizontal meshes (*i.e.* 8 elements) were added in the range AB when calculating the contact for $\alpha_o = 60°$.

Table 1. Material constants for annealed oxygen-free copper.

Symbols	Values	Unit
A_o	7.24×10^{12}	
D_o	0.1×10^{-4}	m^2s^{-1}
b	2.56×10^{-10}	m
n	7.72	
Q_c	146.7	$kJ\ mol^{-1}$
T_m	1356	K
$G(T)$	$a_G(T/T_m)^2 + b_G(T/T_m) + c_G$,	Nm^{-2}
where $a_G = -4.7\times10^{10}$, $b_G = 1.3\times10^{10}$, $c_G = 4.6\times10^{10}$.		
σ_Y	$[G(T)/G(T_o)]\Gamma(T)\sigma_{Yo}$,	Nm^{-2}
where $\sigma_{Yo} = 380$ MPa at $T_o = 300$ K [17]		
$\Gamma(T)$	$a_o(T/T_m)^2 + b_o(T/T_m) + c_o$,	
where $a_o = -8.8\times10^{-1}$, $b_o = 3.47\times10^{-1}$, $c_o = 9.6\times10^{-1}$.		

The bonding conditions were $T = 673$ K and $P = 100$ MPa. Half the asperity width, L_{oo}, was 10 μm, and the surface asperity angle, α_o, was changed from 15 to 60°. The interfacial extension, ε_I, was calculated from the extension strain of each segment of finite elements along the faying surface (in line OAB in figure 2). During that stage, the initial-contact-induced extension was ignored although the initial contact ratio, S_o ($= X_o/L_o$), was in the range 13–21% for $\alpha_o = 15$–$60°$; that is, ε_I, as defined here, means the post-initial-contact expansion strains of the void surface or contact interface.

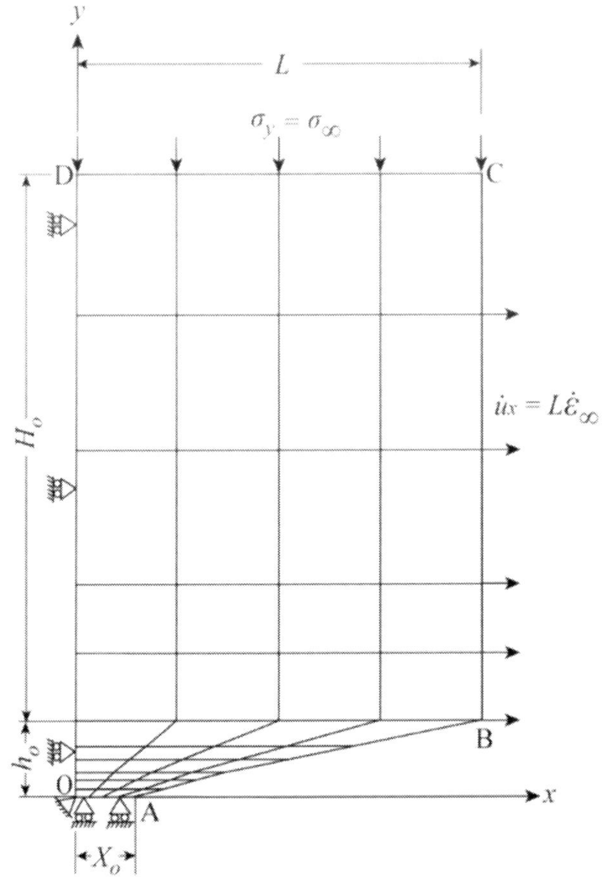

Figure 2. Mesh pattern of finite elements used to simulate interfacial contact ($L = L_o$ at $t = +0$).

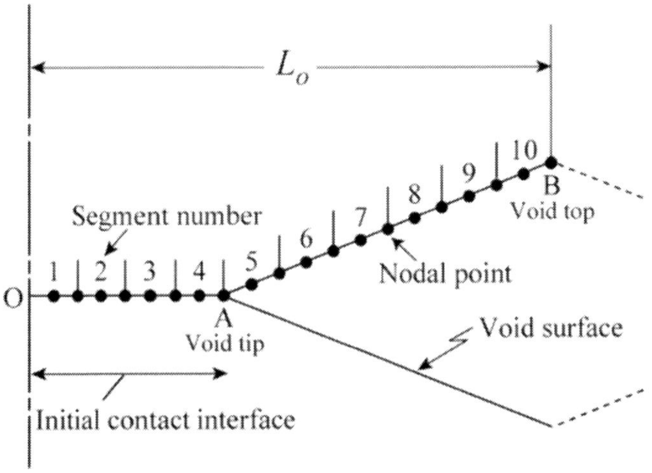

Figure 3. Definition of segment number along interface and void surface (at $t = +0$).

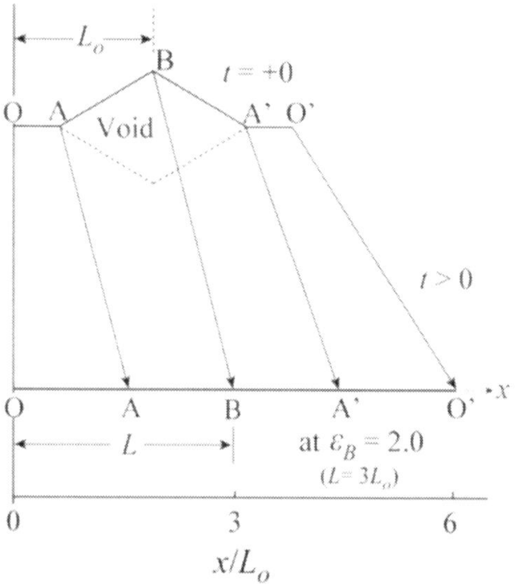

Figure 4. Schematic illustrating interfacial extension when $\varepsilon_B = 2.0$ and $L = 3L_o$.

Figure 3 illustrates the segment number, N_s, and the nodal points of finite elements along the contact interface, OA, and the void surface, AB. As the mother material (*i.e.* bulk) deforms, the void shrinks and disappears and then full contact is achieved. The contact ratio, S, is then 100%. The contact process was simulated up to the bulk strain $\varepsilon_B = (L-L_o)/L_o = 2.0$.

Figure 4 illustrates interfacial extension and void crushing. The change in the positions of points O, A, B, A', and O' is illustrated. Points A and B are called the 'void tip point' and 'void top point', respectively.

3. Results and discussion

When surface asperities exist on the joining surfaces, the contact is classified as either Type I or Type II interfacial deformation, as shown in Figure 5 [9]. In Type I, the contact is dominated by void tip folding, as shown in figures 5 (a-1–3), when $\alpha_o < 30°$. Element E_A touches the bond line (x-axis). However, the contact is mainly produced by the extension of the initial contact area when α_o is large enough. Element E_B in figure 5 (b-1) extends to that in figure 5 (b-2). In Type II, void top folding occurs at the void surface adjacent to the top point. Type I and Type II interfacial deformation largely affect the local ε_i.

Figures 6, 7, and 8 show the results calculated for the change in the interfacial extension under ε_B (*i.e.* elongation in the x-direction) for α_o = 15, 45, and 60°, respectively. Each segment number corresponds to that in figure 3. The red marks represent ε_i at $\varepsilon_B \approx 0.01$. The blue marks represent $\varepsilon_B \approx 0.03$. The outline marks represent the points on the void surface. The solid marks, on the other hand, represent the points on the contact interface. The stars represent S = 100%. Segments N_s = 1–4 are on the initial contact interface. The extensions at N_s = 4 are very high in figures 6 and 7. The original void surface shrinks once, despite $\varepsilon_B > 0$. The fine dotted lines mean that the extension is negative. In figure 8, for $\alpha_o = 60°$, segments $N_s > 6$ are not shown because they are compressed to -0.4 at S = 100% (*i.e.* $\varepsilon_B \approx 0.09$) and are crushed to about -0.7 at $\varepsilon_B = 1$. In other words, the void surface ABA' in figure 4 is squashed and concentrated for $\alpha_o = 60°$. Figure 8, on the other hand, shows segments N_s = 1–4 exhibit very high extension ($\varepsilon_i \approx 9$ at $\varepsilon_B = 1$).

Figure 9 shows the fractional metallic contact area, M_a, at $\varepsilon_B = 2$ for α_o = 15–60°. Points O, A, B, A', and O' correspond to those in figure 4. M_a is calculated from Y in equation (1). $Y = 2/3$ and M_a = 0.44 are obtained at $\varepsilon_B = 2$ when $\alpha_o = 0°$ and $G_d = 0$. Figure 9 (a) shows that $M_a \approx 0.45$ when $\alpha_o = 15°$, which is nearly equal to that of a flat surface, although points A and A' show the peak of $M_a = 0.54$. Figure 9 (b) shows that even when $\alpha_o = 45°$, M_a is rather low around the void top point, B, *i.e.* the metallic bond is not always facilitated around B, even when the surface asperities become coarse.

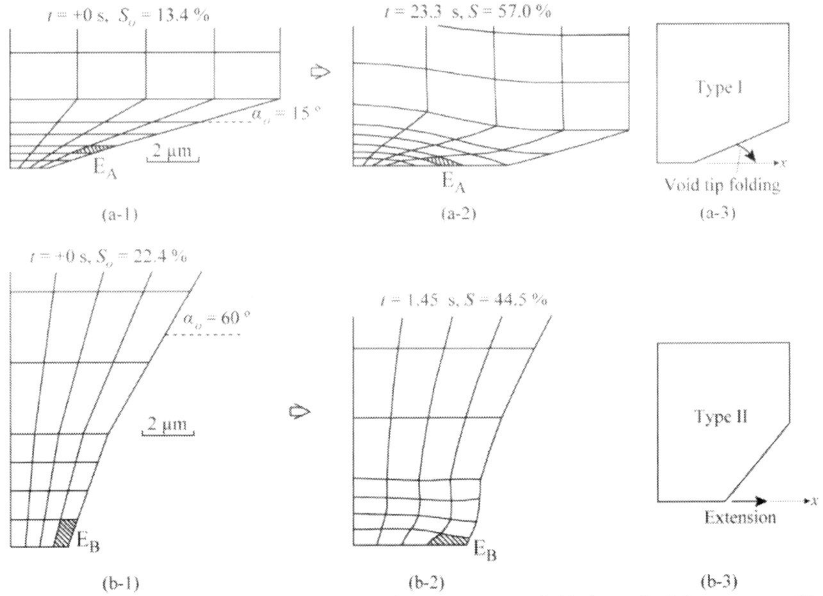

Figure 5. Change in mesh patterns and Types I and II interfacial contact. Bonding conditions were T = 973 K and P = 30 MPa. (a-1) Initial mesh pattern for α_o = 15°, (a-2) deformed mesh pattern for α_o = 15°, (a-3) schematic illustrating Type I (folding mechanism), (b-1) initial mesh pattern for α_o = 60°, (b-2) deformed mesh pattern for α_o = 60°, and (b-3) schematic illustrating Type II (extension mechanism).

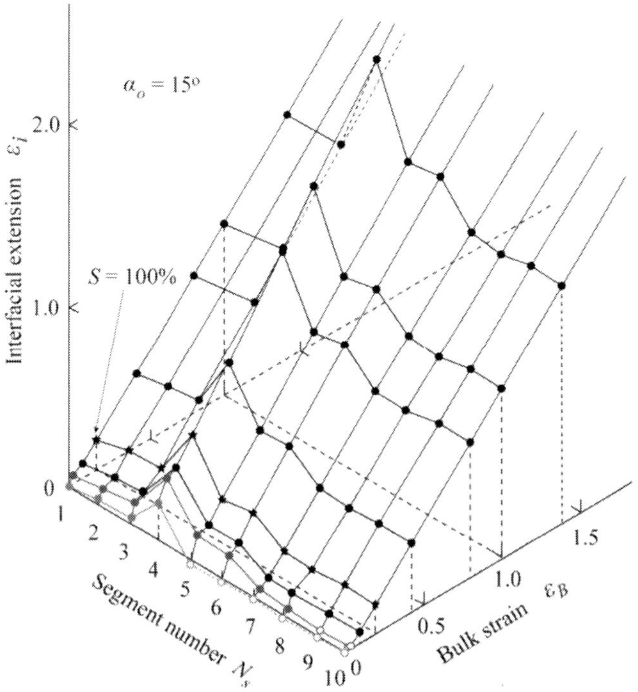

Figure 6. Bulk-strain-dependent interfacial extension calculated for α_o = 15°.

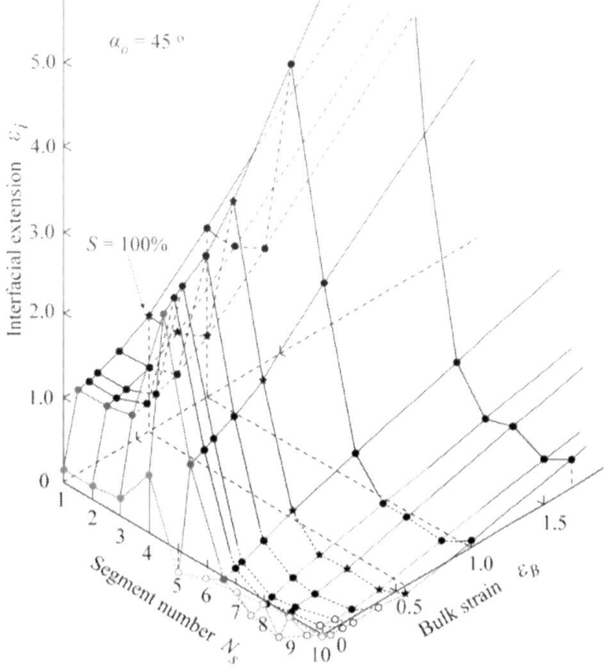

Figure 7. Bulk-strain-dependent interfacial extension calculated for $\alpha_o = 45°$.

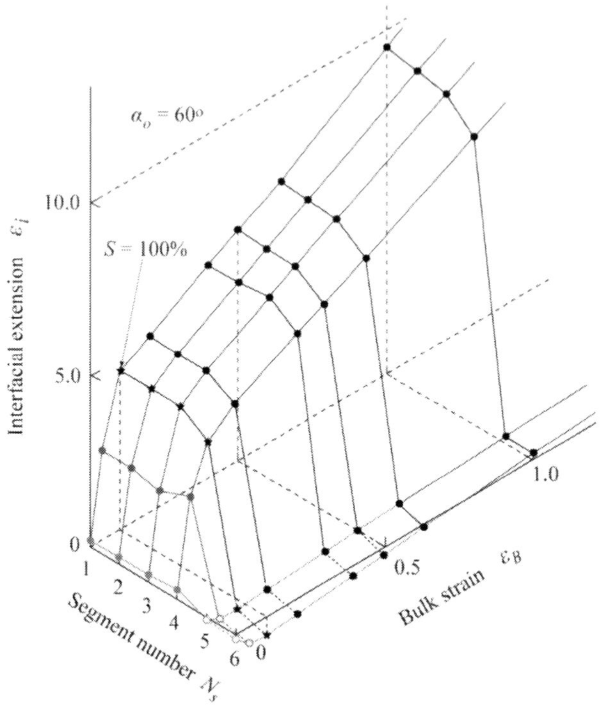

Figure 8. Bulk-strain-dependent interfacial extension calculated for $\alpha_o = 60°$.

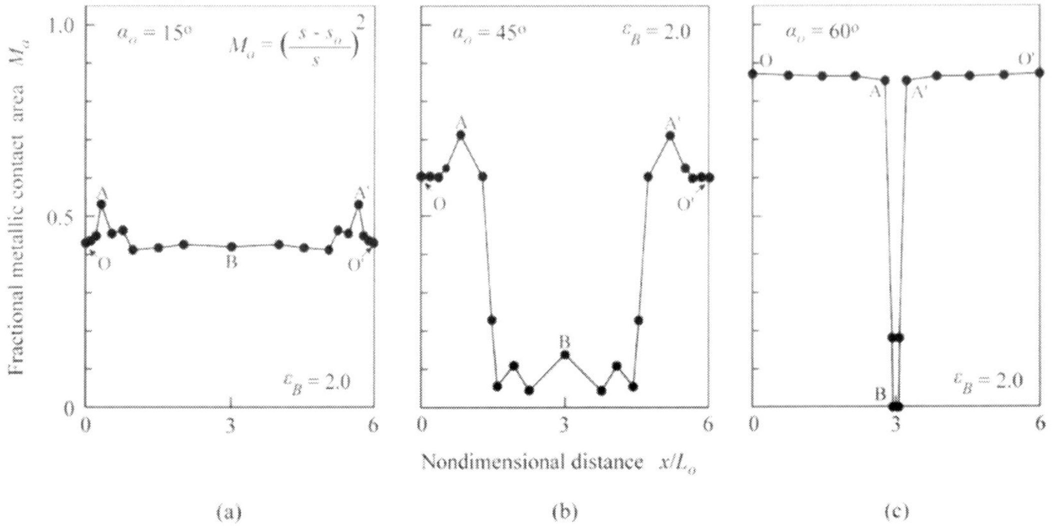

Figure 9. Fractional metallic area, M_a, calculated at $\varepsilon_B = 2.0$ when α_o = (a) 15, (b) 45, and (c) 60°. Surface oxide film is assumed to be brittle ($G_d = 0$).

However, as α_o increases to 60° (figure 9 (c)), M_a becomes very high (> 0.8). The interfacial extension at $\varepsilon_B = 2$ for $\alpha_o = 60°$ is much greater than that at $\varepsilon_B = 4$ for the flat surface (*i.e.* for $\alpha_o = 0°$). When the $\alpha_o = 60°$ faying surface is pressed, the surface oxide film is squashed by interfacial extension and concentrated to B, as shown in figure 9 (c) (i.e. $M_a = 0$). Reiterating, the Type I folding mechanism becomes dominant as the surface asperities become small, inhibiting large local extension and keeping M_a low. The Type II contact mechanism, on the other hand, facilitates metallic bond formation, resulting in high M_a. A large asperity angle is effective for breaking surface oxide films because large asperities produce the Type II mechanism.

4. Experimental study

We used 5-ppm-oxygen copper produced from a cold-drawn rod. The diameter and height of the bonding specimen were 10 and 12 mm, respectively. The faying surfaces were machined using a precision lathe to produce triangular ridges arranged at regular intervals. The asperity angle was changed by regulating the tip angle of bit; that is, three asperity angles were produced: α_o = 10–15, 40–45, and 55–60° corresponding to $L_{oo} \approx 140$ μm, surface roughness (*i.e.* height, h_{oo}) ≈ 32 μm; $L_{oo} \approx$ 90 μm, $h_{oo} \approx 85$ μm; and $L_{oo} \approx 600$ μm, $h_{oo} \approx 950$ μm, respectively, although h_{oo} was different because of the precision of the lathe machine. The specimen was preheated in dry air at 473 K for 1200 s before pressure bonding, producing a few-hundred-nm-thick surface oxide film on the faying surface [4]. Similar specimens were then pressure welded. Specimens were also subjected to bondless contacting, during which the specimen was set between a die and a pressing tool to observe the oxide film distribution (*i.e.* the surface exposure), and 20-μm-thick molybdenum (Mo) foil was inserted between the bonding tool and the specimen [10]. The mean surface roughness of the Mo foil was <0.1 μm. The specimen was pressed under a 10^{-3} Pa vacuum at designated temperatures in the range 273–673 K.

The reduction in ε_H (*i.e.* the compression ratio) of the specimen along the *z*-axis (*i.e.* vertical direction) was defined by $\varepsilon_H = \Delta H/H_o = (H-H_o)/H_o$, where H_o and H represent the heights of the initial and pressed specimens, respectively. The specimen was pressed at $\Delta H/H_o = -0.5$ under axisymmetric deformation although the results were calculated based on plane strain. Hence, the specimen should be pressed such that Y is the same under axisymmetric and plane strain for a flat faying surface.

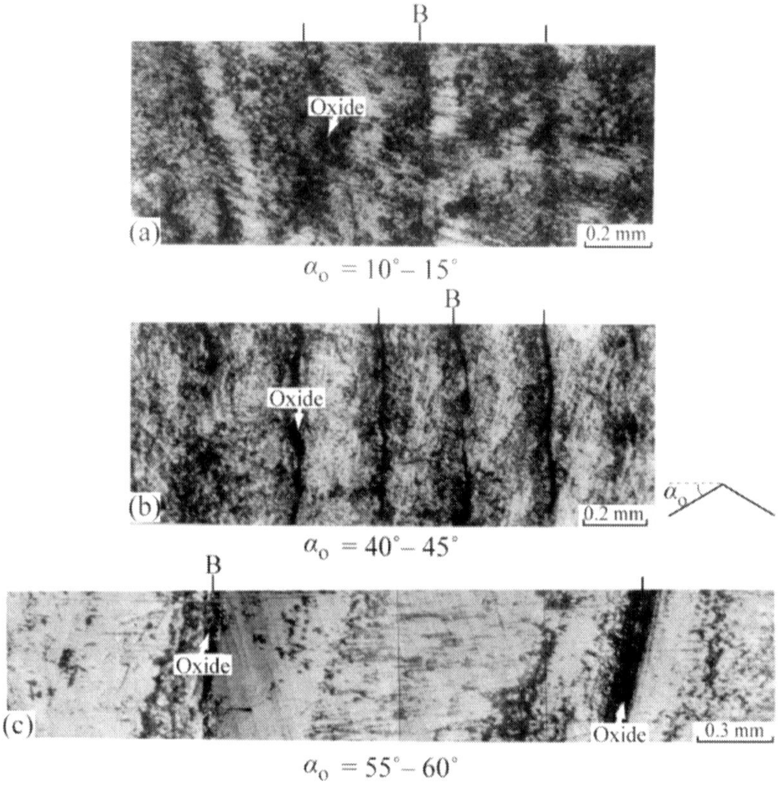

Figure 10. Optical micrographs showing distribution of broken-up oxide films when α_o = (a) 10–15, (b) 40–45, and (c) 55–60°.

The specimens were bonded such that the area expansion, $\varepsilon_A = (A-A_o)/A_o$, was comparable with that shown under plane strain, where $A_o = \pi R_o^2$ represents the initial bonding area, and R_o represents the initial radius of the specimen. A is obtained from $A = \pi R^2$, where R represents the radius of the pressed specimen, as measured at the joining surface; $\varepsilon_A = 1.0$ is equivalent to $\varepsilon_B = 1.0$ when $\varepsilon_H = -0.5$. If the faying surface is flat and the surface oxide film is brittle enough ($G_d = 0$), $Y = (\pi R^2 - \pi R_o^2)/(\pi R^2) = 0.5$ when $\varepsilon_A = 1.0$. The experiment was performed to determine whether the effects of the asperity angle on the interfacial extension and the distribution of the oxide area in the pressed specimens were consistent with the numerically predicted effects. Hence, the experimental and calculated data were compared for the same $\varepsilon_A = 1.0$ ($\varepsilon_H = -0.5$).

Figure 10 shows optical micrographs of the copper side of the contact interfaces between the copper specimen and Mo foil when the foil was removed from the copper specimen. The black (brown) area shows broken oxide film, and the bright (yellow-white) area shows the interface-extension-produced metallic area. The micrographs were obtained during the pressing tests conducted at room temperature because the broken oxide films partially stuck to the Mo foil with increasing pressing temperature. Pressing temperatures in the range 273–473 K did not significantly affect the distribution morphology of the broken oxide film when the bulk strain (or compression ratio) and the asperity angle were kept constant. In figure 10, $\varepsilon_H = -0.5$ but $\varepsilon_A = \varepsilon_B = 1$ under plane strain. The place marked by 'B' in each micrograph corresponds to the void top point, B. Further, figure 10 shows that α_o largely affects the distribution of the broken oxide film. In figure 10 (a), the black-brown oxide area is wholly distributed, as suggested by the calculated result shown in figure 9 (a), although a bright

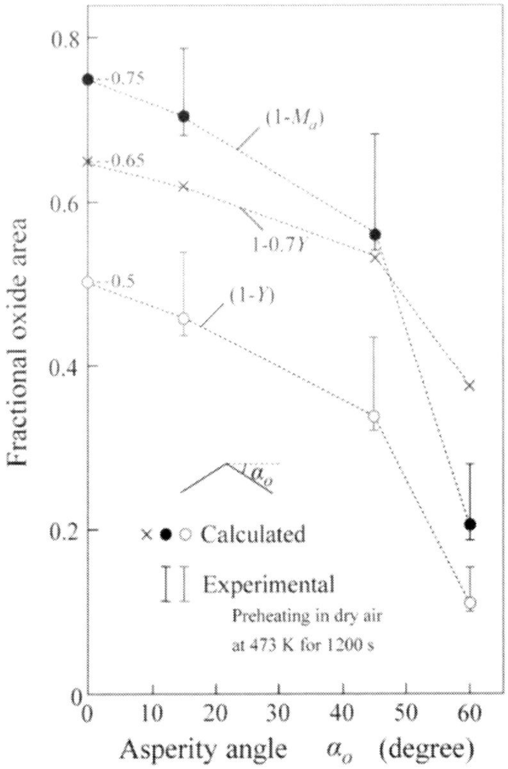

Figure 11. Influence of asperity angle on fractional oxide area.

(metallic) area is observed in places because the surface roughness is irregular. Figure 10 (b), on the other hand, shows a periodic black stripe whose period length = ~2L. Figure 10 (c) shows that the black stripe becomes very sharp when α_o = 55–60°. Traces of the interface-expansion-torn-up oxide film are observed in places. Interfacial extension breaks up the oxide film and gathers it at the void top point.

Suppose that the oxide film is only broken by diffusionless plastic deformation. The fractional oxide film area (1-M_a) at the interface of solid-state pressure-welded bonds is larger than the black area in the micrographs in figure 10 because the micrographs correspond to only one side of the bond interface. If the oxide film is brittle enough (*i.e.* $G_d = 0$), (1-M_a) = (1-Y) on one side and (1-M_a) = (1-0.7Y) if $G_d = 0.3$ [5]. If A_{ox} and A_M represent the black-brown and white-yellow areas in figure 10, respectively, then $(1\text{-}M_a) = 1\text{-}(A_M/A)^2 = 2(A_{ox}A_M/A)\text{-}(A_{ox}/A)^2$, where A, A_{ox}, and A_M represent the total, oxide, and metallic areas, respectively, and $A = A_{ox}+A_M$.

A_{ox} and A_M were separated using image processing. The 0–255 binary representations were used to convert the original colour images into grayscale ones. The black area was obtained as a threshold less than $\theta^* = 60$, *i.e.* it was binarized using $\theta^* = 60$. For example, the A_M/A was about 0.85–0.90 for α_o = 55–60°. Although exactly quantifying (1-M_a) was difficult as A_M/A naturally decreased with increasing θ^*, the tendency of the change in (1-M_a) could be taken into account. Figure 11 shows (1-M_a) and (1-Y), as estimated from the image processing. The red marks represent (1-Y). For the flat surface (*i.e.* α_o = 0°), (1-M_a) = 0.75 at the bond interface and (1-Y) = 0.5 at one side surface. The calculated results shown in figure 11 and the experimental results are consistent, suggesting that G_d is nearly zero for a thick copper oxide film, although the experimental error bars are wider than the circles representing the calculated results. The roughness of surfaces showing large asperity angles breaks oxide films and reduces (1-M_a) from 0.75 to 0.2 when $\varepsilon_H = -0.5$.

5. Conclusion

The interfacial contact of a joining surface containing surface asperities has been discussed. The contact is controlled by void surface folding and interfacial extension. If surface asperities exist on the faying surface, the interfacial extension is not uniform and can be much greater than the bulk extension. Interfacial expansion is facilitated and the oxide film can be easily broken when $\alpha_0 > 50°$. The initial void surface exhibits negative extension before full contact. The surface oxide film is gathered to original void top points. Consequently, surface asperities can produce clean metallic surface area much better than flat surface extension can.

References

[1] Takahashi Y 2003 *Ceramic Trans.* **138** 29-59
[2] Takahashi Y and Inoue M 2002 *ASME J. of Electronic Packaging* **124** 27-36
[3] McQueen H J 2012 *Canadian Metallurgical Quarterly* **51** 239-249
[4] Takahashi Y, Nakamura T and Nishiguchi K 1992 *J. Mater. Sci.* **27** 485-498
[5] Zhang W and Bay N 1997 *Weld. J.* **76** 417s-420s
[6] Maeda M, Kitamori S and Takahashi Y 2013 *STWJ* **18** 103-107
[7] Takahashi Y, Koguchi T and Nishiguchi K 1993 *Trans. ASME, J. Eng. Mater. and Technol.* **115** 150-155
[8] Takahashi Y, Koguchi T and Nishiguchi K 1993 *Trans. ASME, J. Eng. Mater. and Technol.* **115** 171-178
[9] Takahashi Y and Tanimoto M 1995 *Trans. ASME, J. Eng. Mater. and Technol.* **117** 330-335
[10] Takahashi Y and Tanimoto M 1995 *Trans. ASME, J. Eng. Mater. and Technol.* **117** 336-340
[11] Furushiro N, Okada Y and Hori S 1980 *J. Soc. Mater. Sci. Japan* **29** 776-781

Solid-state bonding of superplastic ultra-high carbon steel and structure steel with an interlayer

Keke Zhang[1], Zhonghao Heng[2], Ranfeng Qiu[1], Hongxin Shi[1] and Yasuo Takahashi[3]
1. School of Materials Science and Engineering, Henan University of Science and Technology, 471023, China
2. Graduate School of Engineering, Osaka University, Japan
3. Joining and Welding Research Institute, Osaka University, Japan

E-mail: zhkekekd@163.com

Abstract. Solid-state bonding between two superplastic steels—ultra-high carbon steel (UHCS) and a structure steel (commonly referred to as 40Cr in china; yet contains only 1 mass% Cr)—was investigated. Industry-grade pure iron sheet was used as an interlayer and the bonding was carried out at the superplastic deformation temperature $T = 750$℃. For a pre-pressing stress of $P = 56.6$ MPa, the initial strain rate was $1.5\times10^{-4}s^{-1}$, and for a bonding time of $t = 10$min, the joint tensile strength reached up to 702 MPa a value 35% higher than what could be obtained without an interlayer. The bonding strength reached the tensile strength of 40Cr base metal through the same thermal-stress cycling. Based on the results, it was concluded that the weldability of 1.6 mass%C-UHCS / 40Cr steel was improved by using a pure iron interlayer.

1. Introduction

Fine-grained ultra-high carbon steel (UHCS) has excellent mechanical properties such as high strength, hardness and wear resistance. It can be hopefully developed as a novel resource-saving and economical engineering material. However, the presence of large volume of carbide and several alloying elements adversely affect its weldability through normal fusion welding methods, thus greatly restricting its application [1, 2].

Fine-grained superplastic materials can be readily bonded in the solid state [3, 4]. By taking advantages of the superplasticity, great progress has been made in the field of solid-state bonding of materials with poor weldability [5, 6]. However, the relatively low deformability and the issue of harmful eutectic structure at the bonding interface due to the alloying elements, prevent obtaining high-quality joints [2, 7, 8]. Developing an effective approach for bonding such a material with poor weldability is, therefore, imperative.

In this study, an appropriate interlayer material was designed for solid-state bonding of superplastic UHCS and a structure steel. The effect of the interlayer on the interfacial behaviour was investigated. This novel diffusion bonding method, based on the superplasticity and the use of an interlayer, is discussed.

2. Interlayer design

The use of an appropriate interlayer material is necessary for obtaining better joints. Previous studies on diffusion bonding of UHCS have elucidated the associated problems that seriously affect the

Content from this work may be used under the terms of the Creative Commons Attribution 3.0 licence. Any further distribution of this work must maintain attribution to the author(s) and the title of the work, journal citation and DOI.
Published under licence by IOP Publishing Ltd

bonding quality, as characterized by the low bonding ratio caused by the poor deformability and the presence of harmful eutectic microstructure at the interface.

The first consideration for a material to be used as an interlayer for solid-state bonding of UHCS is that it should be an iron-based material in order to match the base metals. Then, it should have low yield strength to facilitate interfacial deformation at both bonding interfaces. Additionally, it also should have only a few alloy elements to prevent the enrichment of alloy element and avoid the formation of detrimental low-melting eutectic structure at the bonding interface. Based on these criteria, industry-grade pure iron was selected as the interlayer material for the present study.

3. Experimental

The materials used in the present study were 1.6 mass%C-UHCS and the structure steel 40Cr, which contains 1 mass% Cr (40Cr is the standard nomenclature for such a steel in China). The chemical compositions are shown in Table 1. The 1.6 mass%C-UHCS was manufactured via vacuum melting, hot rolling, and a spheroidizing treatment in that order, and was comprised of fine globular cementite and ferrite. The 40Cr steel, obtained via two quenching in salt bath at 820℃ followed by a low-temperature tempering treatment at 200℃, consisted of small lath martensite. Both fine-grained steels could show good superplasticity. As the interlayer is required to be as thin as possible, a sheet of pure iron was rolled to a thickness of 50 μm. The microstructure of the steels and the interlayer were as shown in Figure 1. The 1.6 mass%C-UHCS and 40Cr steel were machined into φ15×25 mm samples.

The bonding faces of the samples and the interlayer sheet were polished and cleaned with acetone. The interlayer was placed between the 1.6 mass%C-UHCS and 40Cr steel samples. Bonding tests were carried out under ambient conditions without any vacuum or gas protection. The bonding temperature T was 750℃, the best superplastic deformation temperature of 40Cr steel. The UHCS was capable of exhibiting superplasticity well at this temperature [2, 5, 6]. The pre-pressing stress was 56.6MPa to have an intimate interfacial contact in the initial bonding stage. After a pre-pressing time of 20 minutes, the samples were bonded under a constant compression speed (initial strain rate $\dot{\varepsilon}_0$ = $1.5\times10^{-4}s^{-1}$) for several values of bonding time t.

Table 1. Compositions of 1.6 mass%C-UHCS and 40Cr steel (mass %).

Steel	Fe	C	Si	Mn	Cr	Al	S	P	Nb	V	Ti
UHCS	Bal.	1.60	0.49	0.44	1.54	1.66	0.0075	0.012	0.021	0.043	0.082
40Cr	Bal.	0.40	0.25	0.6	1.0	-	-	-	-	-	-
Pure iron	Bal.	0.02	0.008	0.02	0.01	-	0.005	0.007	-	-	-

Figure 1. Microstructures of fine-grained steels and pure iron interlayer. (a) globular pearlite in 1.6 mass%C-UHCS. (b) martensite in 40Cr steel. (c) cold-rolled microstructure of the pure iron interlayer.

The joint was machined into 15 × 5 × 3mm tensile sample. The tensile test was performed at a crosshead speed of 0.3 mm/min at room temperature to evaluate the mechanical properties of joint. The microstructures and the fractures of joints were observed by scanning electron microscopy (SEM). The contents of alloying elements were analyzed by energy dispersion spectrum (EDS).

4. Results and discussion

Figure 2 shows the bonding strength of 1.6 mass%C-UHCS joints, depending on bonding time t for the two cases of bonding with and without an interlayer ($\dot{\varepsilon}_0 = 1.5\times10^{-4}s^{-1}$, $T = 750$ ℃). As seen in Figure 2, the joints with the pure iron interlayer exhibit higher strength than the direct bonding joints for the same bonding condition.

For direct bonding, the joint strength increased with bonding time and reached a maximum when the bonding time was 15min. As the bonding time increased further, harmful interfacial diffusion intensified, thus reducing the bonding strength [8]. By comparison, when bonding time was 10min, the maximal strength of joints with the pure iron interlayer reached 702 MPa, which is 35% higher than that of the directly bonded joints. The bonding strength can potentially reach the strength of base metal, which is the 40Cr steel, when it undergoes the same thermal cycle. As the bonding time increased further, the bonding strength reduced slightly due to grain growth and the deterioration in the microstructure of the interlayer. The pure iron interlayer significantly improved the bonding strength and high-quality joints were obtained with relatively short bonding times. These results are consistent with what was expected. The interlayer design and bonding method for superplastic 1.6 mass%C-UHCS are experimentally proved effective.

Figure 3 shows the microstructure of joints obtained through direct bonding and bonding with the interlayer. As shown in Figure 3(a), many voids are observed at the bonding interface. In contrast, there are few defects at both the bonding interfaces, as shown in Figure 3(b).

In the case of direct bonding, the unbonded areas are caused by the low deformability of the 1.6 mass%C-UHCS. Although the superplastic deformation of base metals contributes somewhat to the interfacial contact, the bonding time is relatively too short to obtain a high bonding ratio for UHCS. However, the bonding strength cannot be improved by just extending the bonding time, for the consequent excessively harmful interfacial reactions can also seriously affect bonding quality. Using a low-yield-strength interlayer affords acceptable joining between the interlayer and the both base metals. Interlayer thickness is about 40 μm after bonding, and a distinct deformation occurs in the pure iron interlayer. This deformation significantly facilitates interfacial contact with both the base metals, thus increasing the bonding ratio.

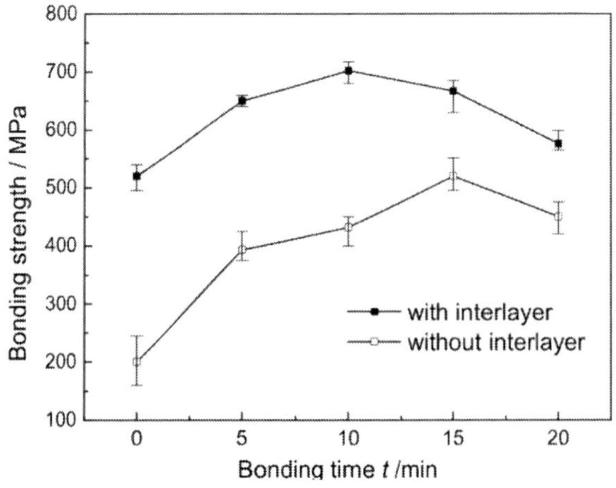

Figure 2. Bonding strength dependence of t ($\dot{\varepsilon}_0 = 1.5\times10^{-4}s^{-1}$, $T = 750$℃).

(a) (b)

Figure 3. Microstructure of bonding joints. (a) direct bonding. (b) with pure iron interlayer. (T = 750℃, t = 10min)

The cold-rolled microstructure of the pure iron interlayer is easily recrystallized during bonding. The concentration gradient between the interfaces makes carbon diffuse from base metals into pure iron interlayer, and consequently, the microstructure and mechanical properties of interlayer are improved as the carbon content increases. High-quality joints of UHCS / pure iron / 40Cr are thus achieved. However, extending the bonding time causes grain growth in the interlayer, resulting in the deterioration of the joint microstructure, and as a result, the bonding strength reduces. This is also reflected in the bonding results in Figure 2.

The fractures of bonding joints, for both the cases of direct bonding and bonding with an interlayer, are shown in Figure 4. As seen in Figure 4(a), numerous intergranular cracks (Figure 4(a) A) can be observed at the interface with 40Cr. EDS analysis shows that alloying element enrichment of Al (1.19 wt%), Si (0.74 wt%) exists in the cracks. The interface and grain boundary are rich in Al, Si because they act as passages for interdiffusion. These cracks are formed near the interface owing to the formation of low-melting phase or eutectic structure under deformation stress and inner stress. Original grinding traces (Figure 4(b) A) and a few metallurgical bonded areas (Figure 4(b) B) can be observed on the UHCS side. The low bonding ratio and the numerous bonding defects seriously reduce the joint strength.

Figure 4(c), (d) shows the fractures of 1.6mass%C-UHCS / pure iron / 40Cr joint. Fracture occurs in the interlayer along the interface and also in some regions of the UHCS base metal (Figure 4(d) A). Pure iron interlayer enhanced interfacial contact, thus causing a significant increase in metallurgically bonded areas (Figure 4(c) A). The fractures in some areas of base metal are also consistent with the results from tensile testing of joints.

Figure 5 shows the distribution of alloying elements in the 1.6 mass%C-UHCS / pure iron / 40Cr joint. The interdiffusion of alloying elements clearly occurs at both interfaces. The Al and Cr content of the interlayer are far below those of base metals. A pure iron interlayer prevents the direct interdiffusion between UHCS and 40Cr, thus preventing the enrichment of the alloying elements, especially those responsible for low-melting-structure (phase)-forming elements. This transition region greatly reduces the formation possibility of harmful structures, and therefore, the excessive interfacial reaction is avoided. No intergranular cracks were observed at the interfaces after bonding with an interlayer. This result is well evidenced by Figure 4. In general, a pure iron interlayer can thus enhance the weldability of superplastic 1.6 mass%C-UHCS.

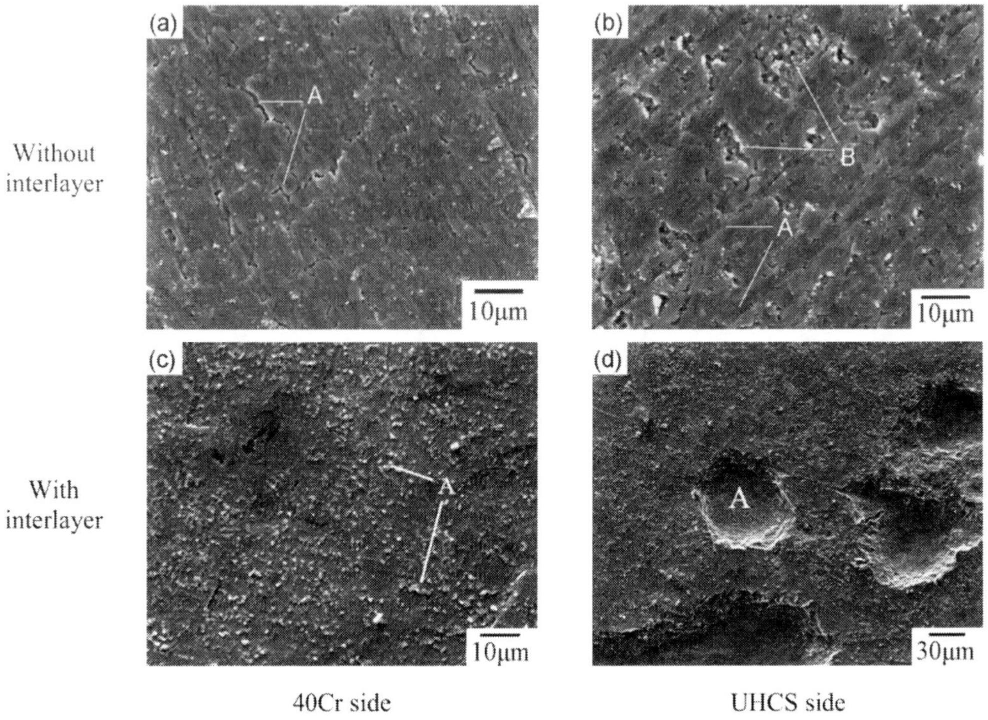

Figure 4. Fractures of bonding joints ($\dot{\varepsilon}_0 = 1.5\times10^{-4}s^{-1}$, $T = 750$℃, $t = 10$min).

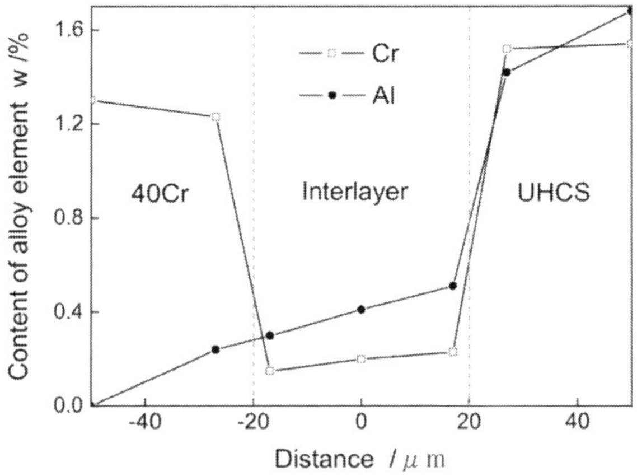

Figure 5. Distribution of alloy elements in the joint. ($T = 750$℃, $t = 10$min)

5. Conclusions

A pure iron interlayer facilitates proper interfacial contact during solid-state bonding of 1.6 mass%C-UHCS and 40Cr steel, reduces required bonding time, and prevents direct interfacial reaction of base metals, thereby preventing the formation of detrimental eutectic phase and structures in the joint. All these contribute to improving joint quality. A pure iron interlayer can be useful in solid-state bonding of dissimilar superplastic steels.

Acknowledgments
The work was supported by the National Natural Science Foundation of China (Grant No. 50774029), Innovation Scientists and Technicians Troop Construction Projects of Henan Province, and Program for Innovative Research Team (in Science and Technology) at University of Henan Province (13IRTSTHN003).

References
[1] Furuhara T, Yamaguchi J, Maki T 2004 *Iron Steel Institute Japan.* **17** 356
[2] Zhang ZL, Liu YL, Zhu JW, Yu G 2008 *Mater. Sci. Eng.***483-484** 64
[3] Maehara Y, Langdon T.G 1990 *Mater. Sci. Eng.* **128** 1
[4] Nieh T.G 1997 *Superplasticity in metals and ceramics* (New York: Cambridge University Press)
[5] Zhang KK, Sun J, Wu ZW, Ma L, Qiu RF, Zhang ZL 2011 *Transactions of the China Welding Institution.* **32** 5
[6] Zhang KK, Yang YL, Wang CS, Li HK, Xue J 2001 *J. Mater. Sci Technol.* **17** 189
[7] Zhang KK, Yue Y, Zhang ZL 2010 *Materials Science Forum.* **287** 177
[8] Zhang KK 2011 *The national natural science funding project concluding report-Grant No. 50774029* (Beijing: National Natural Science Foundation of China)

Identification of predominant interfacial contact mechanisms in solid-state bonding of superplastic high carbon steel

Zhonghao Heng[1], Masakatsu Maeda[2] and Yasuo Takahashi[2]

1. Graduate School of Engineering, Osaka University, 2-1 Yamadaoka, Suita, Osaka 565-0871, Japan
2. Joining and Welding Research Institute, Osaka University, Japan

E-mail: takasy@jwri.osaka-u.ac.jp

Abstract. Interfacial contact mechanisms and the superplastic behaviour of fine-grained high carbon steel during solid-state bonding were investigated. According to the results, typical superplastic deformation occurred at the bonding interface under conditions of superplasticity-causing temperature (T = 1003~1053K), bonding pressure (P = 34~44.3MPa), and bonding area ratio (S = 30~50%). The stress exponent, n, was about 2.25 while the activation energy, Q, was about 189 kJmol^{-1}, both indicating and confirming superplastic solid-state bonding. Predominant bonding mechanism controlled by superplastic deformation was experimentally identified.

1. Introduction

Superplastic materials exhibit high deformability and strong diffusion properties under appropriate conditions of temperature and pressure [1]. One of the important properties of fine-grained superplastic materials is that they can often be readily bonded in the solid state, and this is particularly true for metals with poor weldability [2]. For instance, concurrent superplastic forming and diffusion bonding (SPF/DB) has been recognized as a viable and highly economical manufacturing technology in the aviation industry.

Current researches on solid-state bonding of superplastic materials primarily focus on the optimization of diffusion bonding parameters [3-5]. Although there is no doubt that superplastic flow can facilitate the early stage of bonding process, the predominant bonding mechanisms, including the effect of superplasticity on enhancement of interfacial contact, are not fully understood yet. The role of superplastic deformation therefore should be clarified to maximize the benefits of its application.

Previous experimental researches on conventional solid-state bonding have elucidated fundamental mechanisms of void shrinkage at the bond interfaces [6-8]. In the present study, the predominant bonding mechanism of interfacial contact process was investigated on the basis of the model proposed by Takahashi [6]. The effect of superplastic flow during solid-state bonding was also studied. This research is of great value in clarifying the predominant bonding mechanisms during different bonding stages and in optimizing the bonding parameters of superplastic materials.

2. Identification method

A new method is required to identify the predominant interfacial contact mechanisms in superplastic materials. Referring to an existing theoretical model [6], Figure 1(a) schematically illustrates the solid-

Content from this work may be used under the terms of the Creative Commons Attribution 3.0 licence. Any further distribution of this work must maintain attribution to the author(s) and the title of the work, journal citation and DOI.
Published under licence by IOP Publishing Ltd

state bonding test of superplastic and non-superplastic metals. The bonding surface of the non-superplastic hard metal is perfectly flat while that of superplastic metal is machined to concentric circular ridges (interval of $2L_0$). The voids shrinkage V_s ($= V_2 - V_1$) in the period of t_s ($= t_2 - t_1$) is shown in Figure 1(b). The geometrical parameters X, w, and h in Figure 1 are, respectively, half of the bonded length of a unit, half the void width, and the void height. The subscript 0 indicates initial value of the respective parameters.

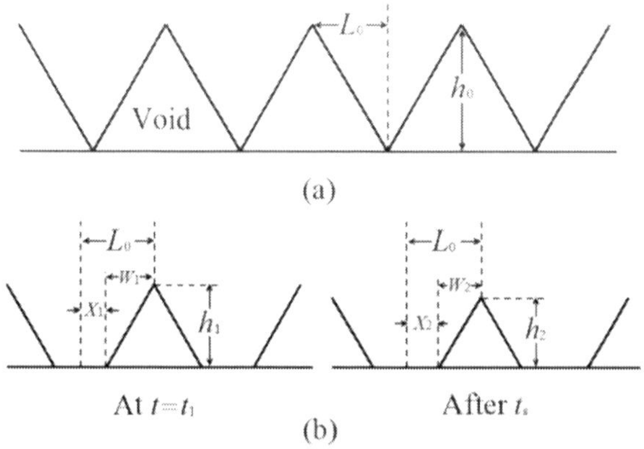

Figure 1. Schematic illustrations for superplastic solid-state bonding. (a) voids at the bonding interface. (b) interfacial contact process.

For solid-state bonding, the interface self-diffusion and volume self-diffusion are predominant when the bonding pressure, P, is low enough. On the other hand, high-temperature deformation becomes more important as P increases [9]. In regard to the bonding conditions used in the present study, the core discussion is focused on the interfacial deformation, that is, interfacial contact controlled by diffusion has been excluded. As the interfacial contact processes occur under conditions of high vacuum and high temperature, effect of surface oxide film were ignored. Interfacial deformation mainly occurs on the side of the superplastic metal under identical bonding conditions, that is, the plastic deformation of the non-superplastic metal need not be considered. Initial contact is made through instantaneous plastic deformation in the early stage of bonding process, in general, when the bonding ratio is less than 20%. In the present study, the set value X_1 is greater than the critical value X_c given by the instantaneous plastic flow, thus, the contribution of instantaneous plastic deformation to the interfacial contact can be ruled out.

Referring to the identification model of void shrinkage due to creep proposed by Takahashi [6], the relation between bonded area growth ΔX and t_s is expressed as

$$\Delta X = \frac{A_c}{kT} \cdot \left(\frac{P}{G}\right)^n \exp\left(-\frac{Q}{RT}\right) \cdot M\{F(L, X, n)\} \cdot t_s, \quad (1)$$

where A_c is a constant when the bonding temperature T is kept constant, P is the bonding pressure, G is the shear modulus, n is the stress exponent, $F(L, X, n)$ is a geometrical function of L, X, and n, Q is the activation energy for creep. ΔV is roughly proportional to ΔX. If ΔX is kept constant, M $\{F(L, X, n)\}$ also becomes constant, and the following relations are expressed as

$$t_s \propto \left(1/P\right)^n \quad (2)$$

when P is constant, and

$$\frac{T}{t_s} G^{n-1} \propto \exp\left(-\frac{Q}{RT}\right) \quad (3)$$

when T is designated properly and kept constant. So, the predominant mechanism during solid-state bonding can be identified by the gradients of $\log t_s$ vs $\log P$ and $\log (T/t_s)$ vs $(1/T)$ plots, that is, the gradients are the stress exponent, n, and the activation energy, Q, respectively. The n value is about 4~6 when power law creep is predominant, whereas the n value is 2~3 and the Q value is close to the interface self-diffusion activation energy Q_b when the superplastic flow is predominant during solid-state bonding [2, 6].

3. Experimental

The material used in the present study was a high carbon steel, SK105. The chemical components and critical points (Ac_1: eutectoid transformation temperature, Acm: temperature of proeutectoid cementite precipitation) are shown in Table 1.

The original steel with its coarse pearlite structure (grain size > 10μm) did not exhibit superplasticity. The superplastic steel with fine-tempered martensite (grain size: 5~8μm) was obtained after an ultra fine heat treatment: circular quenching from 1053K thrice, followed by tempering at 473K [10]. Microstructures of both steels are shown in Figure 2.

Table 1. Chemical components and critical points of SK105.

Steel	Basis element					Critical point [K]		
	C	Mn	Si	Cr	Ni	Ac_1	Ac_3	Acm
SK105	0.92	0.64	0.28	0.18	0.09	1003		1073

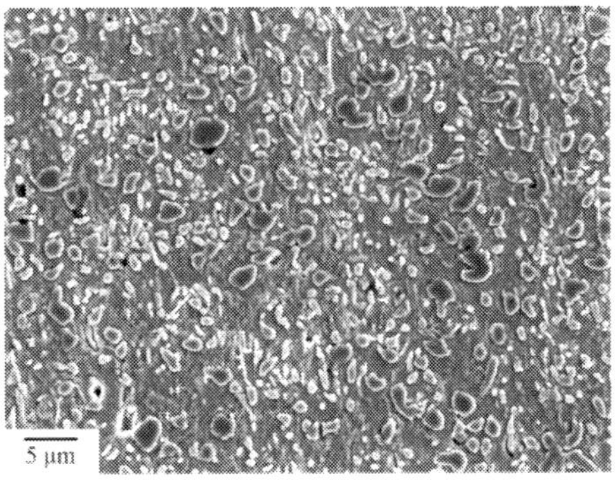

(a)

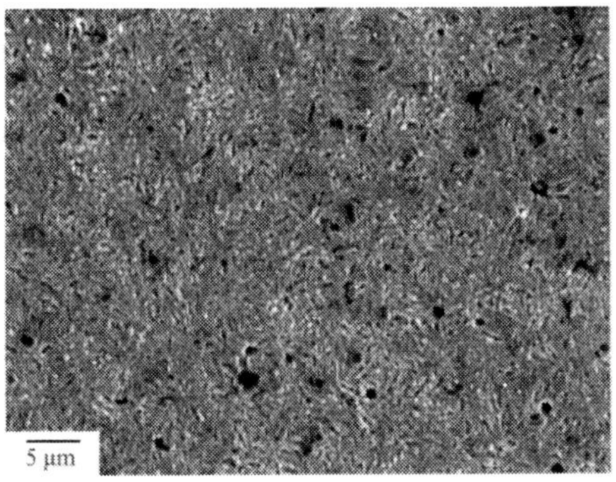

(b)

Figure 2. Microstructure of SK105. (a) pearlite in original steel. (b) tempered martensite in superplastic steel.

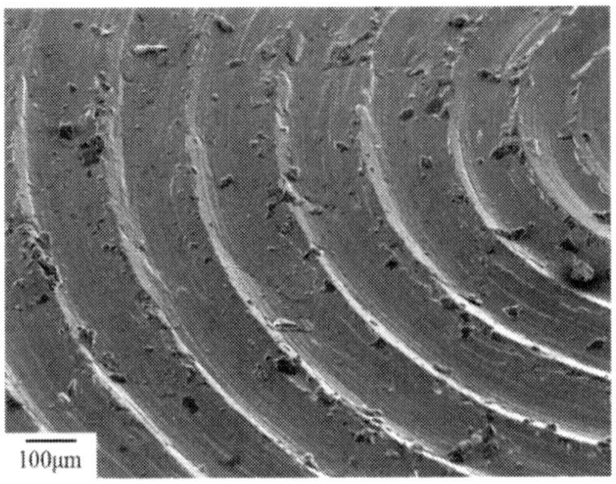

Figure 3. Bonding surface of superplastic steel.

The bonding surface of superplastic steel with concentric circular ridges was lathe-machined. As shown in Figure 3, the height was measured to be about 150 μm, and half the distance L_0 between the ridges was about 80μm. Bonding specimens were cylindrical in shaped of dimensions Φ10mm×12mm.

The bonding tests of the superplastic and original steel were carried out under several conditions for estimating the time t_s required to attain a certain increment ΔS (= $\Delta X/L_0$) in percentage bonded area. The vacuum was 10^{-4} Pa and the bonding temperature was in the range of SK105 superplastic temperatures (1003~1053K) [11]. The range of bonding pressure was 25~60MPa. The pressure was applied in the form of a uniaxial compressive stress. The specimens were heated using a molybdenum heater and cooled after the bonding pressure was removed.

Choosing an appropriate ΔS is necessary for identifying the predominant mechanisms in the middle stage of solid-state bonding. The change in ΔS during cooling was ignored. The estimation of bonded area was inaccurate if the set value of ΔS was less than 10%. Therefore, ΔS = 20% was adopted in the present study. Also, the bonded area percentage S_1 was set at 30% to rule out the influence of instantaneous plastic deformation. The bonded area increment was estimated by the scanning electron microscope (SEM) observation of the fractured surface after bonding. It was difficult to get the time t_s directly, therefore, t_s was estimated by interpolating the data of several bonding tests. Three repetitions of bonding tests under each condition were carried out and the mean value obtained were used to ensure the accuracy of ΔS estimation.

4. Results and discussion

The joints of superplastic and original steel were broken in liquid nitrogen, so, the interfacial deformation that occurred during breaking was ignored. Figure 4 shows the bonding process and surface fractures at the bonded interface in different stages. The shrinkage of voids is relatively uniform. The rugged and bright areas are the bonded zones; the dark areas are the void surfaces. When the bonding time is 0 s, the initial state, S, is 0%. S_1 = 20% is set as the stage 1, for a bonding time of t_1. When the bonding time $t_2 = t_1+t_s$, the bonded area ratio increment ΔS is set to 20% at this stage.

The bonding pressure dependence of t_s at 1023K is shown in Figure 5. From equation (2), the stress exponent n value is calculated to be approximately 5.74 at low bonding pressure (P < 30MPa), therefore, the power law creep is the predominant mechanism. This result is also consistent with the value of n=6 provided by Takahashi [6]. The n value decreases gradually as the pressure increases, as a result of a change of the predominant mechanism of interfacial contact. The n value is about 2.3 for the bonding pressure P = 37~43MPa. The strain rate sensitivity index, m (the reciprocal of n value), is

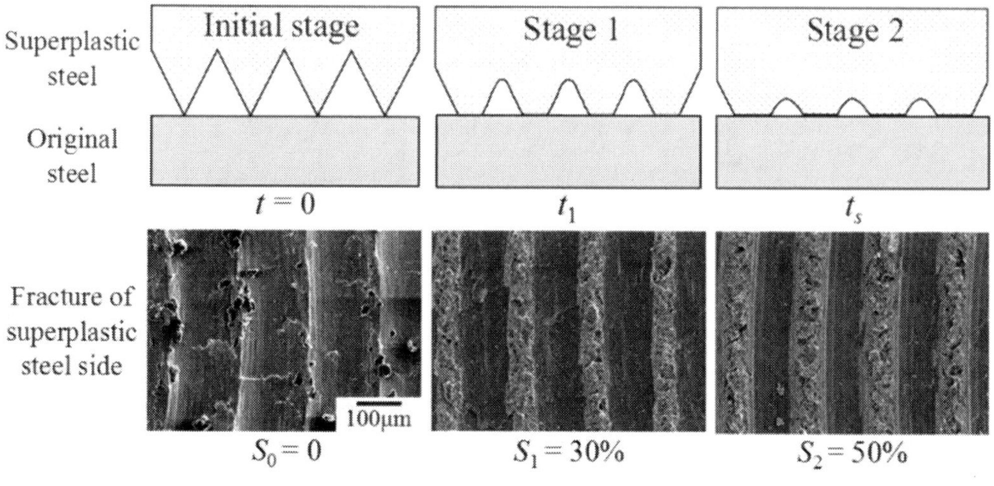

Figure 4. SEM observations of fractured surfaces in different stages.

about 0.43. This suggests that superplastic deformation predominantly contributes to the interfacial contact process. In other words, superplastic flow-based deformation occurs at the bonding interface at this stage. The n value increases with the further increase in bonding pressure and is larger than 3 when $P > 43$MPa, illustrating the change in the predominant bonding mechanism. At such pressure, the creep mechanism becomes increasingly influential and the effect of superplastic flow is weakened significantly. The predominant mechanism thus in effect comes back to exhibit power law creep.

If the n value lies between the values associated with these two fundamental mechanisms, a combined effect of the two mechanisms can be expected to occur. By extending the characteristic curves of the different mechanisms, the intersection point is obtained. This point can be considered as a transition point of these two fundamental mechanisms. Thus, the bonding pressure range can be divided into three regions: P = 34~44.3MPa when the surperplastic flow is predominant; P >44.3MPa or P < 34MPa when the power law creep is predominant. The predominant mechanisms during the bonding process can be identified thus.

This result, P = 34~44.3MPa, also correlates well to the superplastic flow stress range of SK105 (35~45MPa) obtained from uniaxial compression tests [10]. It should be noted that the pressure applied to the bonding interface is larger than that of the base metal. However the compression deformation at the bonding interface is different from that of uniaxial compression, as the interface tangential friction can have an impact on the interfacial contact process. This agreement of pressure is for reference only. The contribution of stress distribution to interfacial deformation needs further analysis.

Figure 6 shows the temperature dependence of T/t_s under 43MPa. The Q value, measured by the log (T/t_s) vs ($1/T$) plot, is about 190 kJmol^{-1} at T = 1003~1023K and 200 kJmol^{-1} at T = 1023~1053K. The Q value increases slightly with increasing T, and that is due to the change in superplasticity at different temperatures. Fine-grained SK105 steel exhibits superplastic behavior when T is in the proximity of the critical temperature Ac_1. Its microstructure is comprised of austenite, γ, and cementite phases when T is slightly higher than Ac_1. This dual-phase fine-grained steel shows superplasticity well. With increasing T, more carbides are dissolved, and the limiting effect of second phases (fine carbide) on grain growth is weakened, resulting in concomitantly reduced stability of grain size. Consequently, superplasticity relatively weakens while the Q value increases slightly.

The Q value during solid-state bonding (P = 43MPa, T= 1003~1053K) is between the activation

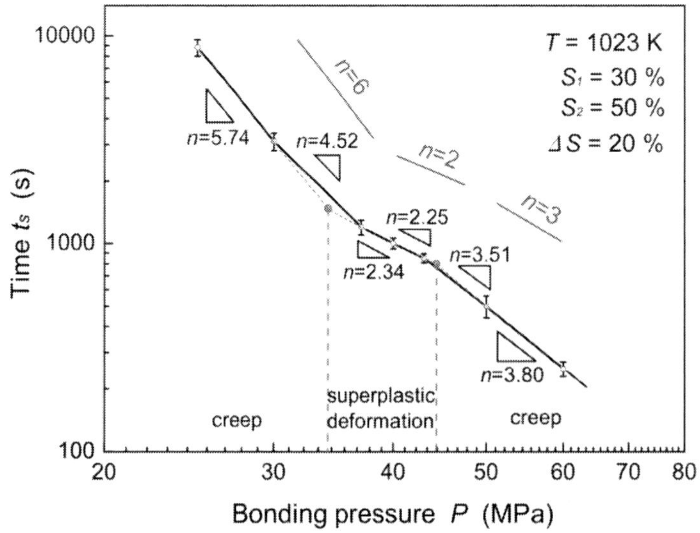

Figure 5. Bonding pressure dependence of t_s.

energy of γ-Fe interfacial self-diffusion (Q_b =159 kJmol^{-1}) and that of volume self diffusion (Q_v =270 kJmol^{-1}) [12]. Additionally, its value is comparatively closer to Q_b. This is reliable evidence of the predominance of superplastic deformation, a grain boundary phenomenon, as the interfacial contact process. The most commonly considered mechanisms for superplastic flow involve grain boundary sliding (GBS) and an accommodation process accompanying GBS [3, 13, 14]. Therefore, the activation energy for superplastic deformation should be equal to that of grain boundary diffusion. This fact is well represented by the results of the present study.

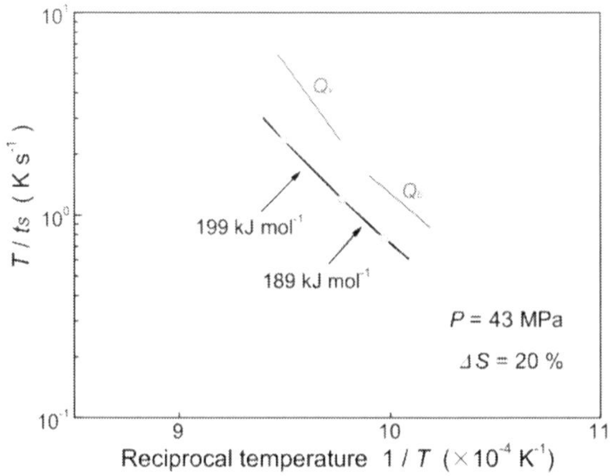

Figure 6. Temperature dependence of T/t_s.

5. Conclusions

It was found that surperplastic deformation takes a dominant role in the interfacial contact process from a bonding area ratio of 30% to 50%, under the conditions of bonding temperature T = 1003~1053K and bonding pressure P = 34~44.3MPa. The predominant mechanism controlled by superplastic flow can be identified using the value of n and Q. Superplastic deformation and its dominant effect are confirmed experimentally. Superplastic bonding was produced under conditions of T= 1003~1053K and P = 34~44.3MPa.

References

[1] Langdon TG 1991 *ICSAM-1991*. 3
[2] Nieh TG 1997 *Superplasticity in metals and ceramics* (New York: Cambridge University Press)
[3] Maehara Y, Langdon TG 1990 *Mater. Sci. Eng.* **128** 1
[4] Huang Y, Ridley N, Humphreys FJ and Cui JZ 1999 *Mater. Sci. Eng.* **266** 295
[5] Somekawa H, Hosokawa H, Watanabe H and Higashi K 2003 *Mater. Sci. Eng.* **339** 328
[6] Takahashi Y, Inoue K and Nishiguchi K 1993 *Acta metall.* **41** 3077
[7] Takahashi Y and Inoue K 1992 *Mater. Sci. Tech.* **8** 953
[8] Hill A and Wallach R 1989 *Acta metal.* **37** 2425
[9] Takahashi Y 2003 *Ceramic trans.* **138** 29
[10] Zhang KK, Yang YL, Liu SZ, Han CX and Xu D 2004 *Acta metall sinica.* **16** 538
[11] Zhang KK, Wang YL, Shi HX, Yu H and Liu S 2009 *Mater. Sci. Eng.* **499** 97
[12] James DW and Leak GM 1965 *Phil. Mag.* **12** 491
[13] Gifkins RC 1976 *Metall Trans.* **7** 1225
[14] Langdon TG 1970 *Phil. Mag.* **78** 689

Interfacial deformation and friction heating in ultrasonic Al ribbon bonding

Yasuo Takahashi[1], Masakatsu Maeda[1], Masaya Ando[2], and Eito Yamaguchi[2]
[1] Joining and Welding Research Institute, Osaka University, Japan
[2] Graduate student, Osaka University, Japan

E-mail: takasy@jwri.osaka-u.ac.jp

Abstract. The interfacial deformation and friction behavior between an Al ribbon and an electric pad (or substrate) during ultrasonic bonding is analyzed, based on numerical simulation and experimental results. The friction heating is estimated by the friction slip work at the bonding interface between the ribbon and pad. The temperature rise of the bonding interface is calculated by the numerical simulation and compared with the experimental results. It is suggested that the electric pad reduces the temperature rise, as compared to the bonding process without a pad. The shear stress at the bonding interface increases as the bonding progresses. The frictional slip due to adhesion increases stress and heats the bond interface.

1. Introduction

Ultrasonic Al ribbon bonding has been applied to power electronics packaging. The ultrasonic vibration applied to the bonding interface produces frictional slip between the Al ribbon and the substrate (or the Al electric pad on the substrate). The frictional slip work heats the bonding interface and causes adhesion at the interface [1, 2]. Because the friction force originates in local interface adhesion, the friction force increases as the interface adhesion increases. In other words, even if poor adhesion occurs, frictional slip can continue. As the adhesion area grows and strengthens the frictional slip will be reduced. If the value of the frictional coefficient is assumed, we can no longer debate the nature of the bonding process (adhesion evolution) because the frictional coefficient is derived from adhesion. Therefore, the vibration amplitude should be given as a boundary condition [3]. The purpose of the present study is to understand the temperature change of the bonding interface and the adhesion evolution during Al ribbon bonding on the basis of experimental and calculated results.

2. Numerical simulation and experimental procedure

A two-dimensional finite element model was adopted [3], as illustrated in Figure 1. The thickness and width of the electric pad were, respectively, 2 μm and 200 μm. The thickness and width of the Al ribbon were 50 μm and 100 μm, respectively. The initial contact width was 100 μm. The deformations of the ribbon and the pad were calculated using the finite element method [3]. The Al ribbon was separated from the electric pad at planes AB and ML. The bonding interface was positioned between planes AB and ML. It was possible for plane AB (the lower surface of the ribbon) to slip on plane ML (the upper surface of the pad). It was assumed that the shear stress τ_{xy} in the x direction (parallel to the bonding interface) and the displacement rate in the y direction were the same at the bonding interface between the lower surface of the ribbon and the top of the pad, i.e., $f_{xr} = f_{xp}$ and $(du/dt)_{yr} = (du/dt)_{yp}$, where f_{xr} and f_{xp} are the nodal forces in the x direction of the lower surface of the ribbon and of the top

Content from this work may be used under the terms of the Creative Commons Attribution 3.0 licence. Any further distribution of this work must maintain attribution to the author(s) and the title of the work, journal citation and DOI.
Published under licence by IOP Publishing Ltd

of the pad, respectively. Also, $(du/dt)_{yr}$ is the displacement rate in the y direction of the nodal point on the lower ribbon surface and $(du/dt)_{yp}$ is that of the top of the pad, where u is the displacement and t is the time. The ultrasonic vibration frequency f_v was 60 kHz. The bonding pressure was 28 MPa. The ultrasonic vibration was applied for a period of 400 ms. The heat input was assumed to be given only by the work associated with the frictional slip at the bonding interface. It was assumed that the local frictional work Q_{in} was the product of the shear force multiplied by the differential displacement rate at the bonding interface, i.e.,

$$Q_{in} = \tau_{xy} \cdot S_b \cdot |(du/dt)_{xr} - (du/dt)_{xp}|, \quad (1)$$

where S_b is the element area on AB, $(du/dt)_{xr}$ is the displacement rate in the x direction of the node on the lower surface of the ribbon, and $(du/dt)_{xp}$ is that of the top of the pad. The value of $(du/dt)_{xr}$ - $(du/dt)_{xp}$ corresponds to the friction slip at each nodal point along AM. The heat loss from the surface was ignored. The heat flow in the z direction was also ignored because the length of the substrate in the z direction was much greater than the thickness of the substrate. The variations in the vibration amplitude A_o and the amplitude ratio R_d (= A_b/A_o) with the time t were given, as shown in Figure 2, taking into account the experimental results [4, 5], where A_b was the average vibration amplitude of the lower surface of the Al ribbon. The variation of A_o was estimated on the basis of the experimental results, but the variation of A_b was assumed from several previously observed patterns [4]. Figure 2 shows the representative patterns for A_b. As seen in Figure 2, the values were changed discretely, resulting in staircase-like profiles, so that the calculation could be simplified. In the first such step, R_d (t = 0–40 ms for Figure 2(a) and t = 0–100 ms for Figure 2(b)) was changed from 1.0 to 0.999. The duration of one vibrational cycle was divided into 8 time increments ($\Delta t = 2.083\times10^{-6}$ s) for calculating the plastic deformation. The heat input Q_{in} was given as an average value for each cycle. The heat flow was calculated using a time step of $\Delta t = 1\times10^{-10}$ s for a few tens of cycles, assuming that the average heat input was given at the interface. The temperature change with time was simulated as detailed

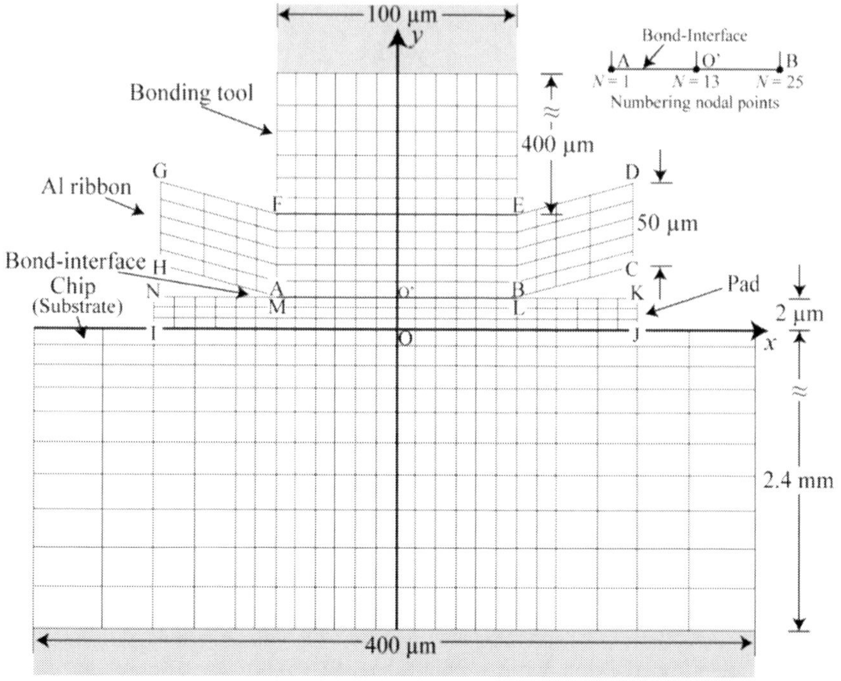

Figure 1. Mesh pattern in cross-section of the Al ribbon, Al pad, and silica (or Si) substrate. In the upper-right part of the figure is an explanation of the scheme used to number the nodal points on the lower surface of the ribbon (AO'B). The z direction is normal to the sheet.

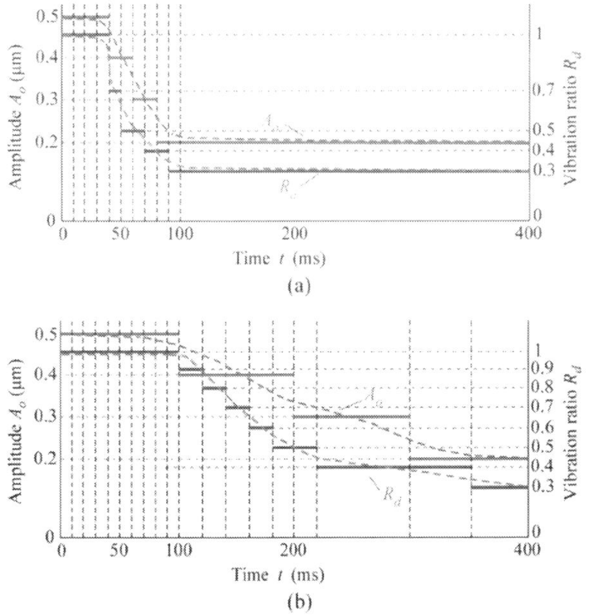

Figure 2. Variations in the vibration amplitude A_o and ratio R_d as boundary conditions. R_d was just an initial average value. The displacement rates on plane AB were changed so that the calculation could converge. (a) Representative vibration change when the Al ribbon is bonded directly to the silica and (b) that of the Al ribbon bonded to the Al pad on the chip (or SiO_2).

Table 1. Materials constants for viscoplastic deformation of Al ribbon and Al pad.

Name	Symbol	Al pad	Al ribbon	Unit
Frequency factor	D_0	0.1×10^{-4}	0.1×10^{-4}	m^2s^{-1}
Constant	A_0	2.28×10^{8}	2.14×10^{5}	
Constant	B	5.60×10^{2}	5.60×10^{2}	
Constant	C	1.00×10^{2}	1.00×10^{2}	
Burgers vector	b	2.86×10^{-10}	2.86×10^{-10}	m
Stress exponent	n	4.4	6.4	
Activation energy	Q	150.7	75.4	kJ mol^{-1}

elsewhere [3] and was compared with the experimental results. The temperature was measured by a micro-thermocouple embedded in the bonding interface. The experimental parameters were $F = 7.0$ N, $P_u = 4.0$ W, and $f_v = 60$ kHz, where F was the bonding force and P_u was the nominal ultrasonic power. The bonding pressure P_B was about 28 MPa because the tool width was 0.25 mm and the thickness was 0.1 mm [3]. The ultrasonic vibration was applied to the specimen for a period of 400 ms. Table 1 shows the materials constants of the Al ribbon and the Al pad for strain-rate-sensitive plastic deformation [1, 3, 6]. The strain rate equation is taken from [4]. The thermal conductivity was assumed to be the same between the Al ribbon and the Al pad. The thermal materials constants were taken from literature [3].

3. Results and discussion

Figure 3 shows the distributions of the displacement rates (calculated results) of the Al ribbon and Al pad (a) for the free-slip condition without vibration and (b) for the frictional-slip condition. The thickness of the pad is magnified by a factor of 10. Ultrasonic vibration enlarges the displacement in the x direction, but the displacement rate of the pad is much smaller than that of the ribbon. As seen in Figure 3, the displacement rate is discontinued at the bonding interface. Interfacial slipping occurs. If

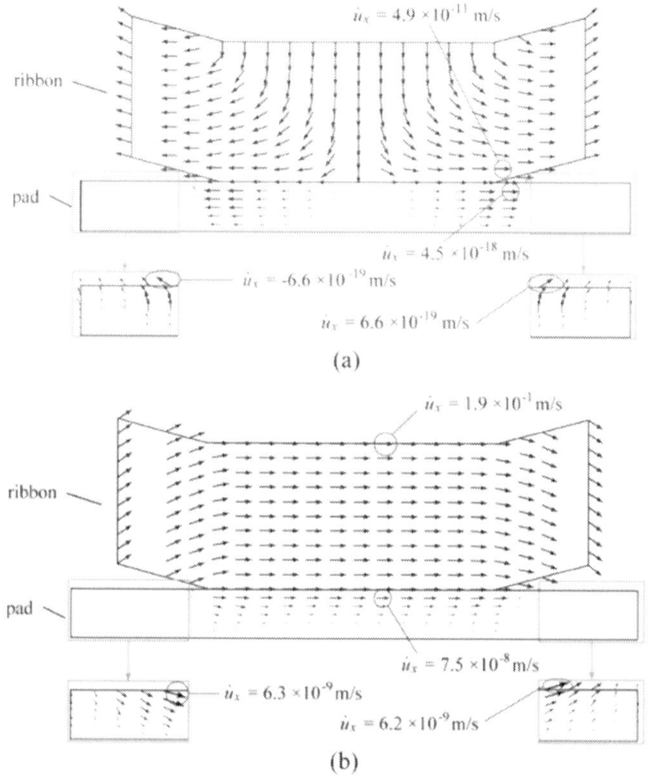

Figure 3. Distributions of displacement rate vectors during ultrasonic bonding at T = 373 K. (a) free slipping without vibration and (b) with ultrasonic vibration (A_o = 0.5 μm, A_b = 0.4995 μm, R_d = 0.999).

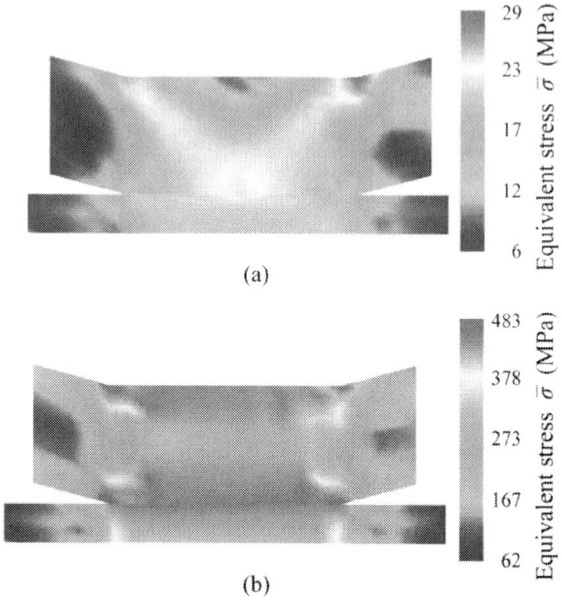

Figure 4. Equivalent stress distributions of the ribbon and the pad at T = 373 K.
(a) Free slip without vibration and (b) with ultrasonic vibration (A_o = 0.5 μm, R_d = 0.9).

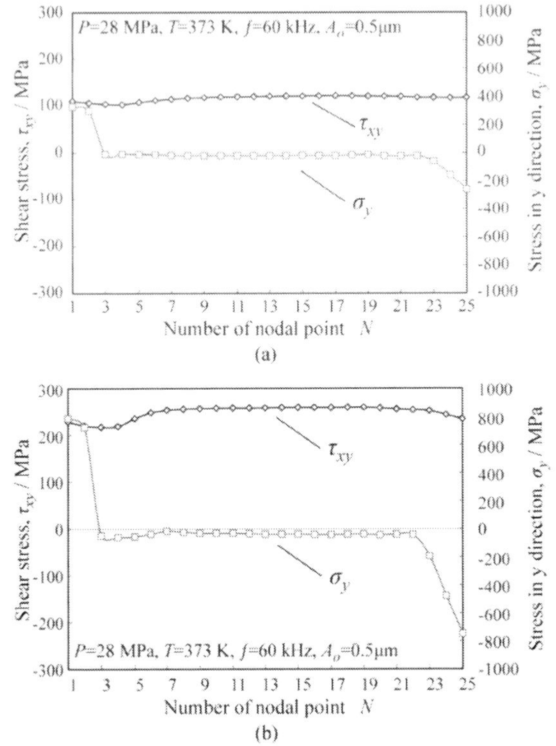

Figure 5. Stress distributions at the bonding interface (ribbon side). (a) $R_d = 0.999$ and (b) $R_d = 0.9$.

the vibration is applied toward the right-hand side, as shown in Figure 3(b), the direction of the displacement at the pad tilts downward along the back side and upward in the front. On the other hand, that of the ribbon tilts downward in the front. Surface folding is produced in the front side, and separation tends to occur in the back side. The displacement rate of the ribbon is much greater than that of the pad.

Figure 4 shows equivalent stress distributions of the ribbon and the pad. Figure 4 (a) shows the case without vibration (the free-slip condition), and Figure 4 (b) shows the case with an ultrasonic vibration of $A_o = 0.5$ μm. If the bonding interface can slide freely, the equivalent stress at the bond-interface is not enhanced at all and not more than the static bonding pressure (28 MPa). However, as seen in Figure 4 (b), the equivalent stress is strikingly enhanced by the ultrasonic vibration even if the vibration ratio R_d is 0.9. It is found that the plastic deformation was extended to the ribbon and pad by the ultrasonic vibration. The stress enhancement is affected by the vibration ratio R_d.

Figure 5 shows the distributions of the shear stress τ_{xy} and the stress in the y direction σ_y at the bonding interface (ribbon side). The abscissa is the numbers of the nodal points N at plane AO'B (the lower surface of the ribbon), as shown in Figure 1. $N = 1$ maps to the nodal point A, $N = 13$ maps to the point O', and $N = 25$ maps to the point B. Figure 5 shows the case when the shear force of the ultrasonic vibration is applied from left to right at the ribbon's top side. As seen in Figure 5, even if $R_d = 0.999$, the shear stress is enhanced, and friction begins to occur. The shear stress is almost uniform because the displacement rate is nearly uniform, but the normal stress is tensile on the left-hand side and compressive on the right-hand side. The friction coefficient becomes very high under ultrasonic vibration even if R_d is between 0.9 and 0.999. Even move the ribbon a little across the adhesive region produces a strong friction force.

Figure 6 shows the evolution of the shear stress at the bonding interface. Figure 6(a) shows the case when the Al ribbon is bonded to silica without pads. Figure 6(b) shows the case when the Al pad exists

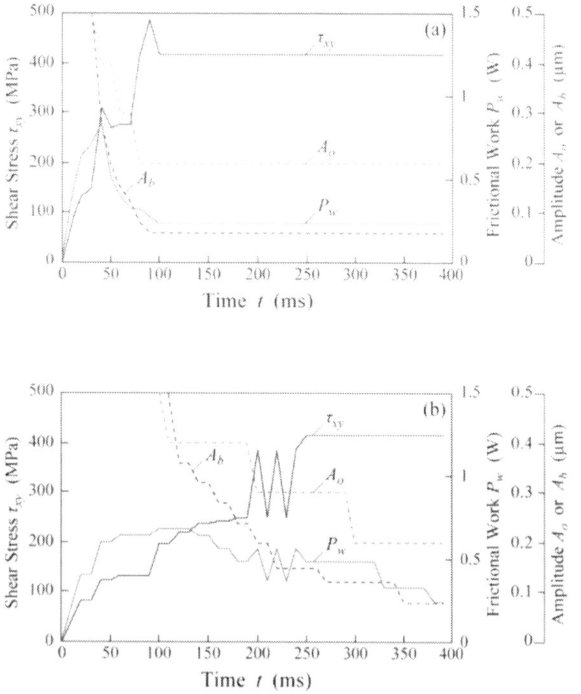

Figure 6. Change in shear stress and frictional work at the bond interface (calculated results) (a) substrate without pads and (b) substrate with pads.

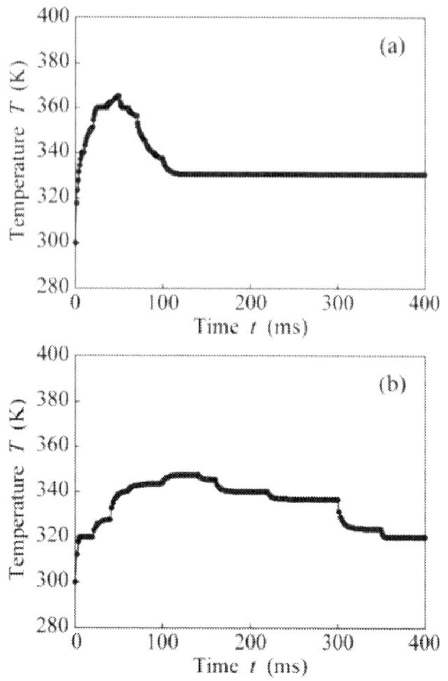

Figure 7. Change in temperature at the bond interface. (a) for a substrate without pads and (b) for a substrate with pads.

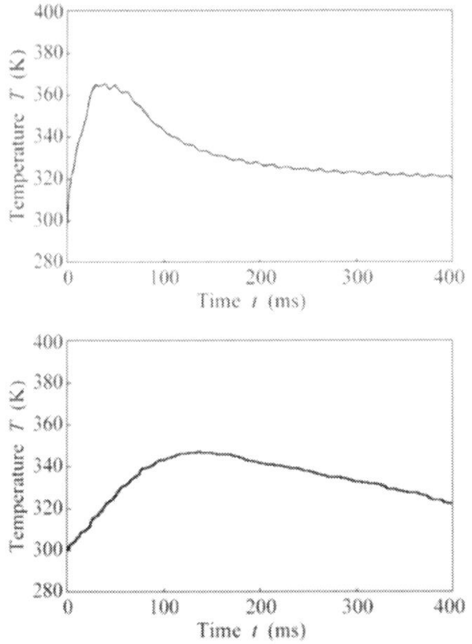

Figure 8. Temperature change at the bond interface with time (a) for a SiO_2 substrate without pads and (b) for the substrate with pads.

on the substrate. The frictional work P_w was computed by summing ΣQ_{in} over the whole bond interface (100 μm × 1 mm). The changes in P_w and τ_{xy} with time are naturally affected by the changes in A_o and R_d, which differ depending on whether the pad exists on the chip. As seen in Figure 6, the frictional stress increases rapidly for the bare chip without pads. The heat input is also large at the start of the bonding process. The adhesion is easily formed by the frictional heating caused by slipping. The evolution of the shear stress is more striking in bare-chip bonding. The shear stress of 400 MPa is much greater than the tensile strength of pure aluminum (≈ 80 MPa). This is due to ultrasonic vibration (high-speed displacement rate) [5]. On the other hand, if the Al pad exists on the chip (substrate), the adhesion is not so easy to form. The vibration amplitude does not decrease so rapidly because the friction is not enhanced, i.e., the heat input is not so large in the beginning. Therefore, the shear force is not so rapidly increased. These behaviors must have an influence on the increasing temperature of the bonding interface.

Figure 7 shows the variation of the interface temperature with time (calculated results). Figure 7(a) shows the temperature change for a chip without pads, and Figure 7(b) represents that of a chip with pads. Because the heat input is large in the beginning of the bonding without pads, the interface temperature rises rapidly, and the temperature begins to drop at $t \approx 50$ ms, as seen in Figure 7(a), but the interface is not cooled down to the room temperature because the ultrasonic vibration is applied for t = 100–400 ms. On the other hand, as seen in Figure 7(b), the interface temperature increases moderately with a gentle slope, and the maximum temperature is not greater than 350 K. After the highest temperature is reached, the temperature decreases more gradually than in Figure 7(a), although the slope has discontinuous points because of the previously mentioned simplification of the numerical simulation. Figure 7 suggests that electric pads can inhibit the increasing of the temperature of the bond interface and yield a moderate temperature change.

Figure 8 shows the experimental results of temperature change at the bond interface with time. Figure 8(a) is for a SiO_2 substrate without pads, and Figure 8(b) is for SiO_2 with pads. If the pads exist on the chip (substrate), the temperature change becomes moderate. The maximum temperature without pads is higher than that with pads. The tendency of the experimental results resembles that of the calculated results in Figure 7. The moderate temperature change means that the frictional work is performed gradually. On the other hand, the rapid temperature increase is due to the large frictional work (heating) produced in the beginning of the bonding process. The existence of the pad restrains the frictional work. The frictional behaviors naturally depend on the surface situation from the view of tribology. It is not easy to determine the change of R_d. Although several patterns were assumed for the ratio R_d, the results can still roughly predict the change of temperature rising and the stress evolution. These properties are very important for understanding the ultrasonic bonding mechanism.

4. Conclusion

Interface deformation and frictional heating have been discussed to understand the increase of the temperature during ultrasonic Al ribbon bonding because the friction coefficient is intimately related with the solid-state adhering behavior. In the present study, the frictional heating behavior was analyzed by changing the ultrasonic vibration amplitude. The ultrasonic vibration behavior largely depends on the existence of the pads. The existence of the pads restrains the evolution of the frictional force (shear stress) and tends to reduce the increase of the temperature. Even if the pad exists on the chip (substrate), the shear stress becomes fifteen times greater than the stress level under only the bonding force condition without ultrasonic vibration. The stress evolution causes a frictional heating, resulting in the increase of the interface temperature.

References

[1] Maeda M and Takahashi Y 2013 *Int. J. Nanotechnol.* **10** 89
[2] Maeda M, Kitamori S and Takahashi Y 2013 *Sci. Technol. Weld. Join.* **18** 103
[3] Maeda M, Yoneshima Y, Kitamura H, Yamane K and Takahashi Y 2013 *Mater. Trans.* **54** 916
[4] Takahashi Y, Suzuki S, Ohyama Y and Maeda M 2012 *J. Phys. Conf. Ser.* **379** 012028
[5] Ando M, Maeda M and Takahashi Y 2013 *Mater. Trans.* **54** 911
[6] Takahashi Y and Inoue M 2002 *ASME J. Electro. Pack.* **124** 27

Stress evolution during ultrasonic Al ribbon bonding

Masaya Ando[1], Kazumasa Takashima[1], Masakatsu Maeda[2] and Yasuo Takahashi[2]

[1] Graduate School of Engineering, Osaka University, 2-1 Yamadaoka, Suita, Osaka 565-0871, Japan
[2] Joining and Welding Research Institute, Osaka University, 11-1 Mihogaoka, Ibaraki, Osaka 567-0047, Japan

E-mail: takasy@jwri.osaka-u.ac.jp

Abstract. The present study reveals the stress distribution in the substrate during ultrasonic bonding. The deformations of the Si substrate, Al ribbon, and Al pad were numerically analyzed using a finite element method. Experimental observation of the interface using a high-speed video camera was also conducted to determine the actual interfacial slip amplitude. This amplitude becomes smaller than that of tool-tip with bonding time. It was suggested from the numerical simulations that frictional adhesion enhanced the friction force, resulting in an increase in the equivalent stress in the ribbon and pad. As a result, very large stresses occur in the substrate during ultrasonic bonding. These stresses evolve with the progress of ultrasonic bonding, i.e., frictional adhesion.

1. Introduction

Ultrasonic bonding of thick Al ribbon is necessary for packaging heavy-duty power electronics devices. However, the under-pad structures of the power electronics devices can be damaged by the large interfacial shear stresses that emerge during ultrasonic bonding [1]. To overcome this problem, it is important to understand and control stress evolution during the bonding process. The stress distribution in an Al ribbon and a pad electrode has been investigated by numerical simulation [2, 3], and this simulation has to be extended to the under-pad structure to understand its influence on failure behavior. The present study reports on the stress distribution in the substrate during ultrasonic bonding.

Many *in-situ* measurements and observations have been made in order to understand the phenomena present during ultrasonic bonding [4-6]. Vibration amplitude at the tool-tip is one of the important parameters related to the state of bonding [7-9]. However, there are no reports about actual interfacial slide. It is supposed that the vibration amplitude is different between the tool-tip and the interface. It is important to understand the origin of this amplitude difference because it is a significant parameter in numerical simulations. Another purpose of the present study is to determine the actual vibration amplitude at the bonding interface during ultrasonic bonding.

2. Experimental procedure

The interface between SiO_2 substrate and Al ribbon during bonding was observed with a high-speed video camera (HX-3, nac Image Technology, Japan). The ultrasonic bonding apparatus used in the present study is schematically illustrated in Figure 1. A transparent SiO_2 substrate was used to observe the interface from the underside of the device. The thickness and width of Al ribbon were 0.2 mm and

Content from this work may be used under the terms of the Creative Commons Attribution 3.0 licence. Any further distribution of this work must maintain attribution to the author(s) and the title of the work, journal citation and DOI.
Published under licence by IOP Publishing Ltd

1.0 mm, respectively. The thickness of the SiO_2 substrate was 2.0 mm. The bonding tool was made of tungsten carbide. The vibration direction was parallel to the longitudinal direction of the Al ribbon. The ultrasonic frequency f was 60 kHz. The bonding force F_b was 7.0 N, which was comparable to the bonding pressure P_b = 28 MPa in the numerical simulation. The ultrasonic power P_u was 3.0 W. The ultrasonic vibration was applied for 400 ms. The frame rate of the video camera was 5×10^5 fps, which is sufficient for tracking the actual motion of the ribbon at the interface during ultrasonic bonding. The video data was analyzed to determine the amplitude at the interface by tracking the motion of a certain point in the interface during bonding. Simultaneously, the vibration amplitude at the tip of the wedge tool was monitored with a laser-Doppler vibrometer.

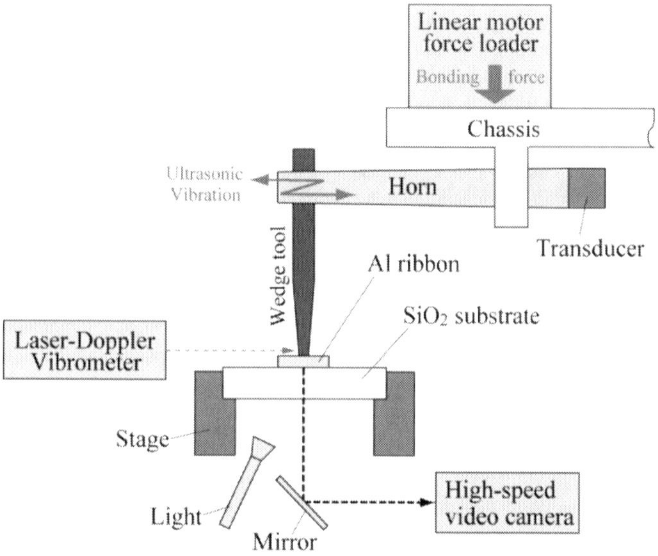

Figure 1. Schematic illustration of the ultrasonic bonding apparatus and measuring system.

3. Numerical procedure

The numerical analysis was performed using a finite element method. The mesh pattern used is illustrated in Figure 2. The thickness of Al ribbon was 200 μm. The thickness and width of the electric pad were 3–10 μm and 500 μm, respectively. The initial contact width was 250 μm. The ultrasonic frequency f was 60 kHz, equal to that in the experimental condition. The bonding pressure P_b was 28 MPa.

The temperatures of the ribbon and pad change with bonding time because of friction heating at the interface, and this temperature change should be considered. However, the numerical calculation will be a very hard task. So, the temperatures of the Al ribbon and the Al pad were assumed to be approximately constant at 373 K in the numerical simulations. This assumption is appropriate for estimating the stress evolution of the substrate during the ultrasonic bonding process and is based on the previous numerical and experimental study [2, 3]. Also, the substrate under the pad will be heated by friction. However, it was assumed that materials constants of the SiO_2 substrate were almost constant in the temperature range of 273–373K. That is, the temperature of the SiO_2 substrate was fixed at 293 K, and we used the elastic constants of SiO_2 at 293 K for the substrate.

Because frictional slip occurs at the interface between the ribbon and pad, the deformations of the ribbon and pad were calculated separately as in a two-body problem [3]. It was assumed that the Al ribbon and Al pad were visco-plastic bodies and that the Si substrate was a perfectly elastic body. The visco-plastic deformation of the Al ribbon and Al pad were calculated according to the previous

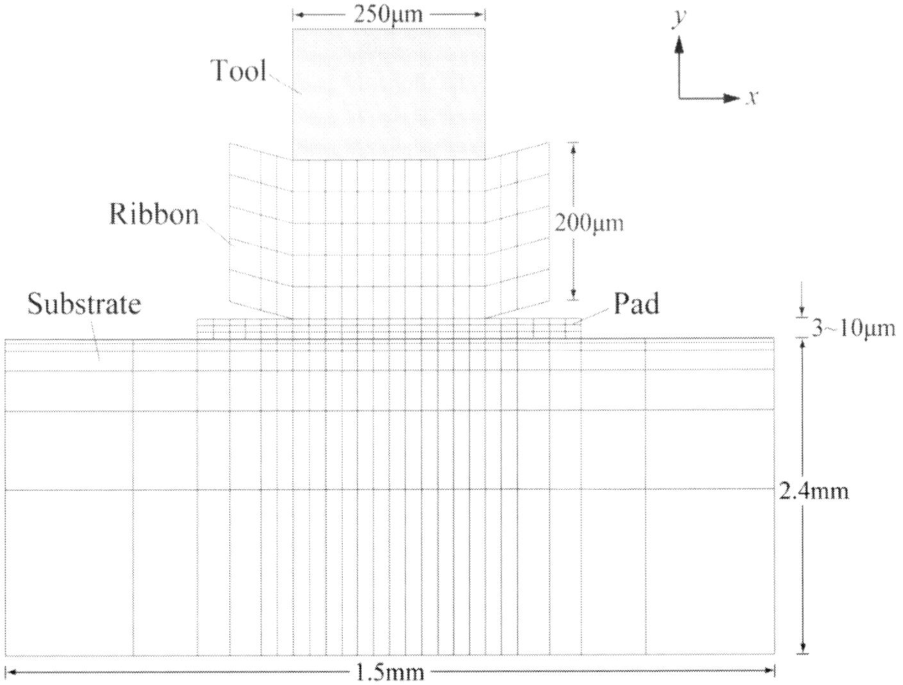

Figure 2. Mesh pattern in cross-section of the tool, Al ribbon, Al pad, and substrate.

studies [2, 3]. It was also assumed that the interface between the pad and SiO2 substrate was fixed without any frictional slip.

The detailed procedure for this calculation is in [2, 3]. However, the size of this model is different from that of the previous study. In the present study, the model is enlarged to fit the experimental size. The vibration amplitude A_o and the amplitude at the bottom of the ribbon A_b was changed based on experimental results. These amplitudes are important parameters which show the state of the bonding interface. The decrease in A_b represents the fact that the constraint on the interface becomes stronger as adhesion progresses.

The elastic deformation of the substrate was calculated using CAE software (Sanseikai, Japan). It was assumed that the force was balanced between the bottom of the pad and the top of the substrate, i.e., the nodal force of pad-bottom was equal to that of substrate-top. The material constants of SiO_2 substrate were taken from the literature [10]. The change in stress in the substrate with time was simulated and compared with the experimental results. The shear stress applied at the interface between the Al ribbon and the SiO_2 substrate was measured using a piezoelectric load cell. The detailed monitoring procedure is described in [1]. Experimental condition was F_b = 7.0 N, P_u = 2.5 W, f = 60 kHz. The ultrasonic vibration was applied for 400 ms.

4. Results and discussion

Figure 3 shows the experimental results for the amplitudes of both the tool-tip A_t and bonding interface A_i over the range of bonding times t of 1–100 ms. The bonding condition was F_b = 7.0 N, P_u = 3.0 W. A_i could not be determined after t = 100 ms because the motion was too small to track given the resolution of the video camera. At t = 1 and 5 ms, A_i is slightly larger than A_t , but this may be because of the measuring accuracy of A_i. It is supposed that the ribbon slides on the substrate almost freely, i.e., $A_i \approx A_t$, at the beginning of bonding. After t = 10 ms, A_t begins to decrease with bonding time. On the other hand, A_i becomes smaller than A_t and the difference between A_i and A_t increases with progress of the bonding process. *In-situ* observations indicate that adhered areas at the interface expand with bonding time [11], and the Al ribbon becomes increasingly constrained.

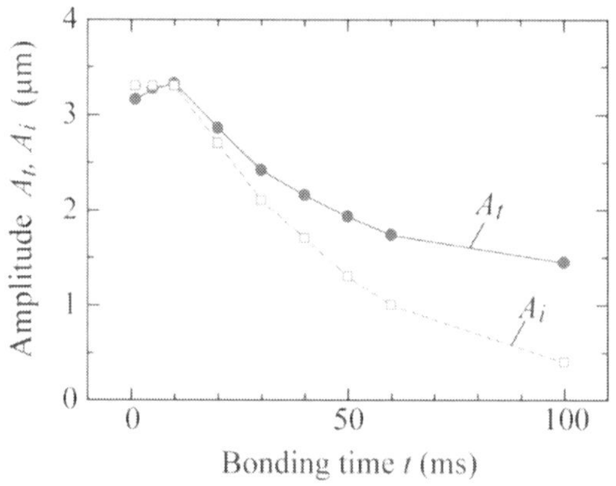

Figure 3. Experimental results for the amplitudes of the tool-tip and bonding interface vs. bonding time. The bonding condition was F_b = 7.0 N, P_u = 3.0 W.

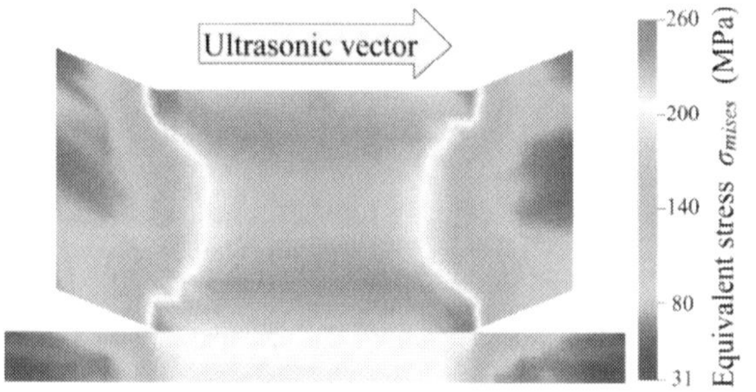

Figure 4. Equivalent stress distribution of the ribbon and pad. The parameters for the calculation are A_o = 3.5 μm, A_b =3.4965 μm, T = 373 K, and t_p = 4 μm.

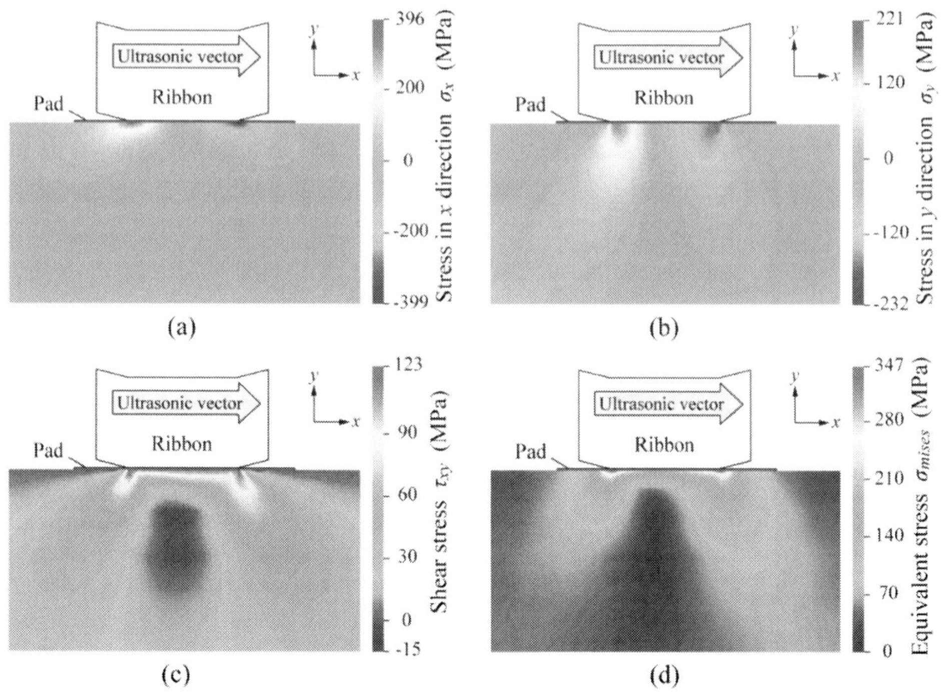

Figure 5. Distribution of (a) σ_x, (b) σ_y, (c) τ_{xy}, and (d) the equivalent stress σ_{mises} in the substrate. The calculations were performed using A_o = 3.5 μm, A_b =3.4965 μm, T = 298 K and t_p = 4 μm.

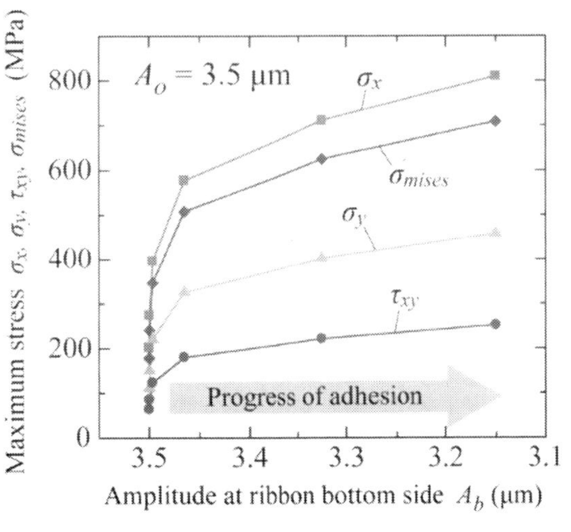

Figure 6. Change in the maximum stresses in substrate σ_x, σ_y, τ_{xy}, and equivalent stress σ_{mises} with the progression of adhesion. The decrease in A_b represents that the constraint of the interface increasing with the progress of adhesion.

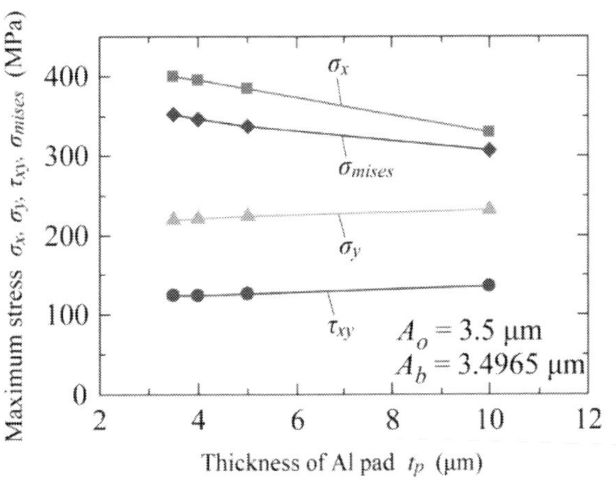

Figure 7. Change of the maximum stress in the substrate σ_x, σ_y, τ_{xy}, and the equivalent stress σ_{mises} depending on the thickness of Al pad.

Figure 4 shows the equivalent stress distribution of the ribbon and pad. The parameters for the calculation were A_o = 3.5 μm, A_b =3.4965 μm, temperature T = 373 K, and thickness of Al pad t_p = 4 μm. The thickness of the pad has been magnified 10 times for illustration purposes. This calculated distribution results when the ultrasonic vibration is applied from the left- to the right-hand side. A_b is 0.999 times as small as A_o. A large equivalent stress develops in the ribbon and pad even though A_b is just slightly smaller than A_o. Figure 5 shows the distribution of (a) the stress in x-direction σ_x, (b) the stress in y-direction σ_y, (c) the shear stress τ_{xy}, and (d) the equivalent stress σ_{mises} in substrate. The calculation parameters were the same as in Figure 4. The x-direction is in parallel to the vibration direction, and the y-direction is perpendicular to x-direction. In Figures 5(a) and (b), positive values indicate tensile stresses and negative value indicate compressive stresses. σ_x emerges just under the edge of the interface; it is tensile at the left-hand side and compressive at the right-hand side. This tensile and compressive σ_x is caused by shear stress in the interface. The distribution of σ_y is similar to that of σ_x but it spreads in the y-direction. Although the ultrasonic vibration is applied in the x-direction, a moment is applied to the ribbon [3]; this moment causes tensile and compressive σ_y. As seen in Figure 5(c), a large τ_{xy} develops under the interface. The interfacial shear stress during bonding produces this large τ_{xy} in the substrate. The maximum σ_{mises} is approximately 347 MPa, which is considerably larger than the bonding pressure.

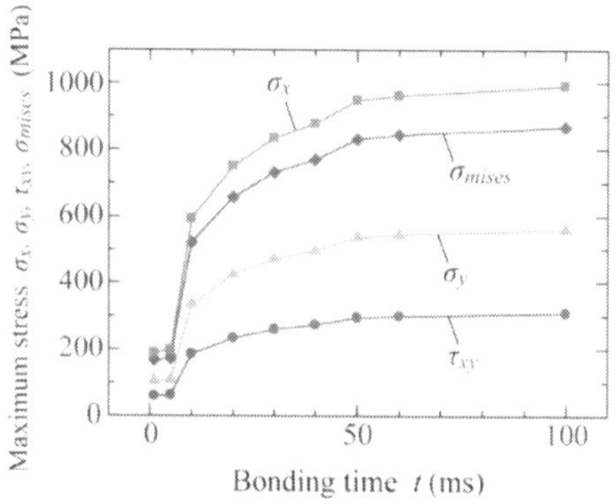

Figure 8. Change of maximum stresses in substrate σ_x, σ_y, τ_{xy} and equivalent stress σ_{mises} with bonding time.

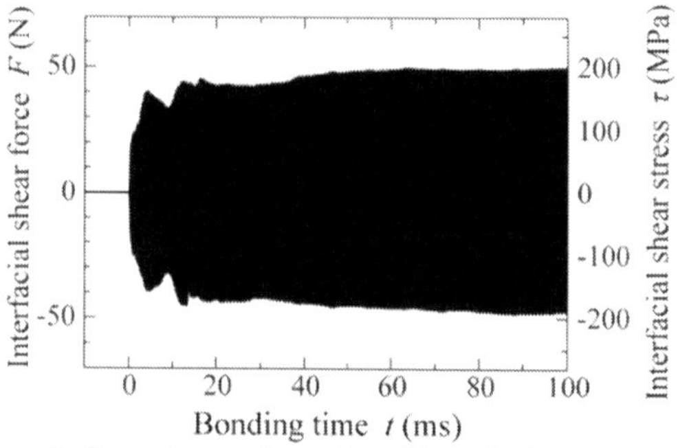

Figure 9. Experimental results of interfacial shear stress. The bonding condition was $F_b = 7.0$ N, $P_u = 2.5$ W.

As described in the numerical procedure, the decrease in A_b represents the enhancement of the constraint of interface with the progress of adhesion. Figure 6 shows the change of the maximum stresses in substrate with the progress of adhesion. The calculation parameters were A_o/μm = 3.5 and 3.5 > A_b/μm > 3.1, T = 373 K, and t_p = 4 μm. That is, A_o was fixed and only A_b was altered in Figure 6 in order to investigate the influence of the difference between A_o and A_b on progress of adhesion. As frictional adhesion increases and the adhesion areas expand in the interface, large stresses develop in the interface [2], and the maximum stresses in the substrate also increase. As seen in Figure 6, the maximum stresses rapidly increase when A_b is just slightly smaller than A_o. This suggests that even though the adhered areas are very small in the bonding interface, stress is significantly enhanced in the substrate.

Figure 7 shows the change in the maximum stresses in substrate with the thickness of the Al pad. The calculation parameters were t_p = 3, 4, 5, 10 μm, A_o = 3.5 μm, A_b = 3.4965 μm, and T = 373 K. σ_x is reduced by increasing the thickness of the Al pad. It is suggested that stress in x-direction is buffered by a thick Al pad. In contrast, σ_y and τ_{xy} slightly increase with increasing t_p. This slight increase could be related to convergence error of the displacement rates within the pad and/or the temperature assumption, but the detailed explanation is not known. It is supposed that σ_y and τ_{xy} also decrease with increasing thickness of pad because a thick Al pad similarly buffers σ_y and τ_{xy}.

Figure 8 shows the change in the maximum stresses in substrate with bonding time t over the range 1–100 ms. The experimental amplitude results were used in these calculations; the values of A_t and A_i (shown in Figure 3) were used as the inputs A_o and A_b in the simulation. However, as described above, A_i is slightly larger than A_t in the initial stages because of the measuring accuracy of A_i. Thus, A_b was assumed to be 0.99999 A_o in the initial stage (t < 5 ms). The thickness of the Al pad was set to t_p = 4 μm for calculating the results in Figure 8. As seen in Figure 8, the stress emerges in the substrate as t

increases from 0 to 5 ms, but it is not so large. From t = 5 to 10 ms, the maximum stresses sharply increase even though the interface seems to keep sliding (A_r is nearly equal to A_t) as shown in Figure 3. After 10 ms, maximum stresses gradually increase and saturate to a constant value. These results indicate that, at first, the ribbon slides on the substrate almost freely. As frictional adhesion emerges, strengthens, and the adhered areas expand, large shear stresses develop in the interface and stress in the substrate increases. Finally, the maximum stresses in the substrate remain steady because the developed adhesion areas strongly constrain the motion of the ribbon. σ_x reaches approximately 1 GPa after progression of ultrasonic bonding, and this large stress may damage the substrate during bonding.

Figure 9 shows experimental results for the interfacial shear stress up until t = 100 ms. The bonding condition was F_b = 7.0 N, P_u = 2.5 W. The shear stress applied at the interface between Al ribbon and SiO_2 substrate was measured with a piezoelectric load-cell. The trends in the experimental results resembles those of the calculation results, as shown in Figure 8, though the initial stage with free slip is very short in the experimental result. The interfacial shear stress is comparable to τ_{xy} in the calculations. τ_{xy} reaches approximately 300 MPa, and the interfacial shear stress reaches approximately 200 MPa. The stresses which occur in the substrate in the numerical simulation may be overestimated because all of ultrasonic vibration energy is assumed to be consumed around the bonding interface area. On the other hand, the experimental result is understood to be an underestimation because there is always energy loss in the measuring system. Although the stress evolution calculated in the present study may be overestimated, the numerical results are valid to some extent.

5. Conclusions

The amplitude of the interface and the stress evolution during ultrasonic Al ribbon bonding have been investigated and discussed. The following results were obtained:

(1) The amplitude of the interface becomes smaller than that of tool-tip, and the difference between these amplitudes increases, with the progression of ultrasonic bonding.
(2) Large stresses can occur in the substrate during ultrasonic bonding, and the stress increases with the progress of adhesion.
(3) Thick Al pads can reduce the stress in the substrate during ultrasonic bonding.
(4) The stress in the substrate increases with bonding time. The stress rapidly increases when adhered areas are formed at the interface. The stress stabilizes after adhered areas have developed.
(5) Even if adhesion occurs during bonding, frictional slip can be produced.

Acknowledgments

The authors express their gratitude to Mr. Akira Hirami and nac Image Technology for their technical support and advice.

References

[1] Ando M, Maeda M and Takahashi Y 2013 *Mater. Trans.* **54** 911
[2] Takahashi Y, Suzuki S, Ohyama Y and Maeda M 2012 *J. Phys. Conf. Ser.* **379** 012028
[3] Takahashi Y, Maeda M, Ando M, and Yamaguchi E 2015 ibid. in press
[4] Maeda M, Yoneshima Y, Kitamura H, Yamane K and Takahashi Y 2013 *Mater. Trans.* **54** 916
[5] Shah A, Rezvani A, Mayer M, Zhou Y, Persic J and Moon J.T 2011 *Microelectron. Reliab.* **51** 67
[6] Seppänen H, Kurppa R, Meriläinen A and Hæggström E 2013 *Microelectron. Eng.* **104** 114
[7] Mayer M, Schwizer J, Paul O, Bolliger D and Baltes H 1999 *Adv. Electron. Packag.* **26** 973
[8] Maeda M, Yamane K, Matsusaka S and Takahashi Y 2009 *Quart. J. Jpn. Weld. Soc.* **29** 138
[9] Gaul H, Shah A ,Mayer M Zhou Y, Schneider-Ramelow M and Reichl H 2010 *Microelectron. Eng.* **87** 537
[10] Toshio H et al. 2012 *Chronological Scientific Tables 2012* (Tokyo: Maruzen) p 391
[11] Oyama Y, Kitamura H, Maeda M and Takahashi Y 2009 *Proc. National Autumn Meeting of Jpn. Weld. Soc. (Tokushima)* **85** 449

Effect of tool geometry on ultrasonic welding process

Tomohiro Sasaki, Yutaro Sakata and Takehiko Watanabe

Graduate School of Science and Technology, Niigata University
8050 Ikarashininocho, Nishi-ku, Niigata-shi, Niigata Japan

E-mail:tomodx@eng.niigata-u.ac.jp

Abstract. Ultrasonic welding of pure aluminum sheets is performed using two weld tools, one with a knurled surface and one with a cylindrical surface. Relative motion behaviors of each weld tool, with respect to the working materials, during ultrasonic welding tests are analyzed using the digital correlation method. Weld microstructure development is investigated on the basis of transitional weld stages in the context of relative motion behaviors. The dominant relative motion is between the two work materials at the beginning of the weld but changes to be the motion between the weld tool and the work material it is in contact with as weld time increases. Thermo-mechanical effects of the relative motion of the weld tool and the work materials, on the development of weld microstructure, are discussed.

1. Introduction

Ultrasonic welding is a solid-state bonding technique that is widely used to bond electrodes to semiconductors and secondary batteries in automobile industry because of its low temperature and high speed capabilities [1,2]. For ultrasonic welding, a high frequency vibration is applied to the part to be welded using a weld tool attached to an ultrasonic horn. The weld, essentially, is accomplished by friction at the weld interface owing to the relative motion of the working materials. Interfacial phenomena, such as oxide film fracture and adhesion [3,4], material flow near the weld interface [5-6], and changes in weld microstructure [7-9] have been metallographically investigated within the context of relative motion at the weld interface. However, relative motion also occurs between the weld tool and the welding material it is in contact with. It has been reported that weld properties strongly depend on weld tool geometry [10] and anvil geometry [11]. These dependencies can be interpreted as resultant phenomena of relative motion between the weld tool and material. However, this relative motion is not completely understood. The present study aims to investigate the effects of tool geometry on relative motion behavior and weld microstructure.

2. Experimental procedure

Sheets of pure industrial aluminum with thicknesses of 0.5 and 0.8 mm were used as test specimens in this study. Weld process observations were performed on aluminum sheet pieces that were cut to be 50 mm long × 10 mm wide and 50 mm long × 10.5 mm wide along their roll directions (Figure 1(a)). In addition, weld specimens with lengths of 100 mm were prepared for joint strength measurements (Figure 1(b)). The weld areas of the specimens were electropolished.

Ultrasonic welding equipment was operated at a vibrational frequency of 15 kHz, and a power of 2.4 kW. The load-free vibrational amplitude was 53 μm (peak to peak). The equipment consisted of a lateral drive system [1] which applied clamping force to the welding part through an ultrasonic horn as shown in Figure 2. The 0.8 mm thick specimen was placed on the anvil (lower specimen), and the 0.5

Content from this work may be used under the terms of the Creative Commons Attribution 3.0 licence. Any further distribution of this work must maintain attribution to the author(s) and the title of the work, journal citation and DOI.
Published under licence by IOP Publishing Ltd

mm thick specimen was stacked on the lower specimen (upper specimen) as shown in Figure 2. Vibration was applied perpendicular to the lengths of the specimens. The lower specimen was clamped so as to suppress slippage across the anvil due to applied vibration. Two types of welding tools, shown in Figure 3, were used in this study. A conventional knurled tip tool (K-tip, Figure 3(a)) with a pitch of 0.8 mm, a height of 0.3 mm, and a groove angle of 90° was used as the basis for comparison. Watanabe *et al.* [10] have demonstrated that using a weld tip with a cylindrical contact surface significantly improves bonding strength. Thus, a tip with a cylindrical surface, having a radius of curvature of 200 mm (C-tip, Figure 3(b)) was used for comparison with the K-tip. The axis of cylindrical curvature was perpendicular to the direction of vibration (x-axis). The welding test was conducted under a constant clamping force of 588 N. In addition, to estimate the temperature near the weld interface, a K-type thermocouple was percussion-welded to the upper specimen as shown in Figure 2. Temperature observations were made simultaneous to high-speed camera observations, discussed presently.

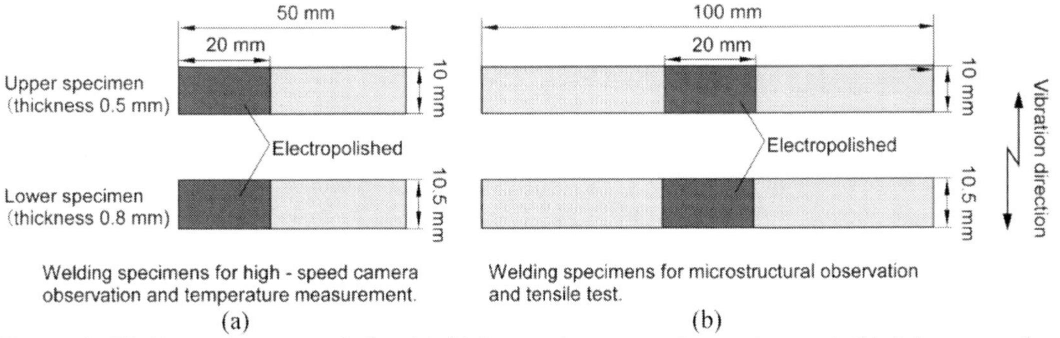

Figure 1. Weld specimens used for (a) high speed camera observation and (b) joint strength measurements.

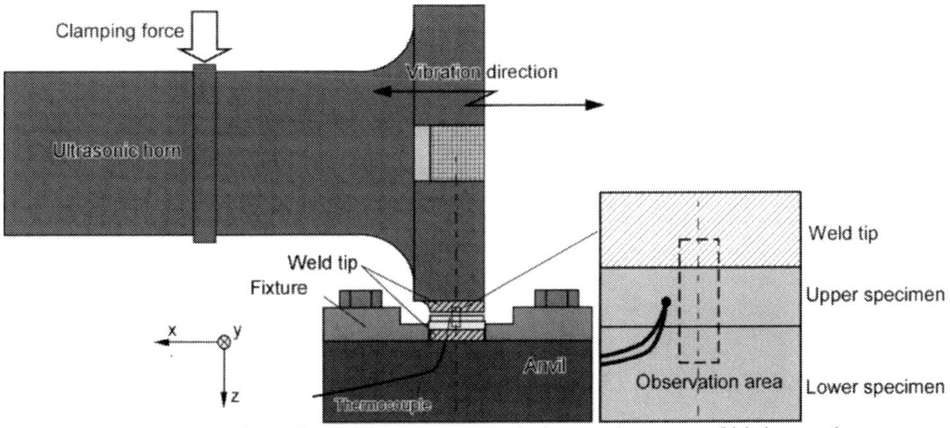

Figure 2. Schematic of welding test apparatus and observation area of high speed camera.

The motions of the weld tools and specimens during ultrasonic welding were observed using a high speed camera. The observation area was a coupon area that included the tip of the weld tool (weld tip), and the upper and lower specimens, as shown by the broken line in Figure 2. The capture speed of the high speed camera was 100,000 frames per second. Weld tip and the specimen displacements in the direction of vibration were calculated using the digital correlation method discussed previously [12]. Three subset areas were defined in the captured image, the weld tip, the upper specimen, and the lower specimen. The coordinate of each subset area, during the welding test, was determined based on its intensity array in an 8-bits digital image. The displacements of the two weld tips and the upper / lower specimens in the vibration direction are denoted, respectively, by $u_{K\text{-}tip}$, $u_{C\text{-}tip}$, u_{upper}, and u_{lower}. Relative displacements between the upper and lower specimens and those between each weld tip and the upper specimen were obtained as follows: $u_{upper} - u_{lower}$, $u_{K\text{-}tip} - u_{upper}$, and $u_{C\text{-}tip} - u_{upper}$.

To evaluate joint strength, welded joints made on 100 mm long specimens were processed into "U-bent shape specimens [7]". The maximum load in the tensile test was defined as the joint strength in this study. For weld microstructure observations, welded specimens were cut at the center of the welded area, parallel to the vibration direction. The cross-sections were then polished with diamond slurry and electrochemically etched with Barker's reagent (1 vol% HBF4 solution).

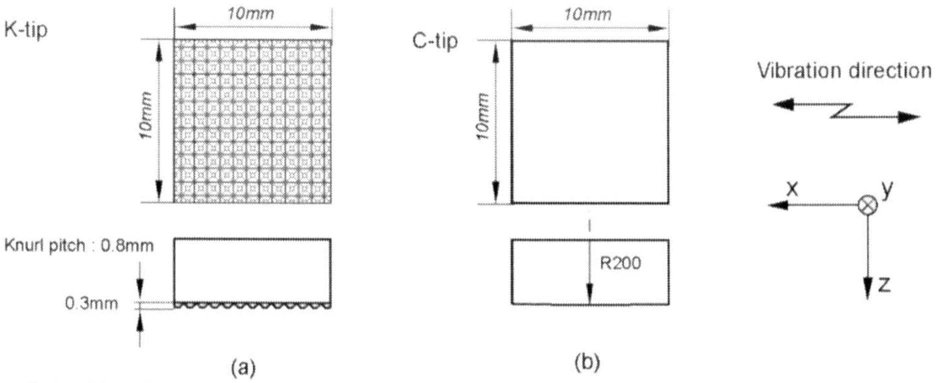

Figure 3. Weld tool geometries used in this study. (a) presents a knurled surface tip (K-tip), and (b) a cylindrical surface tip (C-tip).

3. Results and discussion

3.1 Motion behavior and weld stages of ultrasonic welding

Figure 4 shows typical C-tip and specimen displacements during an 800 ms welding test. The three transitional stages indicated in Figure 4 as I, II, and III were established based on features of the displacement curves. Similar weld stage progressions have been observed in welding tests that used K-tips [12]. In the first stage ($t_w < 50$ ms, weld stage I) upper specimen displacement, u_{upper}, rapidly increases to approximately 150 μm more than the vibratory amplitude of the ultrasonic horn. This indicates that vibration causes the upper specimen to slip macroscopically. This macroscopic slippage is likely due to the induction of a bending moment by the clamping force and asymmetric shape of the ultrasonic horn. The slippage distances of C-tips tend to be smaller than those of K-tips, which demonstrate slippage distances of approximately 250 μm [12]. In the second stage ($t_w > 50$ ms, weld stage II), u_{upper} showed displacement. Finally, in the third stage ($t_w > 300$ ms, weld stage III), the welding parts began to be plastically compressed by the clamping force. In this stage, the observation face expanded in the out-of-plane direction, hence u_{upper} and u_{lower} show apparent increases in Figure 4. Displacement analysis was impossible for weld times larger than 600 ms because the captured images were out of focus. Figure 5 shows how temperature changed over the course of a 600 ms welding test and indicates the time ranges of the weld stages, as determined from high-speed camera observations. Temperature monotonically rose during welding, reaching a maximum of approximately 230 °C at the beginning of weld stage III. This temperature should be high enough to initiate aluminum recrystallization. Although the measured temperature indicates an average value for the welding part, it is the compressive deformation associated with the temperature rise of weld stage III that is considered to be responsible for the softening of the upper specimen.

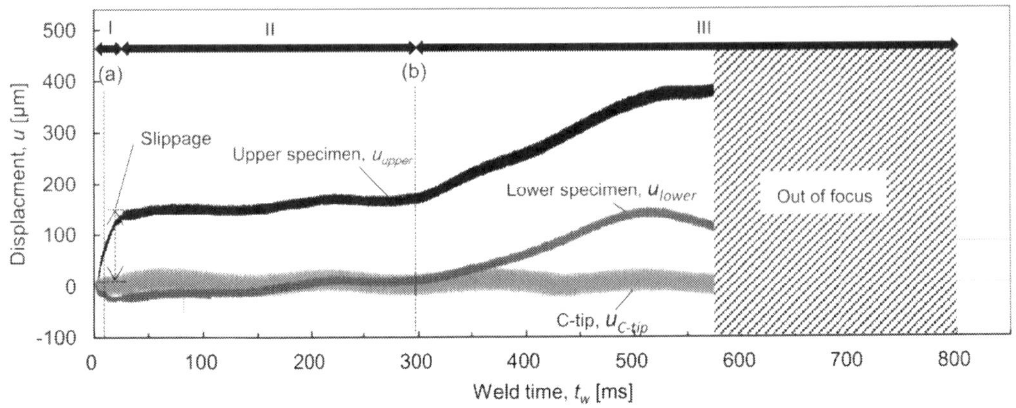

Figure 4. Displacement in the direction of vibration during ultrasonic welding with a C-tip.

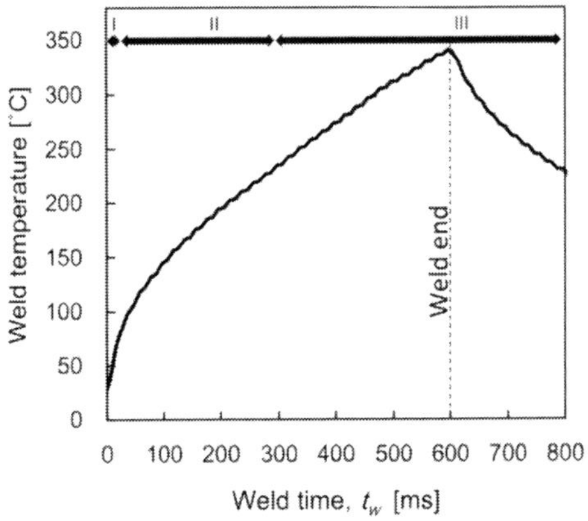

Figure 5. Weld temperature versus time during a 600 ms welding test.

Figures 6 and 7 show the relative motion behaviors, with respect to weld times, that correspond to weld stages I and II. The plots shown in Figures 6 and 7 present relative displacement data, with respect to the upper specimen, for of each weld tip, $u_{K\text{-}tip}$ - u_{upper} and $u_{C\text{-}tip}$ - u_{upper}, and that of the lower specimen, with respect to the upper specimen, u_{upper} - u_{lower}. At the beginning of the weld, in the case of the K-tip (Figure 6(a)), the majority of the relative motion occurred between the specimens, u_{upper} - u_{lower}, which vibrated extensively. Smaller amplitude vibration is also observed in the $u_{K\text{-}tip}$- u_{upper} plot of Figure 6(a). The u_{upper} - u_{lower} vibrational amplitude decreased to 2–4 μm during weld stage II using the K-tip, as shown in Figure 6(b). Meanwhile the vibrational amplitude of $u_{K\text{-}tip}$ - u_{upper} became dominant, *i.e.* relative motion occurred predominantly between the weld tip and the upper specimen. It was found that during weld stage I, partially bonded regions formed at the contact surface and suppressed the relative motion of specimens, u_{upper} - u_{lower}. In contrast, compared to that of the K-tip, the vibrational amplitude of u_{upper} - u_{lower} of the C-tip was relatively small at the beginning of the weld (Figure 7(a)). The cylindrical surface without knurled edges should suppress penetration of the weld tip into the upper specimen and result in greater relative motion of $u_{C\text{-}tip}$ - u_{upper} at the beginning of the weld. However, the vibrational amplitude of $u_{C\text{-}tip}$ - u_{upper} during weld stage II (Figure 7(b)) was smaller than that of $u_{K\text{-}tip}$ - u_{upper} (Figure 6(b)). This may be related to the plastic deformation of the lower specimen, described in the following section.

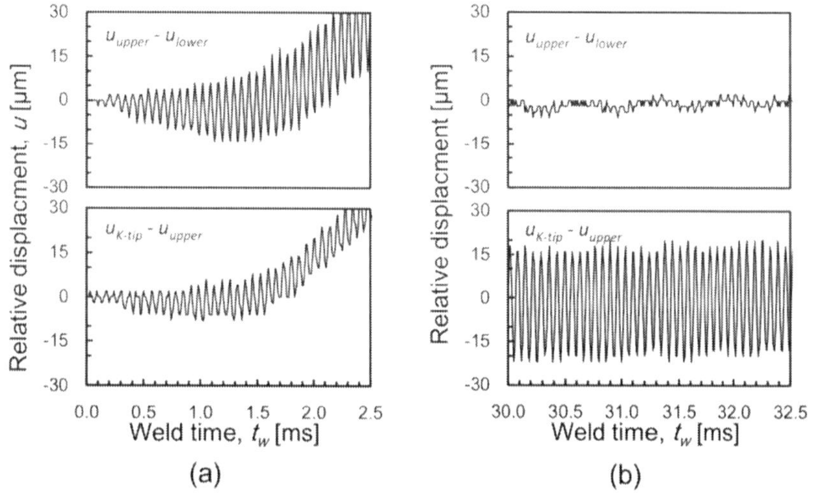

Figure 6. Relative motions during ultrasonic welding using a K-tip. (a) illustrates the motion observed during weld stage I, and (b) illustrates motion observed during weld stage II.

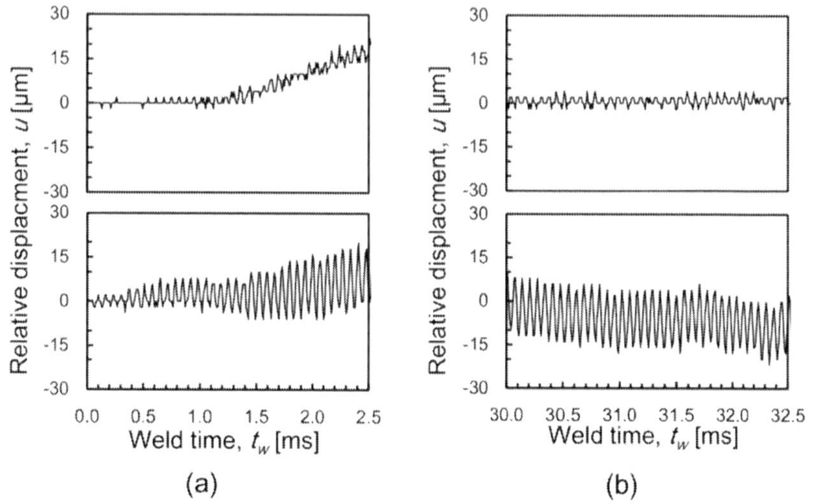

Figure 7. Relative motions during ultrasonic welding using a C-tip. (a) illustrates the motion observed during weld stage I and (b) illustrates the motion observed during weld stage II.

3.2 Microstructures of weld formations

Figure 8 shows cross-sectional micrographs of K-tip welded specimens. Trapezoidal indentations formed by the penetration of knurled edges were observed on the welding tip side of the upper specimens. Indentation depth increased with increasing weld time and led to thinning of upper specimens. When t_w = 20 ms, which corresponds to weld stage I, using the K-tip (Figure 8(a)), almost no change in the microstructure around the weld interface was observed, while the upper side of welding part in contact with the knurled edges was plastically deformed by edge penetration. As shown in Figure 8(b) (t_w = 300 ms), penetration depth reached approximately 300 μm, equal to the height of the knurl, during weld stage II. Figures 8(e) and (f) show high magnification micrographs of the areas indicated by rectangle 1 in Figure 8(b) and rectangle 2 in Figure 8(e), respectively. Figure 8(e) (severe plastic deformation zone) shows that the microstructure between the knurled edges was severely deformed. Figure 8(e) shows a partially bonded region at the weld interface, indicated by a broken line. The severe plastic deformation zone started to spread toward the weld interface at t_w = 550 ms (Figure 8(c)), which corresponds to the beginning of weld stage III, when using the K-tip. Furthermore, at t_w = 600 ms (Figure 8(d)) the severe plastic deformation zone had expanded into the lower specimen and the weld interface had become unclear. The weld temperature has been found to rise to approximately 300 °C during weld stage III when using a K-tip [12]. As shown in Figure 6, relative motion occurred predominantly between the weld tip and the upper specimen after weld stage II. Friction at the weld tip / upper specimen interface and plastic deformation of the upper specimen due to vibration are the most likely major causes of temperature rise during weld stage II. In contrast, Figure 8(c) indicates that the partially bonded region observed at the weld interface was relatively small. It is concluded that heating, due to the relative motion of the weld tip and upper specimen causes the welding part to soften and results in the formation of the severe plastic deformation zone.

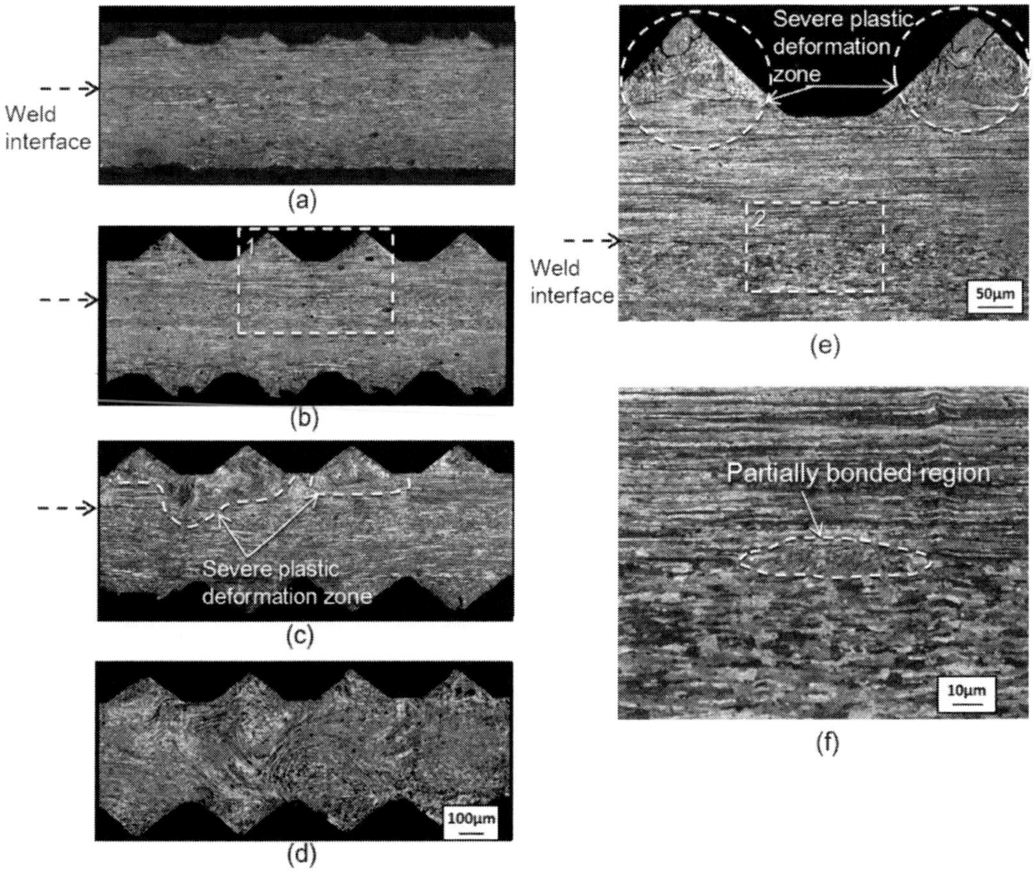

Figure 8. Cross-sectional images of joints welded using a K-tip. Images were obtained at t_w = (a) 20 ms, (b) 300 ms, (c) 550 ms, and (d) 600 ms. (e) and (f) are high-magnification micrographs taken from the region near the indentations indicated by rectangle 1 in (b) and the bonded region indicated by rectangle 2 in (f), respectively.

In contrast, Figure 9 shows cross-sectional micrographs of joints welded using the C-tip. In the case of the C-tip, the start of weld stage III is observed earlier. The severe plastic deformation zone has already been generated in the upper side of the welding area at the beginning of weld stage III (Figure 9(b), t_w = 300 ms). In addition, the entire welding region, except for the severe plastic deformation zone, was also deformed, giving it a wavy shape. The smaller vibrational amplitude of the weld tip, relative to the upper specimen, observed in weld stage II when using the C-tip probably arose from the fact that plastic deformation of the welding region, including the lower specimen, assisted the motion of the upper specimen. The microstructure of the welding region was more dramatically disturbed as weld time increased (Figure 9(c) and (d)). The changes observed in weld microstructures during weld stage III, Figure 8(d) and Figure 9(d), are mainly attributed to thermo-mechanical effects caused by horizontal vibration and clamping force. It is concluded that the cylindrical surface of the C-tip facilitated the temperature rise caused by weld tip / upper specimen relative motion.

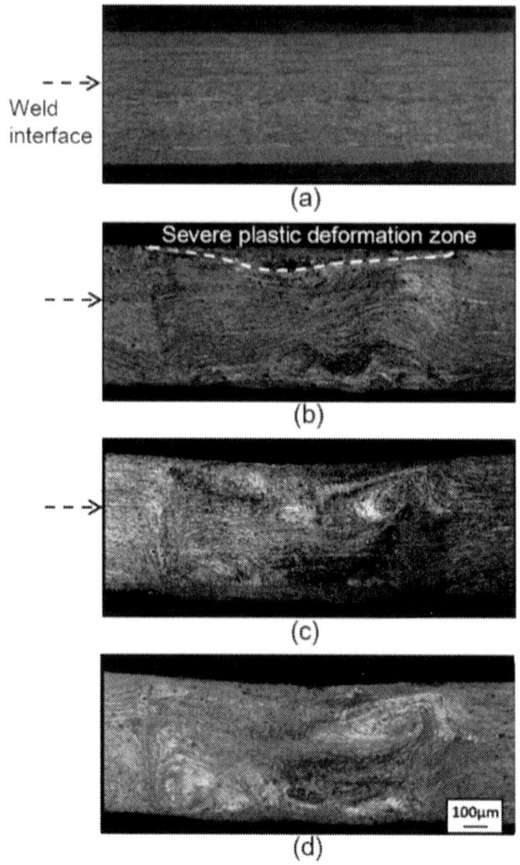

Figure 9. Cross-sectional images of joints welded using a C-tip. Images were obtained at t_w = (a) 20 ms, (b) 300 ms, (c) 550 ms, and (d) 600 ms.

3.3 Joint strength

Figure 10 shows how joint strength changed with weld time and indicates the three weld stages observed in the relative motion behaviors of each weld tip at the top. Measurement results for the K-tip were obtained in a previous work [12]. Joint strengths were relatively low for weld times corresponding to weld stage I but increased monotonically, reaching maximum values during weld stage III. Photographs of the fractured specimens, obtained after tensile tests, are shown in Figure 11. At weld times corresponding to weld stages I and II, fractures occurred at the weld interface, as shown in Figures 11(a) and (b). Dispersed white areas, which indicate partially bonded regions, can be seen in the image of a fracture surface presented Figure 11(a). These areas expand as weld timed increase (Figure 11(b)). The increase in joint strength observed in weld stage II of the K-tip curve can be attributed to the expansion of partially bonded regions. In weld stage III, the fracture mode changes to "button fracture" in which the bonded part of the upper specimen remains attached to the lower specimen, as shown in Figure 11(c). The microstructure of the area surrounding the weld interface is intensely disturbed by the relative motion of the weld tip and upper specimen during this weld stage. Disturbance of the microstructure of the weld interface may contribute to the expansion of the bonded area and, hence, lead to increased joint strength during weld stage III. However, longer weld times, t_w > 700 ms, lead to compressively deformed welding parts resulting from the clamping force. This phenomenon decreases joint strength, which is due to thinning of the welding part, as shown in Figure 11(d). In addition, uneven surfaces resulting from indentation by knurled edges of the K-tip also result in decreased joint strength and increased joint strength variance. It was found that the strengths of joints produced using the C-tip were higher and varied less than those produced using the K-tip. In the case of the C-tip, concentrated bonded regions were observed at the centers of fracture surfaces (Figure 11(e)). In addition, button fractures were already observed late in weld stage II (Figure 11 (f)). The cylindrical surface of the C-tip not only suppresses penetration by the weld tip, it also leads to the formation of concentrated bonded regions because of the high pressures it produces in the contact area at the beginning of the weld. These effects may expedite the onset of weld stages II and III. We attribute the higher joint strengths obtained from the C-tip to its smaller penetration and more significant effect on microstructure.

4. Conclusion

Ultrasonic welding of pure aluminum sheets was performed using two weld tips, a K-tip and a C-tip, and the relative motion behaviors between the weld tip and the specimen, and between the specimens were analyzed. The weld stages defined using features of the displacement curve of the C-tip were similar to those of the K-tip. The relative motion between the weld tip and the upper specimen became dominant after weld stage II (weld time t_w > 50 ms). The relative motion between the weld tip and the upper specimen of the C-tip showed a larger vibration at the beginning of the weld compared to that of the K-tip, thus, resulting in greater weld microstructure disturbance during weld stage III. It was found that the difference in weld tool geometry explored affected microstructure growth through differences in relative motion.

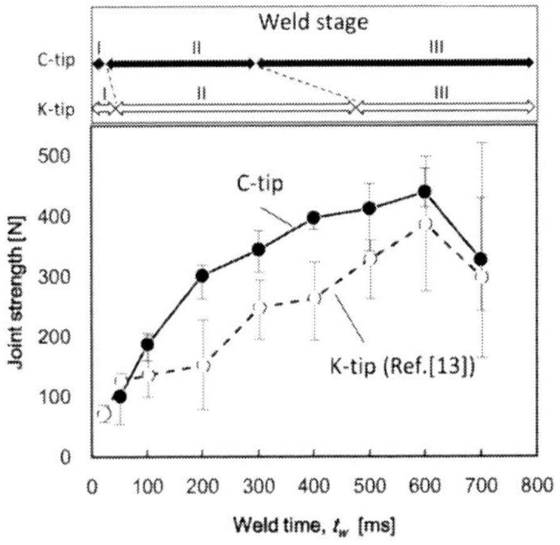

Figure 10. Evolution of weld stage and joint strength with weld time.

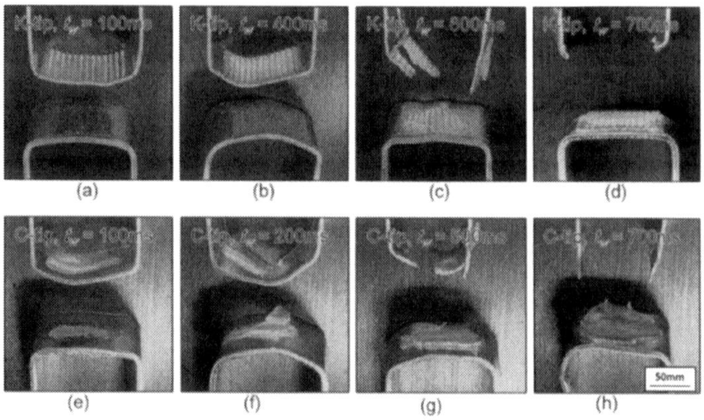

Figure 11. Photographs of fractured specimens. (a) - (d) are joints produced using the K-tip, and (e) - (h) are joints produced using the C-tip.

References

[1] R L O'Brien 1991 *Ultrasonic Welding, in Welding Handbook vol.2 8th edition* (Miami: American Welding Society) pp 784

[2] Hetrick E, Jahn R, Reatherford L, Ward S, Skogsmo J, Devine J, Graff K and Geherin R 2005 *Welding Journal* **84** 26-31

[3] Watanabe T, Yoneda A, Yanagisawa A, Konuma S, and Ohashi O 1999 *Quart. J. Jpn. Weld. Soc.* **17-2** 234-42 (in Japanese)

[4] Takahashi Y and Inoue M 2002 *J. Electron. Packag.* **124** 27-36

[5] Elangovan S, Semeer S, Prakasan K 2009 *J. Mater. Proc. Technol.* **209** 1143-50

[6] Maeda M, Yoneshima Y, Kitamura H, Yamane K and Takahashi Y 2013 *Mater.Trans.***54-6** 916-21

[7] Watanabe T, Ito H, Yanagisawa A, Hiraishi M 2008 *Quart. J. Jpn. Weld. Soc.* **26-2** 117-23 (in Japanese)

[8] Bakavos D and Prangnell P B 2010 *Mater. Sci. Eng. A* **257** 6320-34

[9] Maeda M, Kitamori S and Takahashi Y 2013 *Sci. Technol. Weld. Join.* **18-2** 103-7

[10] Watanabe T, Miyajima D and Yanagisawa A 2009 *Quart. J. Jpn. Weld. Soc.* **27-1** 7-12 (in Japanese)

[11] Jahn R, Cooper R and Wilkosz D 2007 *Metall. Mater. Trans.* **A38-3** 570-83

[12] Sasaki T, Watanabe T, Goto T and Hosokawa Y 2012 *Quart. J. Jpn. Weld. Soc.* **30-2** 125-32 (in Japanese)

IOP Conf. Series: Materials Science and Engineering **61** (2014) 012007 doi:10.1088/1757-899X/61/1/012007

Effect of surface roughening of aluminum plates on the strength of bonds formed between aluminum and polyphenylene sulfide by thermosonic bonding

Kiyokazu Yasuda and Ryo Saito

Division of Materials and Manufacturing Science, Osaka University
2-1 Yamadaoka, Suita, Osaka 565-0871, Japan

E-mail: yasuda@mapse.eng.osaka-u.ac.jp

Abstract. Thermosonic bonding of aluminum on polyphenylene sulfide was carried out in order to examine the effect of surface roughening of aluminum on the joint strength. Repeated chemical treatment of aluminum by immersion in aqueous sodium hydroxide and hydrochloric acid solutions increased its surface roughness (Ra ~25 μm) and surface area (~445% increase). Consequently, the bonding strength (~1.8N in average) was enhanced through anchoring effects.

1. Introduction

Polyphenylene sulfide (PPS) is one of the most promising heat- and chemically resistant super-engineering thermoplastics. Adhesive bonding is generally adopted for the joining this polymer to surface-treated metals for use in automotive and electronics industries [1]. The use of adhesives, however, has certain problems, such as low working efficiency, detrimental effect on environment and workers, and unreliability of joint over a long term due to chemical degradation. Considering these issues, we conducted thermosonic bonding, whose primary joining principle is the anchoring effect obtained by plastic resin deformation and its penetration into metal surface pits. For this reason, chemical etching in alkaline/acid solutions is one of the simpler choices of metal surface roughening. For example, upon immersion in an active base of pH 13, such as an aqueous solution of sodium hydroxide, the dissolution reactions at the surface of an aluminum plate occur as described by the following reaction equations.

$$Al_2O_3\ (s) + 2NaOH(aq.) + 3H_2O \rightarrow 2Na[Al(OH)_4]\ (aq.) \quad (1)$$

$$2Al\ (s) + 2NaOH(aq.) + 6H_2O \rightarrow 2Na[Al(OH)_4]\ (aq.) + 3H_2\ (g) \quad (2)$$

Since the product of both reactions (1) and (2), sodium aluminate, is soluble in the basic solution of sodium hydroxide [2], aluminum, even in the presence of its stable passivating native oxide layer, is etched uniformly to a matte finish, thereby introducing a high density of nucleation sites for pits on the surface [3]. When subsequently immersed in a solution of hydrochloric acid, aluminum is etched, as described by the following equation.

$$2Al\ (s) + 6HCl\ (aq.) \rightarrow 2AlCl_3\ (aq.) + 3H_2\ (g) \quad (3)$$

Content from this work may be used under the terms of the Creative Commons Attribution 3.0 licence. Any further distribution of this work must maintain attribution to the author(s) and the title of the work, journal citation and DOI.
Published under licence by IOP Publishing Ltd

This reaction occurs rather preferentially at the pitting nucleation sites related to surface flaw distribution [4], and proceeds anisotropically, resulting in the formation of rectangular etch pits. When again subjected to the NaOH solution, the step described by eq. (2) occurs again, and the rectangular etch pits grow rather rapidly to take on round-shaped crater forms through rapid anodic dissolution due to the exposure of the cathodic second phase (iron-rich precipitates in case of A1050) on the aluminum surface [5]. By repeating the alternating alkaline/acid immersions, anomalous roughening, characterized by a very large increase in the size of the aluminum surface craters, can be produced. In this study the effect of surface roughening of aluminum plates on the strength of the bonds formed between aluminum and PPS by thermosonic bonding was investigated.

2. Experimental procedure

Surface roughening of aluminum was carried out as pretreatment under the following three conditions: (treatment C1) immersion in sodium hydroxide (5 wt% aq. solution) for 120 s at 40 °C and then in hydrochloric acid (3 wt% aq. solution) for 300 s at 40°C; (treatment C2) dipping in sodium hydroxide (5 wt% aq. solution) for 150 s at 40 °C after C1; and (treatment C3) dipping in sodium hydroxide (5 wt% aq. solution) for 300 s at 40 °C after C1. After each treatment, a pair comprising an aluminum plate (A1050P H24, 5 mm square and 0.5 mm thick) and a PPS plate (1 mm in thickness) was used for the joining process. Physical properties of PPS are shown in Table 1. After subjecting to the surface treatments, the aluminum plates were bonded on to PPS plates using an ultrasonic bonder by solid-state bonding. The bonding tool head vibrates horizontally at a frequency of 50 kHz, and is capable of performing thermo-compression as shown in Figure 1. The bonding conditions were fixed as follows: temperature of 200 °C, pressure of 12 MPa, bonding time of 2.5 s, and an ultrasonic vibration amplitude of 6 μm. The tool vibration amplitude was controlled through closed-loop feedback during the bonding cycle. Shear tensile testing of the joined specimen was conducted for the evaluation of the bond strength at a crosshead speed of 0.1 mm/s. The tensile direction was set parallel to the direction of ultrasonic vibration.

Table 1. Physical properties of polyphenylene sulfide.

Melting point (°C)	Glass transition temperature (°C)	Specific gravity	Rockwell hardness	Tensile strength (MPa)
250	87	1.50	R118	130

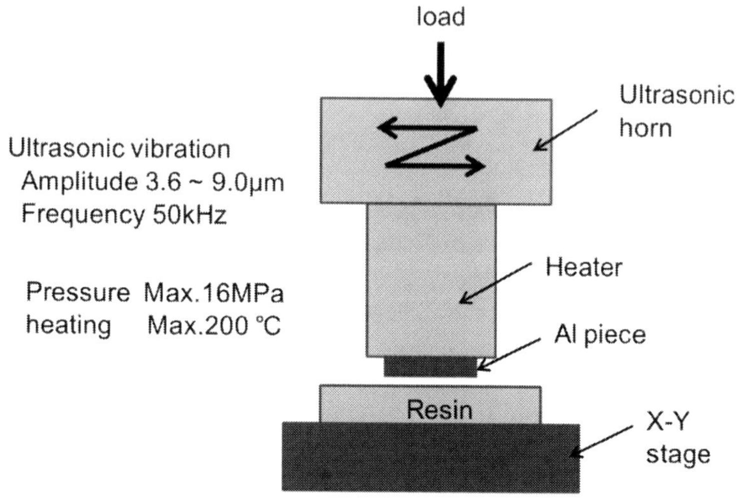

Figure 1. Schematic illustration of the thermosonic bonding tool, which vibrates horizontally and is capable of performing thermo-compression.

Quantitative topological characterization of the aluminum surface was conducted by scanning laser microscopy (Keyence, VK-5800). Surface roughness *R*a (μm) and surface area *S* (μm^2), as observed in the images, were calculated using the system software provided with the microscope from the measured topographical array data. The percentage increase in surface area, α, was evaluated using the following equation.

$$\alpha\,(\%) = (S_1 - S_0)/S_0 \times 100\,, \qquad (4)$$

where S_1 and S_0 are the surface areas of aluminum after and before chemical etching, respectively.

3. Results and discussion

Aluminum sample surfaces after the chemical etching treatments are shown in Figure 2. Surface evaluation results of etched aluminum are presented in Table 2. For C1 treatment, small pits with a low density (pit ratio 13.8%) were formed. The pits were either small rectangles or rectangular aggregates with sharp edges. However, the majority of almost surface area remained unaffected, showing a flat topology with rolling scratches similar to surface of as-received aluminum. On the other hand, hemispherical pits with a high density (pit ratio > 80%) were formed after both C2 and C3 treatments. The average pit diameter of pits was several tens of micrometers, comparable to the surface roughness *R*a. Moreover, the percentage increase in surface area, α, increased linearly with *R*a up to a maximum of 445% (α ≒ 20.3 ×*R*a - 54.7). It is thus concluded that the aluminum surface roughness (*R*a~25 μm) can be increased with keeping almost linear relation to the increase in surface area by simple repeated alkaline/acid chemical etching.

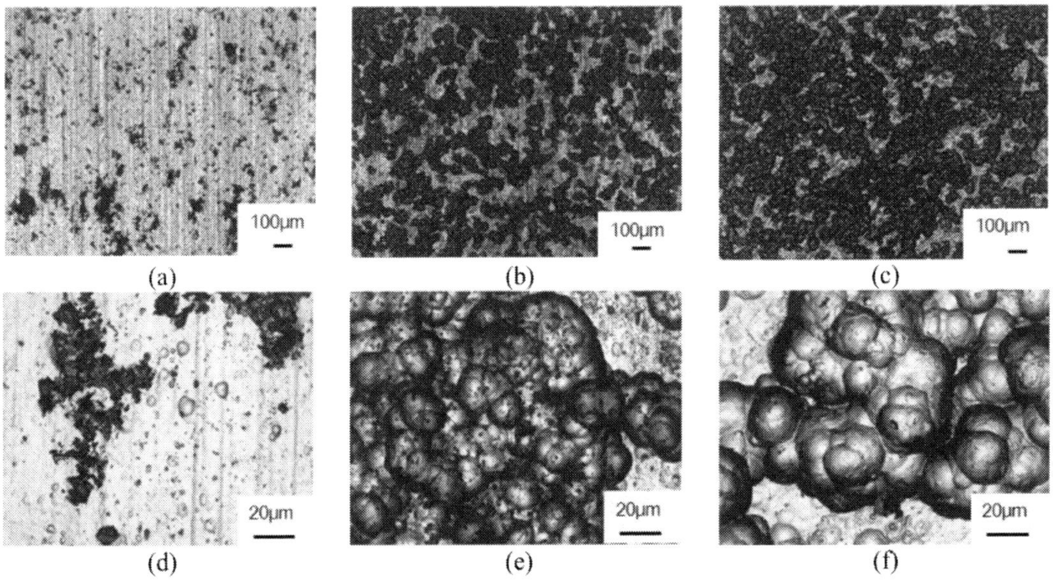

Figure 2. Aluminum surface etched by the following chemical treatments: C1 ((a), (d)); C2 ((b), (e)); and C3 ((c), (f)).

IOP Conf. Series: Materials Science and Engineering **61** (2014) 012007 doi:10.1088/1757-899X/61/1/012007

Table 2. Characterization results of aluminum surface after chemical treatments.

Treatment	Surface roughness *R*a (μm)	Percentage increase in surface area, α	The average diameter of the pit (μm)	Pit ratio in the observed range (%)
As received	3.2	0	-	-
C1	4.8	38	-	13.8
C2	15.5	292	12-14	80.9
C3	25.5	445	20-25	93.5

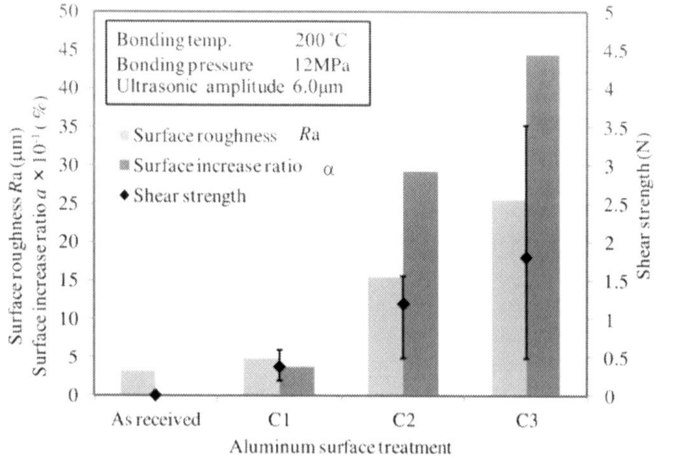

Figure 3. Effect of chemical treatments on the surface roughness *R*a, surface increase ratio α, and shear strength.

The effect of the chemical treatments on the shear strength of the bonded joints is shown in Figure 3. In the figure, *R*a (light gray) and α (dark gray) are also represented as solid bars. The joints made using specimens of as-received aluminum and those subjected to C1 treatment, both of which have small values of *R*a and α, showed a small average shear strength value of less than 0.5 N. Because of the small density of the etch pits, as shown in Figure 2(a), it is thought that sufficient anchoring onto aluminum pits could not be achieved. It should also be pointed out that using too large an ultrasonic amplitude (6 μm) compared to the surface roughness (3.2 μm) was ineffective for proper bonding. In such cases of a large difference between amplitude and roughness, it is very probable that the resin flows in and out of the small aluminum pits repeatedly, and consequently, no stable anchor bonds are formed, this effect produced being much like the grinding of PPS. In contrast to the C1 treatment, the C2 and C3 treatments caused an increase of the average shear strength, approximately 1.2 N and 1.8 N, respectively. It is assumed that the increase of the average shear strength is attributable to the surface roughening of aluminum to a surface roughness value far greater than the amplitude of the ultrasonic vibration. Nevertheless, the minimum values of shear strength obtained through C2 and C3 treatments merely reached approximately 0.5 N, which is almost similar to the average value obtained through C1 treatment. Since no interfacial interatomic chemical reaction or inter diffusion can be expected from such low-temperature thermosonic bonding [6], the variations in bonding strength is solely attributable to the variations in the 'interfacial topology' between aluminum and PPS.

In order to elucidate the origin of such bonding characteristics, the bonded aluminum and PPS plates of a few bonded samples were separated, and the previously bonded areas were studied. As shown in Figures 4(a) and (b) for the case of C3 treatment, it was confirmed that very similar surface topology was observed on both aluminum and PPS surfaces at the micrometer scale. It clearly shows that the localized deformation of PPS at the aluminum pits was achieved through ultrasonic vibration amplitudes smaller than pit sizes even under low bonding pressures (12 MPa) below the melting point of PPS. The horizontal vibration of the pit peripheral ridges might have caused localized plastic deformation of the resin. In such a case, the thermosonic energy (the sum of ultrasonic energy and thermal energy) could have been effectively absorbed only by the PPS surface, leading to bonding. However, it should be mentioned that there existed a certain amount of topology mismatch between PPS and aluminum. The PPS image was not sharp, and its surface was not accurately identical to that of aluminum, especially around the smaller pits across the rugged area (light gray zone). This observation leads to the inference that the existence of non-bonded gaps at the interface is the potential reason for the variations in the bonding strength.

The cross-sectional optical microphotograph of a bonded specimen subjected to C3 treatment is shown in Figure 5. The resin remains adhered to the aluminum surface despite the rough interface. The increase in aluminum surface roughness and the resultant increase in the interfacial area between aluminum and PPS together caused the average bonding strength to increase. However, non-bonded small voids at the interface were also observed, such as the one shown by white arrow in the figure. This incomplete intrusion of resin might be another cause of the variations in the shear strength in the case of C2 and C3 treatments, since the existence of voids at the interface increases the local stress, and initiates cracks during shear testing. It is suggested that the wider variation of the shear bonding strength was caused by the uncontrolled variation in the non-bonded interface defects. Further study on the effects of surface pretreatment through more controlled roughening by an electrochemical approach as well as on the effects of the therrmosonic process parameters shall be undertaken for improving bonding quality.

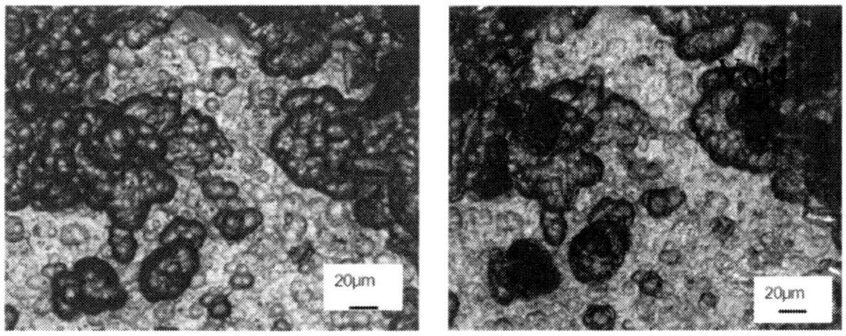

Figure 4. Bonded C3-treated pair of (a) aluminum surface and (b) PPS (mirror images).

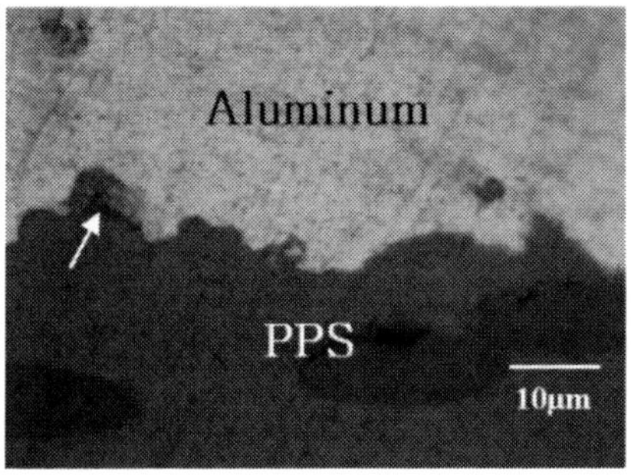

Figure 5. Cross-sectional optical microphotograph of the bonded interface of aluminum (C3 treatment) and PPS. A pit (white arrow) remains as an interfacial void.

4. Conclusions

Thermosonic bonding of aluminum on PPS resin was examined through the use of a horizontally vibrating tool head. The aluminum surface treatment by the repeated alternated immersion in NaOH and HCl aqueous solutions increased the surface roughness (*Ra* ~25μm) with a linear increase in surface area (α ~445%) through the formation of a high density of micro pits. The average bonding strength increased up to 1.8 N with increasing surface roughness. The application of ultrasonic vibration was effective due to the thin plastic, but not severe, deformation and anchoring of PPS. The non-bonded zone caused by low roughness (< ultrasonic vibration amplitude) and incomplete anchoring are supposed to be the causes of the wide scatter of the bonding strength.

References

[1] Anagreh N, Dorn L and Bilke-Krause C 2008 *Int. J. Adhes. Adhes.* **28** 1-2 16
[2] Moolenaar R J, Evans J C and McKeever L D 1970 *J. Physical Chemistry* **74** 20 3629
[3] Wu X and Hebert K 1996 *J. Electrochem. Soc.* **143** 1 83
[4] Richardson J A and Wood G C, 1970 *Corrosion Science* **10** 313
[5] Koroleva E V, Thompson G E, Hollrigl G and Bloeck M 1999 *Corrosion Science* **41** 1475
[6] Inoue M, Mitsuoka T, Kikuzawa Y and Takeuchi H 2013 *Surf. Interface Anal.* **45** 1 198

IOP Conf. Series: Materials Science and Engineering **61** (2014) 012008 doi:10.1088/1757-899X/61/1/012008

Effect of material flow on joint strength in activation spot joining of Al alloy and steel sheets

Goro Watanabe, Yasuhiro Yogo and Hisaaki Takao
Toyota Central Research and Development Laboratories, Inc.
41-1, Yokomichi, Nagakute, Aichi 480-1192, Japan

E-mail: gwatanabe@mosk.tytlabs.co.jp

Abstract. A new joining method for dissimilar metal sheets was developed where a rotated consumable rod of Al alloy is pressed onto an Al alloy sheet at the part overlapped with a mild steel sheet. The metal flow in the joining region is increased by the through-hole in the Al sheet and consumable Al rod. The rod creates the joint interface and pads out of the thinly joined parts through pressing. This produces a higher joint strength than that of conventional friction stir spot welding. Measurements of the joint interface showed the presence of a 5–10 nm thick amorphous layer consisting of Al and Mg oxides.

1. Introduction

In order to reduce CO_2 emissions and minimize energy and resource consumption, the automotive industry is increasingly being pressured to reduce the weight of vehicle bodies [1-2]. In order to do so, extending the application range of light metal materials such as Al alloys and high-tensile-strength materials is essential. In multi-material design, joints between dissimilar materials are unavoidable. When an aluminum alloy and steel are joined by conventional fusion joining such as resistance spot welding, brittle intermetallic compounds easily form at the joint interface [3-6]. In addition, the costs associated with mechanical joining methods such as riveting are high [7].

Hence, growing interest in friction stir spot welding to join metals in the solid phase without melting has recently led to its application to dissimilar joining (Al/Fe) [8]. However, achieving sufficient joint strength is of concern because of joint concavity, which tends to occur in correspondence with the tool shape, and the consequent reduction in sheet thickness [9]. Therefore, we propose using activation spot joining (ASJ) to maintain the joint thickness during the joining process and increase the joint strength; we report here on the joining of Al alloy and mild steel sheets using this method. We measured the static strength, interface morphology, and other properties of the resulting joint.

2. Experimental procedure

Friction-process equipment (owned by Kosei Aluminum Co., Ltd.) was used for the joining to control the rotational speed and tool position precisely. The chemical compositions of the workpiece materials are listed in table 1. As shown schematically in figure 1, ASJ was implemented using an Al alloy rod (A2017, 10 mm diameter) to maintain the joint thickness. The rod was pressed and rotated into an Al alloy sheet to simultaneously produce friction and an Al deposit. The deposit was removed after a

Content from this work may be used under the terms of the Creative Commons Attribution 3.0 licence. Any further distribution of this work must maintain attribution to the author(s) and the title of the work, journal citation and DOI.
Published under licence by IOP Publishing Ltd

given holding time to complete the spot joint between the 1 mm thick Al alloy (A6061-T6) sheet and 0.8 mm thick mild steel (SPC270) sheet. The resulting joint strength was assessed in terms of the tensile shear strength (TSS) and cross tensile strength (CTS). The rotational speed N, plunge speed v, plunge depth PD, and hold time t were varied in the ranges listed in table 2 to determine the optimum process conditions. Through-holes with diameters of 0 (no hole), 5, 6, or 7 mm were drilled in the Al alloy sheet prior to joining to investigate the joint strengthening effect. The joint interface morphology was observed using transmission electron microscopy (TEM) and other analysis methods.

Table 1. Chemical composition of workpiece materials (mass%).

Material	Si	Fe	Cu	Mn	Mg	Cr	Zn	Ti+Zr	C	P	S	Al
Rod (A2017)	0.50	<0.70	4.00	0.70	0.60	<0.10	<0.25	<0.20	-	-	-	bal.
Specimen (A6061)	0.59	0.38	0.26	0.03	0.96	0.25	0.02	0.04	-	-	-	bal.
Specimen (SPC270)	0.002	bal.	-	0.205	-	-	-	-	0.019	0.0104	0.005	-

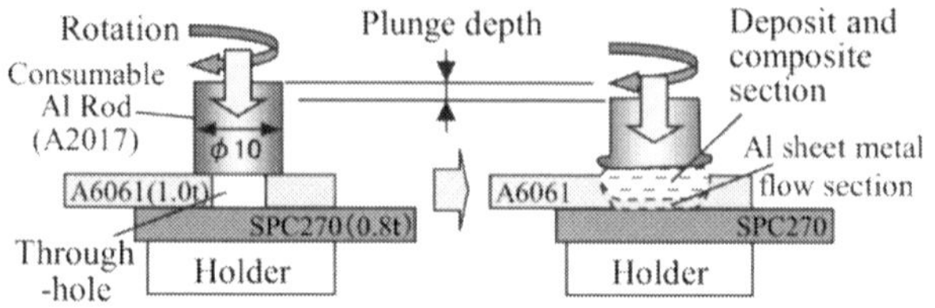

Figure 1. Schematic illustration of activation spot joining using friction surfacing.

Table 2. Spot joining conditions.

Tool rotation	N (rpm)	1000–3000
Tool plunge speed	v (mm/s)	0.5
Tool plunge depth	PD (mm)	0.8–1.4
Hold time of tool	t (s)	1, 2
Through-hole diameter	dia. (mm)	0, 5, 6, 7

3. Results and Discussion

Figure 2 shows the effect of the hole diameter on the joint strength. When the hole diameter was 5–7 mm, the TSS was 4–4.4 kN, which is twice the value of 2 kN for the joint with no through-hole. The TSS with holes was also higher than the value of 3.7 kN for the conventional joint strength as reported in the literature [8]. The CTS for a through-hole diameter of 7 mm was less than 1 kN.

Figure 3 shows the fractured steel and Al alloy sheet surfaces after TSS measurements for the case with the through-hole diameter of 7 mm. In all cases, interfacial fracture modes occurred with or without the through-hole. At the micrometer level, the fracture mode for the Al–steel joint was ductile fracture.

Figure 4(a) shows the cross-section of the joint interface with the 7 mm diameter through-hole. Regions of high elemental copper content were preferentially etched with Keller's reagent. The elemental copper from the Al rod mixed into and permeated the Al sheet metal. As shown by the electron probe micro analyzer (EPMA) element mapping in figures 4(b)–(d), little or no copper was present near the surface of the steel sheet. This indicates that the Al rod did not contact the steel layer. Based on these results, we conclude that the Al sheet metal around the hole flowed into the center, closed the hole, and thus formed a joined layer with the surface of the steel sheet.

The inferred plastic flow behavior is presented in Figure 5, which shows the time-dependent change when the rod was pushed in.

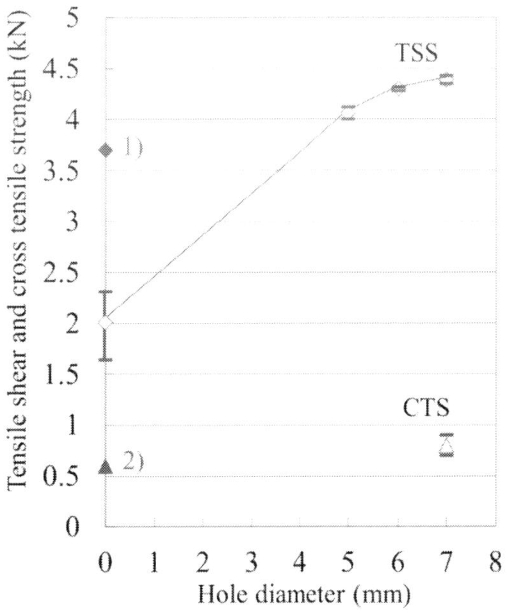

Figure 2. Results of tensile strength test: (1) Al/Fe-TSS (3.7 kN) with tool diameter of 13 mm [8]; (2) Al/Fe-CTS (0.6 kN) with tool diameter of 10 mm [9].

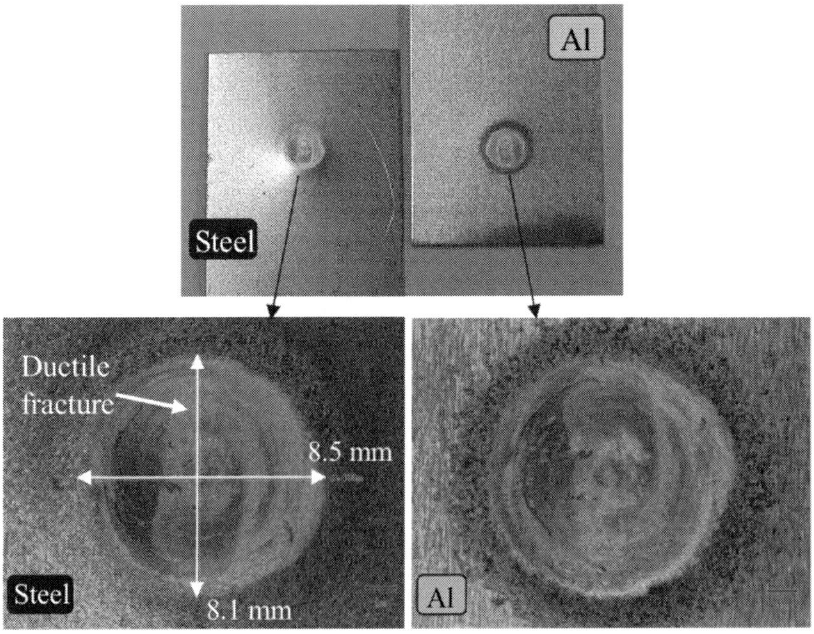

Figure 3. Fractured steel and Al alloy surfaces after TSS measurements of hole with 7 mm diameter.

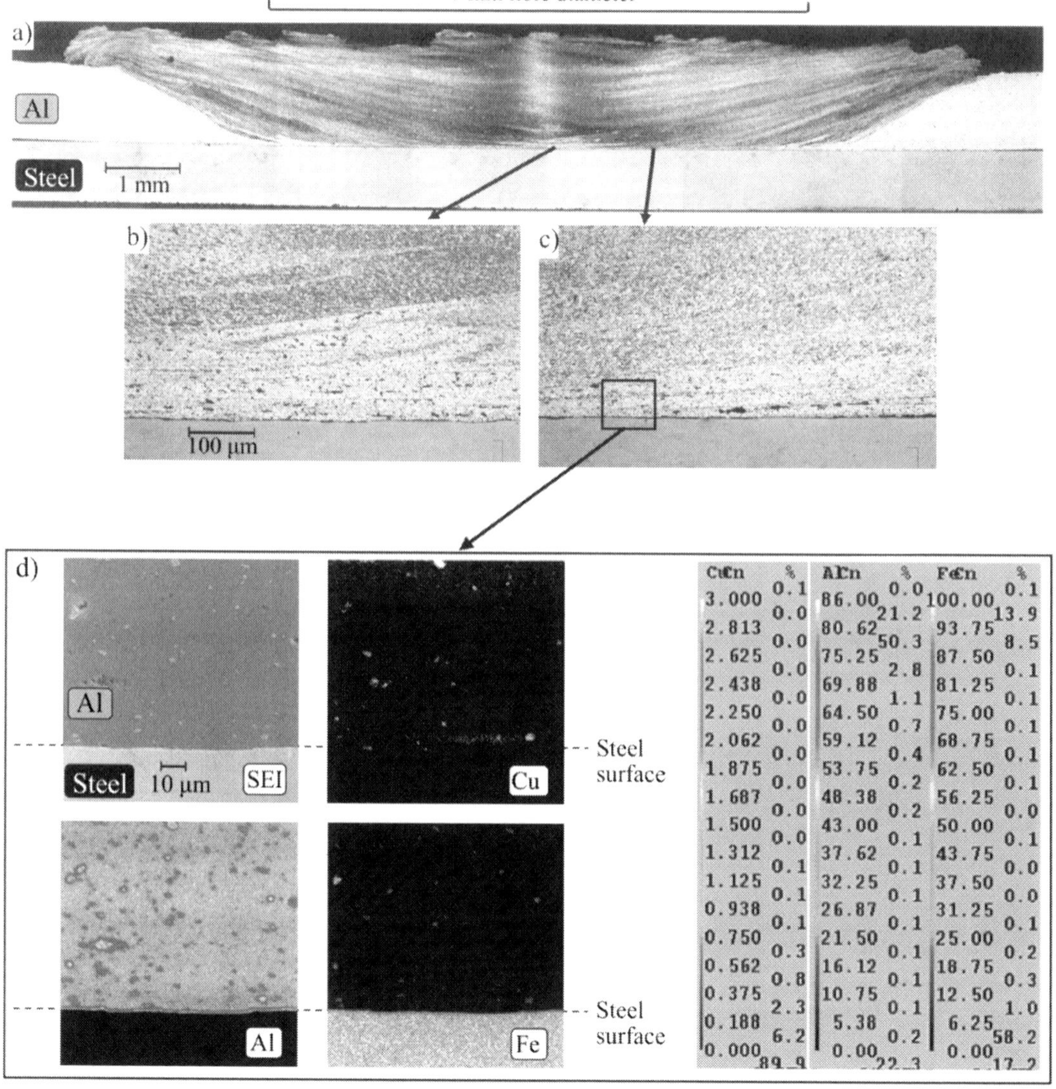

Figure 4. (a–c) Cross-sectional observations and (d) EPMA element mapping of joint interface. Hole diameter of 7 mm.

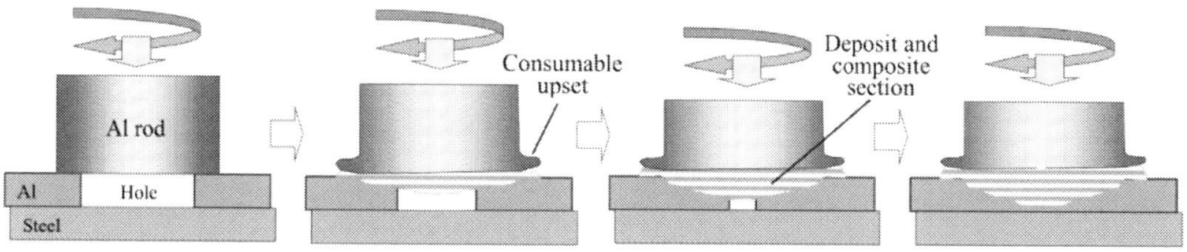

Figure 5. Inference of plastic flow behavior for activation spot joining.

To examine this inference, we performed a simulation using a simple model with a non-consumable steel rod, as shown in figure 6. The simulation was performed using the finite element method with the software FORGE for a 10 mm diameter rotating steel rod pressed down onto an Al alloy sheet with a 7 mm diameter through-hole. The simulation conditions are listed in table 3. Note that the non-consumable steel rod was a rigid body and that the Al alloy sheet was a viscoplastic body. We assumed no adhesion between the materials and $\mu = 0.3$.

Figure 7 shows the simulation and experimental results for the change in the through-hole morphology. As the plunge depth increased, the Al sheet metal tended to flow into the center and close the hole. This was because the material flowed towards areas where it was not restricted.

A simulation was also conducted without a hole; the results revealed key differences between the flow patterns with and without holes. In the simulation, we designated seven points on the Al sheet around the hole perimeter and traced their movement during the joining process with and without a hole, as shown in figure 8. The trajectory of point 6 (red line) was as expected. With the hole, it moved through three and a half revolutions. Without the hole, it stopped moving after one and a half revolutions. This clearly shows the effect of the through-hole on the generation of a dynamic plastic flow. The exposure of the newly formed Al surface at the interface was also evident.

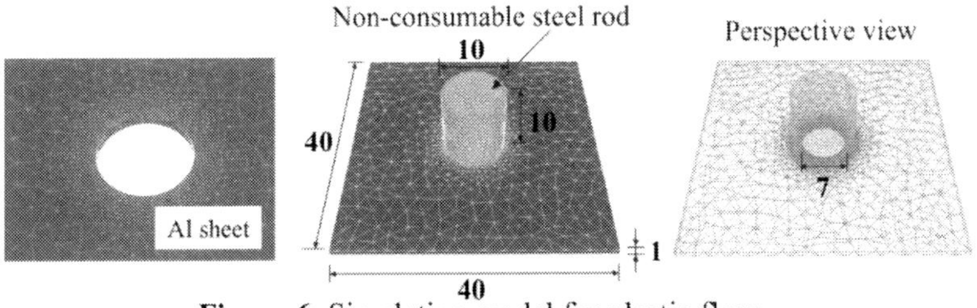

Figure 6. Simulation model for plastic flow.

Table 3. Simulation conditions.

Non-consumable steel rod	Material model	rigid body
	Initial temperature	20 ℃
	Plunge depth	0.8 mm
	Plunge speed	0.5 mm/s
	Rotation	1500 rpm
Al alloy sheet	Material model	viscoplastic body
	Deformation resistance	database of comparable A6061
	Initial temperature	20 ℃
Boundary condition	Coulomb friction	$\mu = 0.3$
	Coefficient of heat transfer	$2\ kW/m^2 \cdot K$

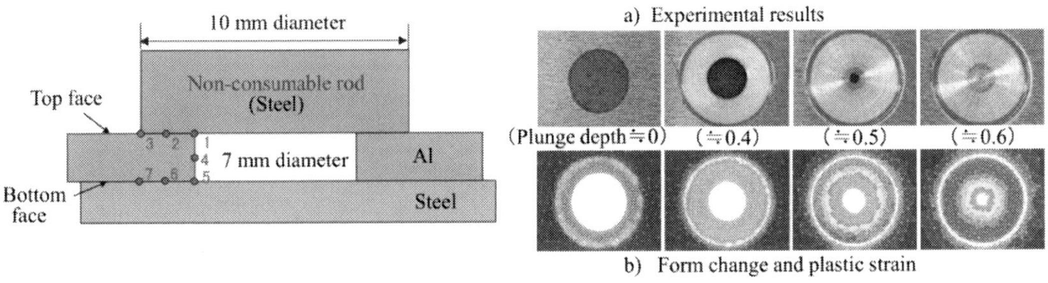

Figure 7. Experimental and simulation results of change in through-hole form.

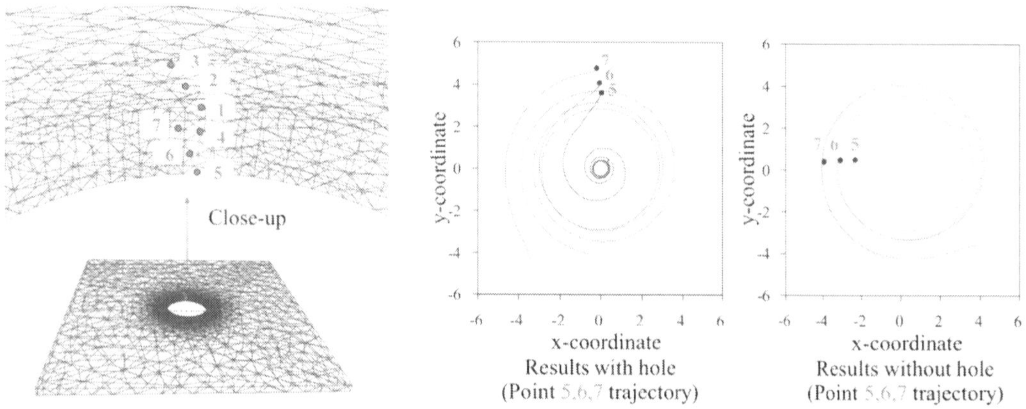

Figure 8. Comparison of Al plastic flows in pressing and rotating process with and without hole.

Figures 9(a) and (b) show the TEM imaging and energy-dispersive X-ray spectroscopy (EDS) element mapping of a joint interface with a 7 mm diameter through-hole. The results revealed the presence of both Mg and Al oxides at the interface, where they formed a 5–10 nm thick amorphous layer. The line analysis in figure 9(c) shows that Mg oxides in particular were deposited at the steel side of the interface.

Based on these results, we inferred the following model for the joining mechanism in the presence of a through-hole, as shown in figure 10: (1) The plastic flow increases the surface area of the Al sheet metal, which closes the hole. (2) A newly exposed Al surface forms, and Mg in the Al alloy sheet migrates to the surface. At the same time, the Al and Mg bond with oxygen on the steel surface. A more negative Gibbs free energy means a more easily oxidized element; thus, Mg, Al, and Fe were inferred to oxidize in that order. (3) This leads to a layer of oxidized Mg and Al and the eventual formation of the interface. Here, Mg and Al oxides were not considered to be present as connected films.

The applied load using the present method was investigated. The results are shown in figure 11. In this experiment, we varied the plunge depth of the consumable Al rod. When the plunge depth was 0.5 mm, the flow of the Al sheet metal from the hole periphery toward its center was quite evident. In this case, the actual plunge depth was a little less than 0.5 mm because of a growing consumable upset at the top of the Al rod (shown in figure 5).

These plots show the vertical load over the course of the joining process. With no through-hole (graph on right), the load rapidly increased to 8 kN in the initial stage. With the through-hole (graph on left), the load stopped rising at about 4 kN. Thus, using a hole reduces the applied load of the process.

4. Conclusions

Based on our results, the proposed method of Al/mild steel joining using a consumable Al rod demonstrates three major advantages:

- The metal flow in the joining region is increased by the through-hole in the Al sheet and consumable Al rod, which increases the joint strength compared to that achieved with existing methods. The TSS for the Al/mild steel joint was 4.4 kN (82 MPa).
- A 5–10 nm thick layer of Mg and Al amorphous oxides forms at the joint interface. The Mg in the Al appears to reduce the oxide film on the steel and contribute to interface activation.
- The applied load of this method is about half that used in existing methods.

IOP Conf. Series: Materials Science and Engineering **61** (2014) 012008 doi:10.1088/1757-899X/61/1/012008

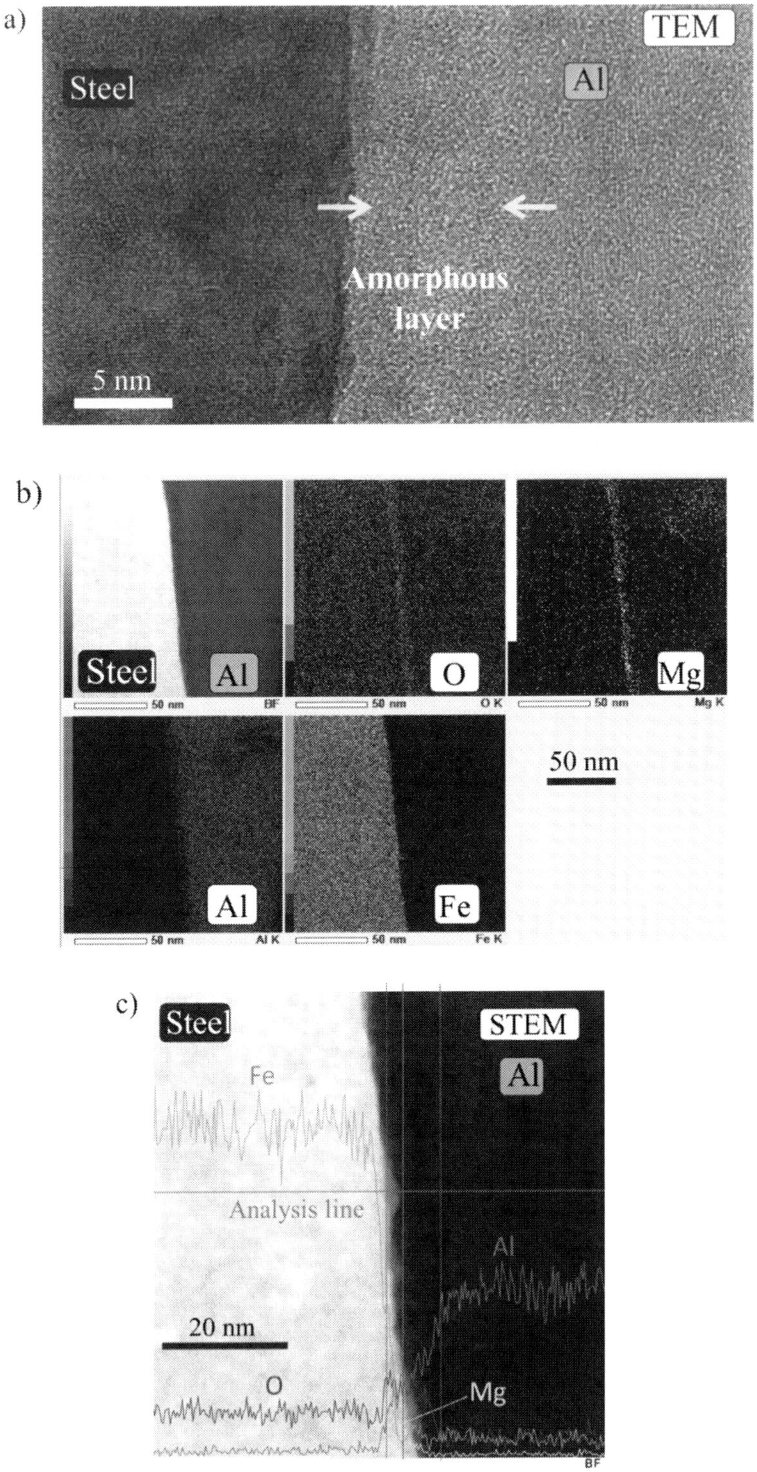

Figure 9. (a) TEM imaging, (b) EDS element mapping, and (c) line analysis of joint interface.

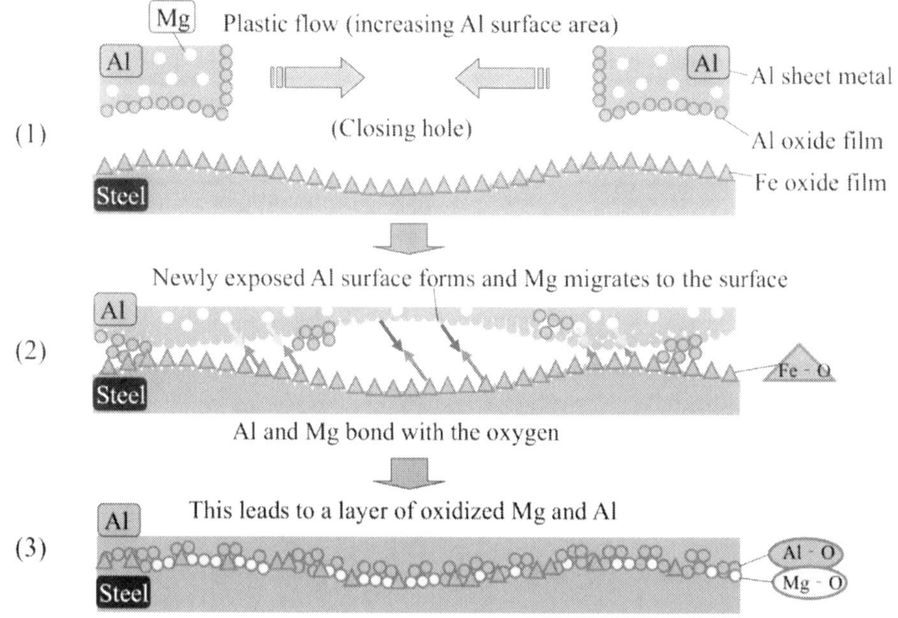

Figure 10. Bonding model of interface for activation spot joining.

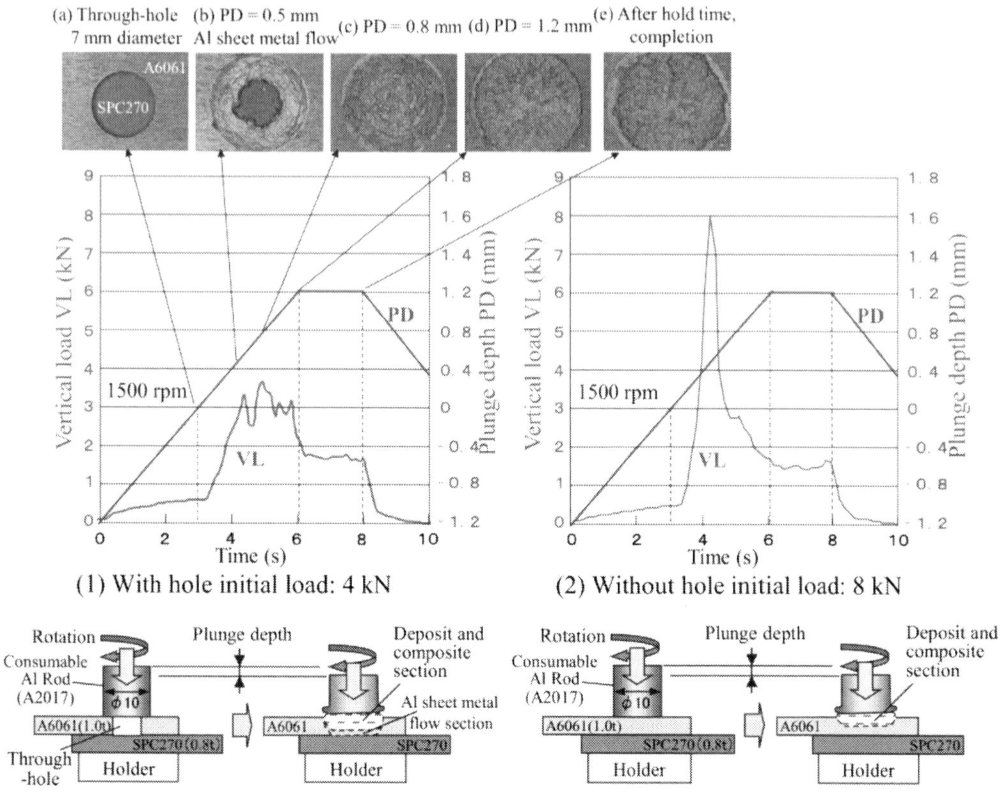

Figure 11. Comparison of changes in vertical load for joining process with and without hole.

Acknowledgments
We thank Professor Masahiro Fukumoto and Associate Professor Toshiaki Yasui at the Department of Mechanical Engineering, Toyohashi University of Technology, for valuable discussions.

References
[1] Furusako S, Watanabe F, Murayama G, Hamatani H, Oikawa H, Takahashi Y and Nose T 2013 *Nippon Steel Tech. Rep.* **103** 69
[2] Figner G, Vallant R, Weinberger T, Schröttner H, Pašič H, and Enzinger N 2008 *Int. Inst. Weld. Doc.* SC-Auto-20-08
[3] Yasuyama M, Ogawa K and Taka T 1996 *Q. J. Jpn. Weld. Soc.* **14(2)** 314
[4] Oikawa H, Saito T, Yoshimura T, Nagase T and Kiriyama T 1996 *Q. J. Jpn. Weld. Soc.* **14(1)** 69
[5] Watanabe T, Doi Y, Yanagisawa A and Konuma S 2005 *Q. J. Jpn. Weld. Soc.* **23(3)** 491
[6] Takeda M, Urushihara W, Matsumoto K and Kato J 2007 *Kobe Steel Eng. Rep.* **57(2)** 69
[7] Sakiyama T, Murayama G, Naito Y, Saita K, Miyazaki Y, Oikawa H and Nose T 2013 *Nippon Steel Tech. Rep.* **103** 93
[8] Tanaka K, Kumagai M and Yoshida H 2006 *J. Jpn. Inst. Light Met.* **56(6)** 317
[9] Miyagawa K, Tsubaki M, Yasui T and Fukumoto M 2008 *Q. J. Jpn. Weld. Soc.* **26(1)** 42

Effect of weld line shape on material flow during friction stir welding of aluminum and steel

Toshiaki Yasui, Naoyuki Ando, Shinpei Morinaka, Hiroki Mizushima and Masahiro Fukumoto

Toyohashi University of Technology, 1-1 Hibarigaoka, Tempaku, Toyohashi, Aichi, JAPAN

E-mail: yasui@tut.jp

Abstract. The effect of weld line shape on material flow during the friction stir welding of aluminum and steel was investigated. The material flow velocity was evaluated with simulated experiments using plasticine as the simulant material. The validity of the simulated experiments was verified by the marker material experiments on aluminum. The circumferential velocity of material around the probe increased with the depth from the weld surface. The effect is significant in cases where the advancing side is located on the outside of curve and those with higher curvature. Thus, there is an influence of weld line shape on material flow.

1. Introduction

Friction stir welding (FSW) is solid-state welding technique in which frictional heating and plastic material flow are induced by a welding tool. It is useful for welding aluminum alloys and is typically applied in the automotive and high-speed railway industries. The FSW of aluminum and steel can achieve high joining efficiency because of the thin intermetallic compound layer (less than 1μm thick) that forms at the weld interface [1-6]. Recently, this method has received attention in the production of hybrid structure components for transportation vehicles, which balances the cost and the structural strength.

Control of material flow behavior is important in FSW to achieve high joining efficiencies. However, the process window for welds between aluminum and steel is smaller than that for welds between aluminum alloys. The material flow changes with welding conditions (welding speed, rotation speed, tool shape, and weld shape) and can generate defects in the weld under unfavorable material flow conditions. Material flow in the FSW of aluminum and steel is different from that of aluminum alloys because of the existence of the rigid steel wall at the weld interface and the weld-shape effect on the material flow and joining efficiency [7]. Thus, material flow during the FSW of aluminum and steel needs to be investigated further. Material flow during FSW has been studied using microstructural observations, X-ray techniques, simulated experiments with simulant materials, and numerical simulation. However, the details of material flow have not been established yet.

Material flow observations made using marker materials or simulant materials are an effective way to understand the material flow visually [8-11]. In this study, the effect of weld line shape on material flow was investigated by evaluating the material flow velocity quantitatively. This simulated experiment was conducted using plasticine as the simulant material. The validity of the simulation was verified by microstructure observations of aluminum with a marker material.

Content from this work may be used under the terms of the Creative Commons Attribution 3.0 licence. Any further distribution of this work must maintain attribution to the author(s) and the title of the work, journal citation and DOI.
Published under licence by IOP Publishing Ltd

2. Experimental procedure

2.1. FSW between aluminum and steel with 2D weld line

The arrangements of the test pieces for the 2D weld line are shown in Figure 1(a). The weld line shape included both straight and curved segments with 30 and 50 mm radii. The thickness of test pieces was 6 mm. The welding tool was composed of a shoulder 20 mm in diameter and a probe with an M4 left-hand screw 4 mm in length, as shown in Figure 1(b). The rotating tool probe is plunged into a test piece on the retreating side (RS) that is made of A6063 or plasticine, as shown in Table 1. In all of the experiments, steel (S45C) was placed on the advancing side (AS).

The rotating tool moves along the weld line, slightly grinds the steel surface on AS, and stirs a test piece on the RS. To understanding the effect of curvature, three types of weld were fabricated, as shown in Figure 2. As for the curve weld line, we need to consider the curvature and arrangement of the test piece. Straight welding is considered as having an infinite radius of curvature ($R = \infty$), as shown in Figure 2(a). Curved welds with a negative radius of curvature ($R < 0$) are defined as having the AS located on the outside of the curve, as shown in Figure 2(b), and curved welds with a positive radius of curvature ($R > 0$) are defined as having the AS located on the inside of curve, as shown in Figure 2(c).

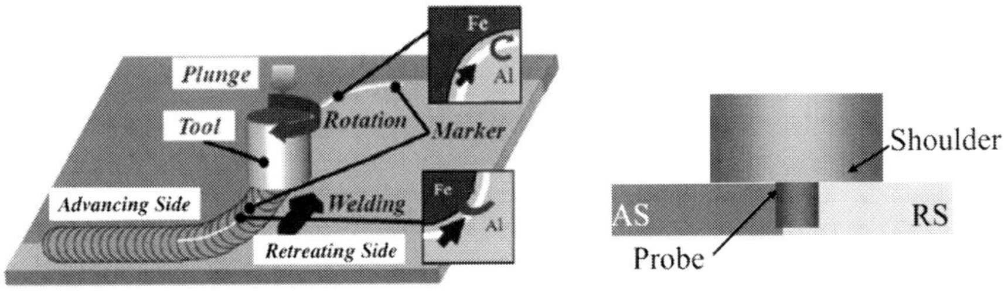

(a) (b)

Figure 1. Schematic drawing of the material flow observation experiment for FSW of steel and aluminum.

Table 1. Test piece and marker materials and geometries.

Test piece (AS)	S45C (t6mm)		
Test piece (RS)	A6063 (t6mm)		plasticine (t6mm)
Marker material	Straight welding	aluminum oxide (6μm)	plasticine (1mm width)
	Curve welding	pure aluminum (ϕ0.3mm)	

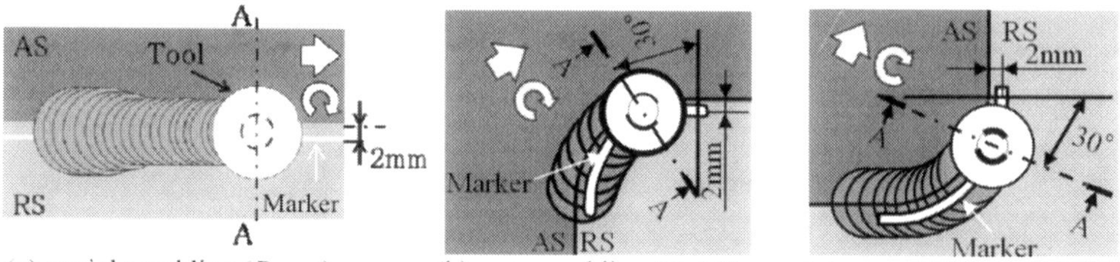

(a) straight welding ($R = \infty$) (b) curve welding ($R < 0$) (c) curve welding ($R > 0$)

Figure 2. Schematic of material flow observation experiments on the FSW of steel and aluminum. The arrows indicate the tool rotation and movement.

In order to visualize the material flow, marker materials, as shown in Table 1, were embedded in the test pieces. In case of straight welds, an aluminum oxide layer was fabricated by a plasma electrolytic oxidation process. A thin aluminum plate was inserted between the marker material and steel for the arrangement of the marker. However, in the curve welding case, it is difficult to fabricate a thin, curved aluminum plate. In this case, vertical holes were drilled along the marker line, and pure aluminum wires were embedded into the holes. Also, different colors of plasticine were layered for the simulated weld experiment. Marker materials were embedded at the tool center position, as shown in Figure 2. The welding tool stopped at the A-A plane and withdrew from the test piece. Material flow around the exit hole of the probe was observed by cutting the weld horizontally at 1, 2, and 3 mm depths from the weld surface. In case of aluminum on the RS, the horizontal cross section was ground and etched with Keller's reagent for marker material observations. The welding conditions are summarized in Table 2.

Table 2. Welding conditions.

Test piece (RS)	A6063		Plasticine	
Straight Welding	No defect	Defect	No defect	Defect
Rotation speed (min^{-1})	2000	2000	500	200
Welding speed (mm/min)	800	100	1100	100
Curvature radius R (mm)	∞ (straight), -50, -30, 30, 50			

2.2. Material flow velocity
Schmidt et al. proposed an evaluation method for the material flow velocity around the tool probe by using a trace of marker material layer [8]. However, it is difficult to introduce marker material layer in FSW because of the difference in properties between aluminum and the marker material. Thus, the material flow velocity was evaluated using the simulated experiments with the simulant materials. A viscoplastic material, plasticine, has been used in the simulated experiment of the metal forming process. Liechty et al. discussed the applicability of plasticine to these simulated FSW experiments [9, 10, 11].

The material flow was observed by using different colored plasticine. Plasticine (Van Aken Int., Inc.) was used as the simulant material, as shown in Table 1, and the circumferential velocity of material flow around the probe was evaluated. Figure 3 shows the model for the velocity evaluation; the derived equation is given by

$$\overline{V}_{\theta_1-\theta_2} = \frac{S_{\theta_1-\theta_2} \cdot U \cdot W_t \cdot \rho_{pixel}}{M_{\theta_1-\theta_2}} \quad (1)$$

where $S_{\theta_1-\theta_2}$ is the measured distance of marker (mm), U is the welding speed (mm/min), W_t is the width of marker material (mm), ρ_{pixel} is the pixel density (pixel/mm^2), and $M_{\theta_1-\theta_2}$ is the number of pixels for measuring the distance.

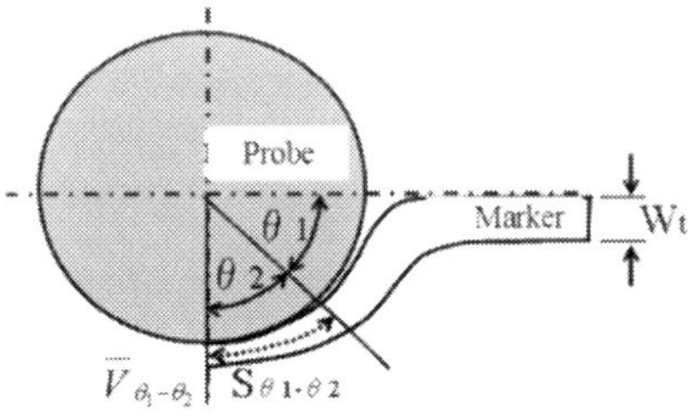

Figure 3. Derivation of the circumferential velocity of material flow around the probe.

3. Results and discussion

3.1. Straight welding

Figure 4 shows the material flow around the probe for straight welding without any defect. The marker materials are observed as a white line for aluminum and a dark strip for plasticine. Both of the marker materials flow toward the back of the probe and reach the weld interface. The distributed area of the marker materials increases with the depth from the surface. Both marker material movements show similar tendencies. From these images of the simulated experiment, the circumferential velocity of the marker material around the probe was evaluated. These velocities are 1212 mm/min at 1 mm depth from the surface and 1380 mm/min for 3 mm depth from the surface. As for the simulated experiment, the inlet velocity of material flow to the probe, i.e. welding speed, is 1100 mm/min and the circumferential velocity of the probe surface is 6280 mm/min. Although the circumferential velocity of the marker material is higher than the inlet velocity, the probe surface moves faster than the material and induces slip between the probe surface and flowing material. Also, the circumferential velocity of marker material at the bottom part is higher than that at the upper part. One reason for this is the difference in the slip condition caused by the difference in temperature around the probe. Another reason is the effect of the downward material flow produced by the probe screw [6].

Figure 5 shows the material flow around the probe for straight welding with defects. The defects are observed at middle and bottom part of the weld interface behind the probe on both aluminum and plasticine. From the marker material movements, this can be attributed to the lack of material flow behind the probe resulting from low heat input. From the simulated experiment, the circumferential velocities of the marker material are 150 mm/min at 1 mm depth from the surface, and 317 mm/min at 3 mm depth from the surface. As for the simulated experiment, the inlet velocity of material flow to the probe is 100 mm/min and the circumferential velocity of the probe surface is 2512 mm/min. The circumferential velocity of the marker material is close to the inlet velocity of material flow to the probe. However, the bottom part of material flow is higher than the inlet velocity. It is considered that the difference in slip condition and the effect of downward material flow by the probe screw produce this effect.

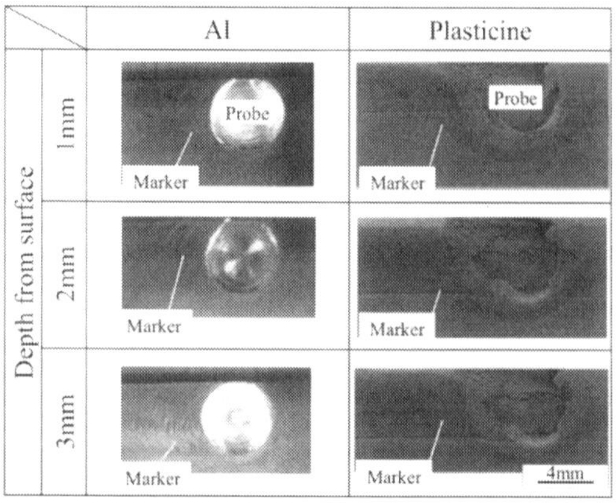

Figure 4. Material flow around the probe during straight welding for no defect.

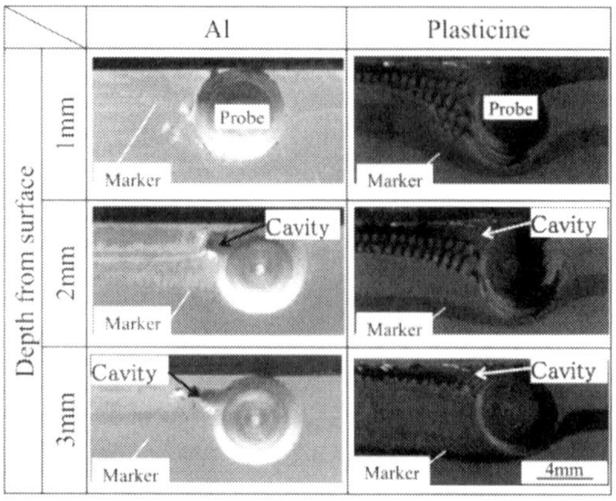

Figure 5. Material flow around the probe during straight welding for defect.

3.2. Curve welding

Figures 6 and 7 show the material flow at positive radius of curvature. The welding conditions were the "no defect" condition of straight welding, above. However, defects are observed at bottom part of the weld behind the probe for both radii. Furthermore, defects are observed at middle part of the weld in the plasticine. The circumferential velocities of marker material ranged from 1600 mm/min to 2200 mm/min. These values exceed the velocities of straight welding.

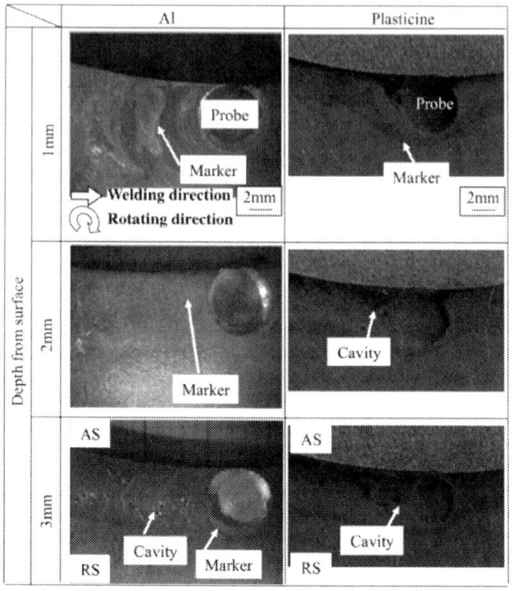

Figure 6. Material flow around the probe at +50 mm radius of curvature.

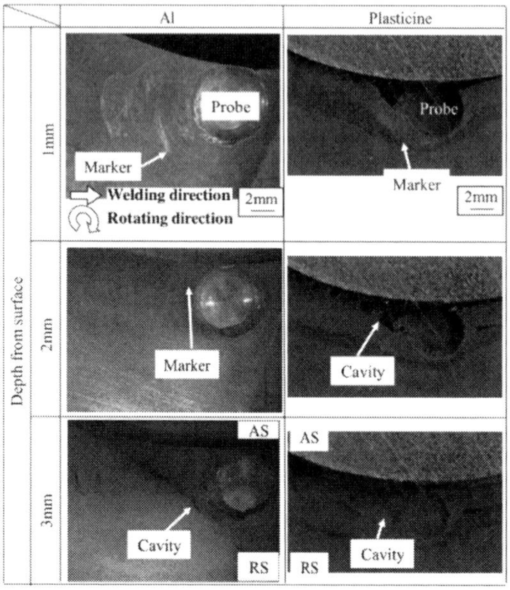

Figure 7. Material flow around the probe at +30mm radius of curvature.

Figures 8 and 9 show the material flow at negative radius of curvature. The welding conditions were the "no defect" condition of straight welding, above. Defects are observed throughout the thickness at the –30 mm radius of curvature for both Al and plasticine. On the other hand, defects are observed in the bottom part of the weld at a –50 mm radius of curvature. From the trace of the marker material for plasticine, the material flow falls short of reaching the weld interface. The circumferential velocities of marker material ranged from 1700 mm/min to 2500 mm/min. The highest velocity was obtained on the bottom of the weld at a radius of curvature of 50 mm.

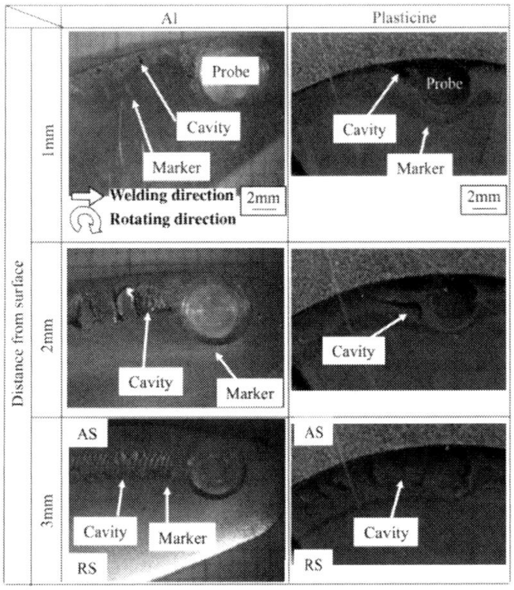

Figure 8. Material flow around the probe at –30 mm radius of curvature.

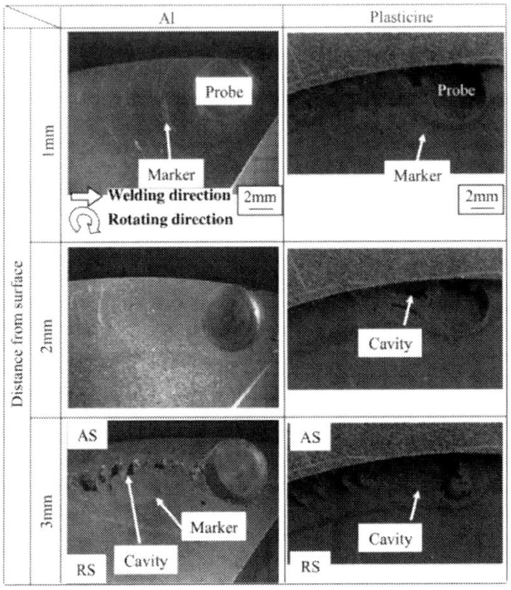

Figure 9. Material flow around the probe at –50mm radius of curvature.

Figure 10 shows the circumferential velocity of marker material around the probe at various curvatures under the "no defect" condition of straight welding. The velocities increase with depth from the surface. The effect is significant at negative curvature and at higher absolute values of curvature ($|R^{-1}|$). This is attributed to the effect of heat input and the travel distance of materials induced by the difference in weld line shape. The heat input depends on the contact area between the tool shoulder and aluminum, and this area differs with the weld line shape. From temperature measurements performed on the probe tip with a thermocouple for the welding of aluminum and steel, the temperature of each weld line at the point of observation is 698.6 K for positive curvature, 697.7 K for straight, and 699.1 K for negative curvature. These differences are small, and it is difficult to estimate the effect of heat input.

As for the travel distance of the marker material for welds with negative curvature, the marker material line in front of the tool curved toward the RS. This means that the marker material flows easily to the back side of the tool because of the short distance. As for the travel distance of marker material for welds with positive curvature, the marker material line in front of the tool curved toward the AS. This means that the marker material flows to the back side of the tool for long distances. The difference in travel distance affects to the volume of stirred aluminum and the overall material flow. Thus, the combined effect of heat input and travel distance affects the velocity difference in welds with different line shapes.

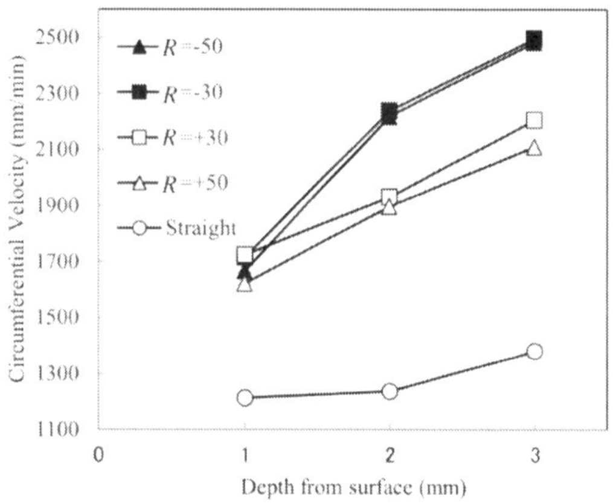

Figure 10. Circumferential velocity of marker material around the probe. (Under the "no defect" condition for straight welding).

4. Conclusion

The effect of weld line shape on material flow during the friction stir welding of aluminum and steel was investigated using marker material experiments and simulated experiments. The validity of the simulated experiments using plasticine was verified by the marker material experiments using aluminum, and the material flow velocity was evaluated for these simulated experiments. The circumferential velocity of the marker material around the probe increased with the depth from the weld surface. The effect is significant at negative curvature and at higher absolute values of radius of curvature. Thus, there is an influence of the weld line shape on the material flow.

Acknowledgement

This study was partially supported by "The Knowledge Hub of AICHI", The Priority Research Project, and by JSPS KAKENHI under Grant Numbers 24560876.

References

[1] Yasui T, Shimoda Y, Tsubaki M and Fukumoto M 2004 *Materials Science Forum* **449-452** 433
[2] Fukumoto M, Tsubaki M, Shimoda Y and Yasui T 2004 *Quarterly J. Japan Welding Society* **22** 309
[3] Yasui T, Shimoda Y, Tsubaki M, Ishii T and Fukumoto M 2005 *Quarterly J. Japan Welding Society* **23** 469
[4] Yasui T, Shimoda Y, Ishii T, Tsubaki M and Fukumoto M 2007 *Quarterly J. Japan Welding Society* **25** 426
[5] Shimoda Y, Tsubaki M, Yasui T, Fukumoto M, Fujita T and Osawa J 2010 steel research international **81** 1108
[6] Shimoda Y, Tsubaki M, Yasui T and Fukumoto M 2011 *Quarterly J. Japan Welding Society* **29** 114s
[7] Ishida S, Tsubaki M, Yasui T and Fukumoto M 2011 *Proc. Japanese Joint Conference for the Technology of Plasticity* **62** 221 (in Japanese)
[8] Schmidt H N B, Dickerson T L, and Hatterl J H 2006 *Acta Materialia* **54** 1199
[9] Liechty B C, Webb B W 2007 *J. Materials Processing Technology* **184** 240
[10] Liechty B C, Webb B W 2008 *Int. J. Machine Tools and Manufacture* **48** 1474
[11] Liechty B C, Webb B W 2008 *J. Materials Processing Technology* **208**, 431

IOP Conf. Series: Materials Science and Engineering **61** (2014) 012010 doi:10.1088/1757-899X/61/1/012010

Effect of formation and state of interface on joint strength in friction stir spot welding for advanced high strength steel sheets

Koichi Taniguchi, Muneo Matsushita, Rinsei Ikeda and Kenji Oi

JFE Steel Corporation, 1 Kawasaki-cho, Chiba-shi, Chiba, Japan

E-mail: ko-taniguchi@jfe-steel.co.jp

Abstract. The tensile shear strength and cross tension strength of friction stir spot welded joints were evaluated in the cases of lap joints of 270 N/mm^2 grade and 980 N/mm^2 grade cold rolled steel sheets with respect to the stir zone area, hardness distribution, and interface condition between the sheets. The results suggested that both the tensile shear strength and cross tension strength were based on the stir zone area and its hardness in both grades of steel. The "hook" shape of the interface also affected the joint strength. However, the joining that occurred across the interfaces had a significant influence on the value of the joint strength in the case of the 270 N/mm^2 grade steel.

1. Introduction

Because it is a solid state joining process, friction stir spot welding (FSSW) is expected to overcome the welding issues for advanced high-strength steel (AHSS) found in conventional fusion welding processes. However, its implementation requires further assurance of the mechanical properties based on a detailed understanding of the welding phenomena.

Studies on the mechanical properties have shown that the factors influencing the joint strength are the stir zone (SZ) area, microhardness of the joint weld, and shape of the interface in the case of a lapped weld joint of two sheets of an identical aluminum alloy [1–3]. Among these, the most influential factor for the joint strength was found to be the SZ area, with a larger SZ area producing a higher joint strength [1]. As a characteristic phenomenon in FSSW, the shape of the interface was deformed by the plastic flow around the tool, which had a significant influence on the joint strength. With the formation of the SZ, the interface was supposed to rise up toward the upper sheet, extend around the SZ, reach the area of the probe, and end up in the form of a hook. This "hook" shape was supposed to cause a failure along its direction and lower the joint strength.

In contrast, sufficient investigations have not yet been performed to determine the factors influencing the joint strength of a lapped friction stir spot weld joint of two sheets of an identical automotive steel, especially AHSS above the 980 N/mm^2 grade. However, some reports have shown that the SZ area and hardness of the joint have effects on the joint strength, just as when using FSSW for aluminum alloys [4–6]. Besides these effects, the influence of the hardness distribution, or the localized strength within the weld, on the joint strength during FSSW should be considered in the case of AHSS, analogous to the trend seen in resistance spot welding (RSW). The higher microhardness of the weld basically increases the joint strength. However, when the microhardness of the weld is unfavorably high, and the joint is loaded in the direction to open the lap sheets, the joint strength may decrease. This is considered to be because the high microhardness in the weld is related to a brittle

Content from this work may be used under the terms of the Creative Commons Attribution 3.0 licence. Any further distribution of this work must maintain attribution to the author(s) and the title of the work, journal citation and DOI.
Published under licence by IOP Publishing Ltd

microstructure, which results in brittle fracture, and the high microstructure in the heat affected zone (HAZ) prevents the plastic deformation, which results in a stress concentration at the weld [7–10].

However, there have been no reports on the influence of the shape and state of the interface on the joint strength with a comparison between conventional mild steel and AHSS up to a grade of 980 N/mm^2. Therefore, the mechanical properties of friction stir spot welds of 270 N/mm^2 grade steel and 980 N/mm^2 grade steel were evaluated, and the influence of the shape and state of the interface on the joint strength was investigated.

2. Experimental procedure

FSSW was used to form lapped weld joints in two sets of 1.2-mm-thick identical steel sheets, as listed in table 1, where TS270 and TS980 represent 270 N/mm^2 grade steel and 980 N/mm^2 grade steel sheets, respectively. A Si_3N_4 ceramic tool was used, which consisted of a flat shoulder with a diameter of 10 mm and a threaded probe with a diameter of 3.8 mm and a height of 1.7 mm. The weld time was 0.5–3.0 s, the plunge force was 6 kN, and the rotation speed was 3000 rpm. During the FSSW, Ar gas was used for shielding the weld.

The cross-sectional microstructures were observed using optical microscopy after etching with a saturated picric acid solution or 3% nital reagent. The projected cross-sectional SZ area, which represented S, was estimated using Eq.(1).

$$S = \pi \cdot \left\{ \left(\frac{Ds}{2} \right)^2 - \left(\frac{Dp}{2} \right)^2 \right\} \tag{1}$$

where Ds is the diameter of the SZ, and Dp is that of the plunged area measured by cross-sectional metallography, as shown in Figure 1.

A tensile shear test and cross tension test were conducted for the obtained weld joints, and the fracture patterns were evaluated by the cross-sectional metallography of the fractured specimens.

Microhardness testing of the weld was carried out using a 1.9-N load and a holding time of 10 s with a 0.2-mm grid spacing in both the through-thickness and transverse directions. The hardness distribution of the welds was plotted as either a contour map or transverse hardness profile for the upper sheet side at 0.3 mm from the surface lapped with the lower sheet.

Table 1. Chemical compositions and mechanical properties of specimens.

	Chemical compositions in mass %			Mechanical Properties		
Steel	C	Si	Mn	YP(N/mm^2)	TS(N/mm^2)	El(%)
TS270	0.002	0.003	0.13	160	300	51
TS980	0.12	1.4	1.9	810	990	18

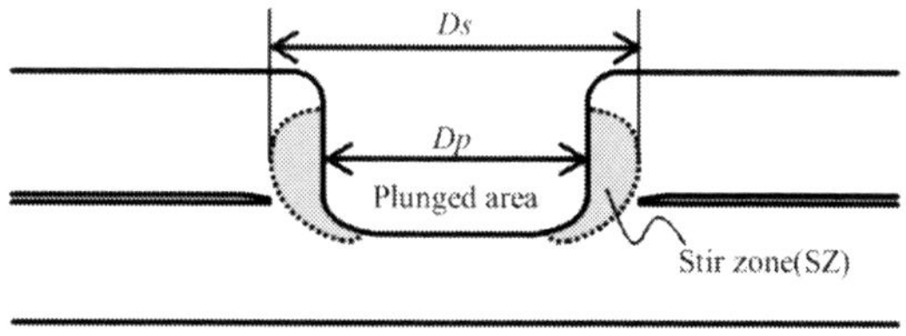

Figure 1. Schematic drawing of parameters for estimation of SZ area.

3. Results

The cross-sectional macrostructures at each weld time are shown in Figure 2. For TS270, the SZ was already formed and two sheets were completely in contact at 1.0 s, and the SZ area gradually increased with the weld time. For TS980, the SZ was formed in 1.0 s despite the lack of full contact between the sheets and increased in area with the weld time. The relationship between the SZ area estimated by Eq.(1) and the weld time is shown in Figure 3. The SZ area basically increased with the weld time in similar ways for both grades. However, that of TS980 was less than that of TS270 at 1.0 s, while that of TS270 was less than that of TS980 after 2.0 s or longer.

The hardened area and its increase were limited to the area around the SZ in TS270, whereas a broad area as wide as the microstructure change showed a large increase in microhardness in TS980. Figure 4 shows the cross-sectional macrostructures etched by the natal reagent and the microhardness distribution based on the values derived from the 0.2-mm grid spacing with a weld time of 3.0 s. In the case of TS270, the hardened area was limited to the area around the SZ, and the values around the interface close to the probe (A1) were almost equal to those of the base metal (BM). However, in the case of TS980, the hardened area extended from the area around the interface close to the probe (B1) to an area at a distance of 0.8 mm (B2). Figure 5 shows the cross-sectional hardness distribution in the weld, measured at the upper sheet side at 0.1 mm from the interface. The highest hardnesses were observed in the SZ areas for both steels, with values reaching Hv195 in TS270 and Hv464 in TS980.

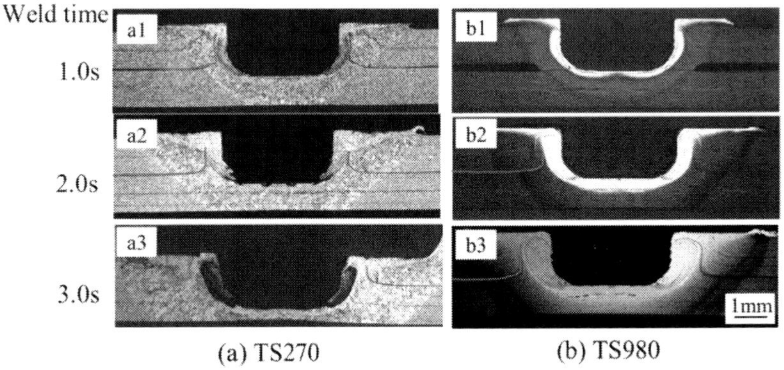

Figure 2. Cross-sectional macrostructures etched by saturated picric acid with respect to weld time, where interface shapes are drawn with curves on left sides.

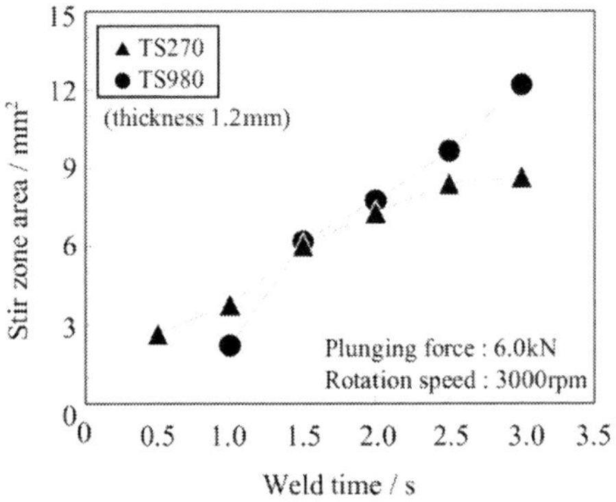

Figure 3. SZ areas of TS270 and TS980 with weld times.

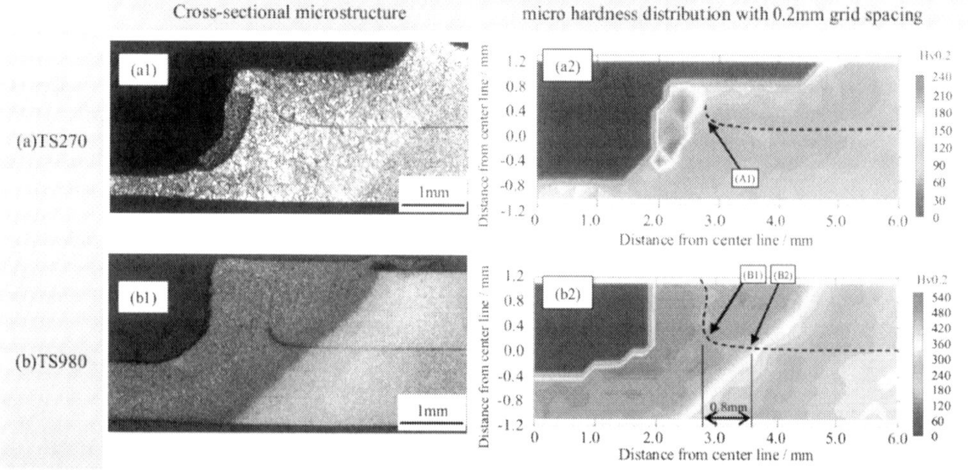

Figure 4. Comparison of cross-sectional microstructure etched by natal regent and microhardness distribution with 0.2-mm grid spacing at weld time of 3 s, where dashed lines represent interfaces in (a2) and (b2).

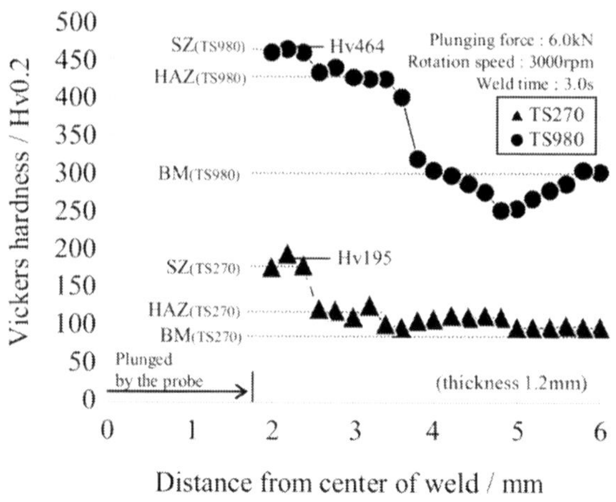

Figure 5. Hardness distributions of TS270 and TS980 on upper sheet side at 0.1 mm from interface with weld time of 3 s, where SZ, HAZ, and BM represent average hardnesses of SZ, HAZ, and BM, respectively.

The value of the average hardness of the HAZ in TS980 was close to that for the SZ. However, that of the HAZ in TS270 showed a low value and was relatively close to that of the BM.

Figure 2 also shows that the interface rose up toward the upper sheets in both cases within 2.0 s or less, which is called the "rising-up" shape in this paper. The interface of TS270 at 3.0 s was also the "rising-up" shape and extended to the surface of the upper sheet. However, the interface of TS980 at 3.0 s was extended around the SZ through the upper sheet and reached the probe, which showed a "hook" shape.

These shape transitions in the interface from "rising-up" to a "hook" were assumed to be caused by the extent of the metal flow, which dragged the lower sheet surface into the upper sheet, formed the "rising-up" shape of the interface, and subsequently drove the "rising-up" interface inward toward the direction of the probe when it was extended close to the shoulder, forming the "hook" shape [11]. Because of the transition, the interface inside the upper sheet was supposed to expand the surfaces of the sheets containing oxides and other inclusions, which resulted in a week interface between the sheets [2].

The joining across the "rising-up" interface could be seen in TS270, whereas the interface was completely separated in TS980, even in the similar "rising-up" shape. The detailed cross-sectional microstructures around the interface with a weld time of 2.0 s are shown in Figure 6 and Figure 7. The interfaces for the two grades seem to have similar "rising-up" shapes in a comparison between (a) and

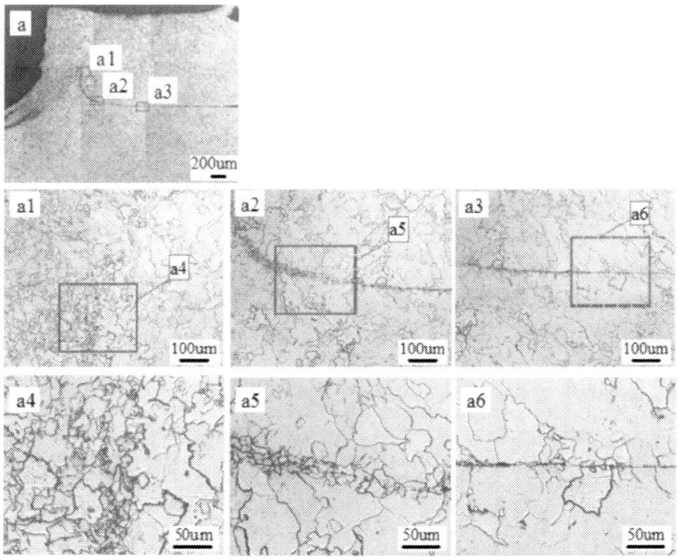

Figure 6. Cross-sectional microstructures of TS270 with weld time of 2 s, where distances from probe are (a1) 0.6 mm, (a2) 1.1 mm, and (a3) 2.0 mm, as shown in (a).

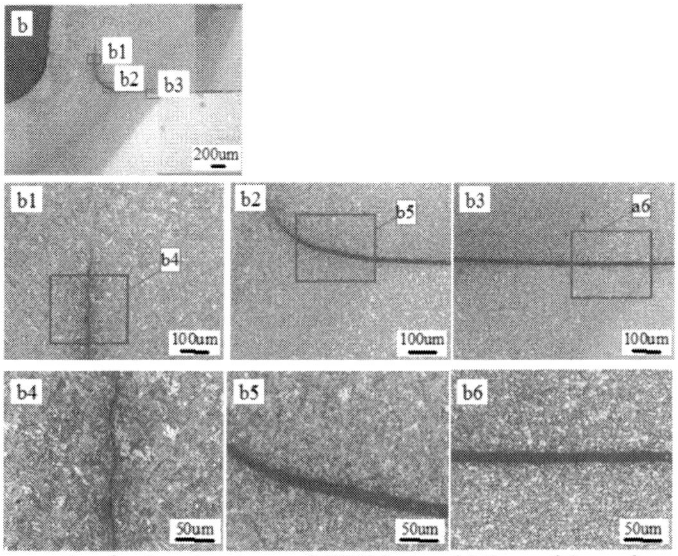

Figure 7. Cross-sectional microstructures of TS980 at weld time of 2 s, where distances from probe are (b1) 0.6 mm, (b2) 1.1 mm, and (b3) 2.0 mm, as shown in (b).

IOP Conf. Series: Materials Science and Engineering **61** (2014) 012010 doi:10.1088/1757-899X/61/1/012010

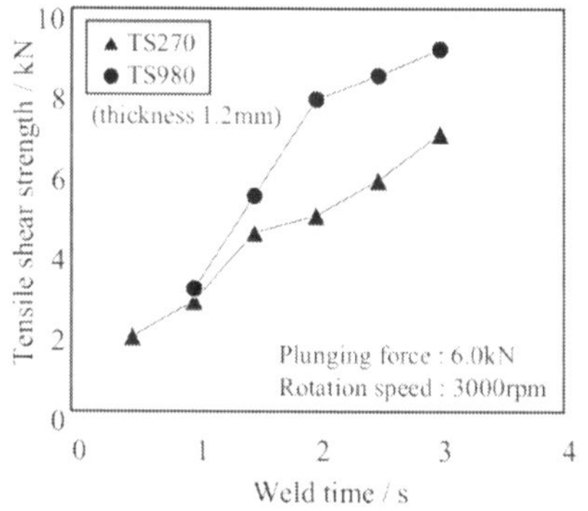

Figure 8. TSS values of TS270 and TS980 with respect to weld time.

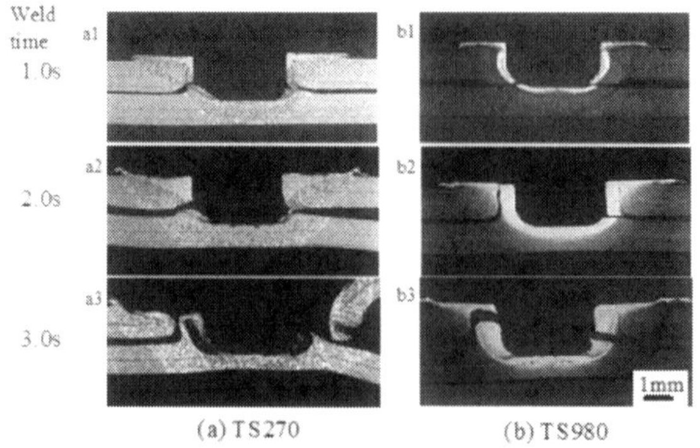

Figure 9. Cross-sectional microstructures of fractured samples after tensile shear test.

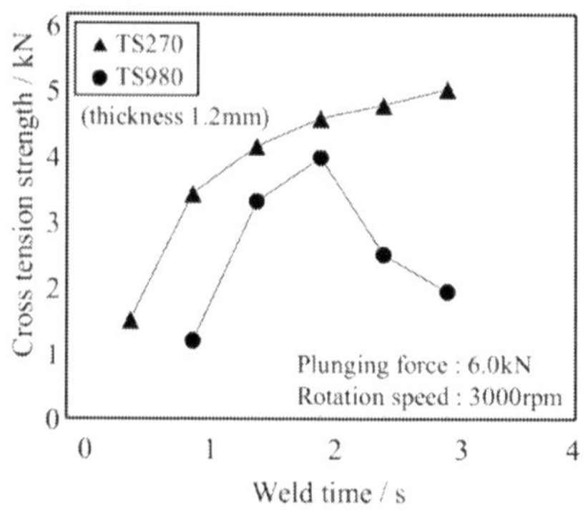

Figure 10. CTS values of TS270 and TS980 with respect to weld time.

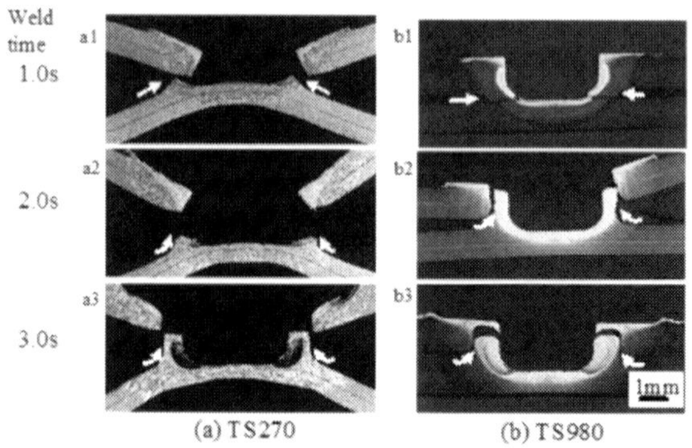

Figure 11. Cross-sectional microstructures of failed samples after cross-tension test, where arrows represent locations of initial cracks and directions suggested by failure mode.

(b). However, at the tip of the "rising-up" interface at 0.6 mm from the probe, the interface seems to be joined as fine ferrite grains in TS270, as seen in (a1) and (a4), whereas an obvious separation can be observed in TS980 at (b1) and (b4). In TS270, the ferrite grains were joined between the upper and lower sheet at 1.1 mm, as seen in (a2) and (a5), and some of them were similarly joined at 2.0 mm, as seen in (a3) and (a6). In contrast, as shown in (b2) and (b5) or in (b3) and (b6), the interface was clearly visible and the microstructures were different across the interface, which indicated that the joining was incomplete in these regions in TS980.

The tensile shear strength (TSS) increased with the weld time in both steels, as shown in Figure 8. Figure 9 shows the cross-sectional microstructure of a fractured specimen after the tensile shear test. At 1.0 s, both steels were fractured across the SZ, as shown in (a1) and (b1). At 2.0 s, the joint of TS270 was also fractured across the SZ, as seen (a2), whereas one side transitioned from the "rising-up" shape of the interface and propagated vertically through the upper sheet in the case of TS980, as seen in (b2). At 3.0 s, in the case of TS270, one side was fractured through the SZ, and the other side was fractured along the "rising-up" interface, as seen in (a3). In the case of TS980, one side was also fractured through the SZ, and the other side was fractured leaving the SZ with a "hook" shape for the interface, as seen in (b3).

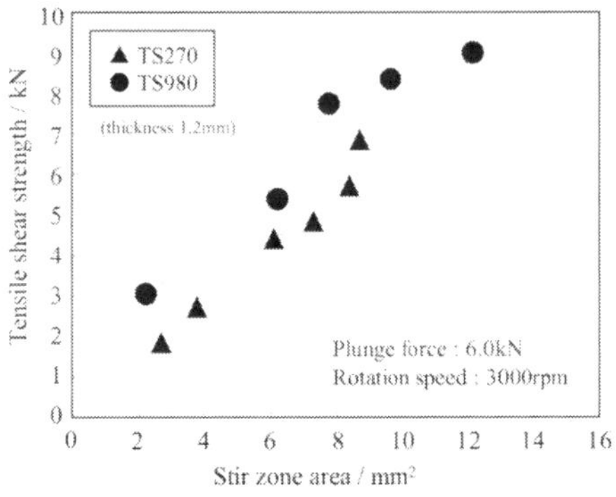

Figure 12. TSS values of TS270 and TS980 with SZ area.

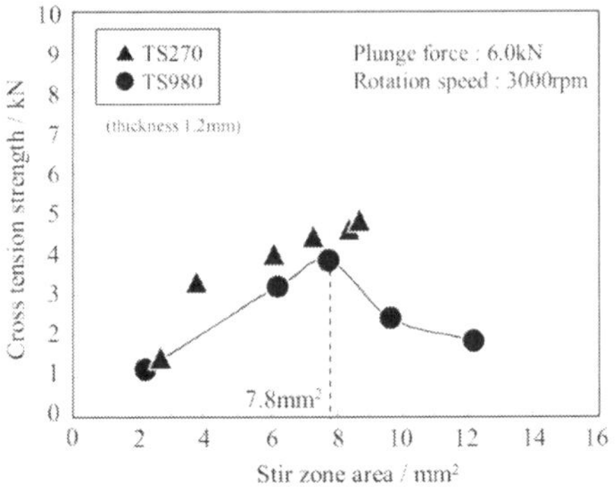

Figure 13. CTS values of TS270 and TS980 with SZ area.

The cross tension strength (CTS) of TS270 increased with the weld time, whereas the CTS of TS980 showed a peak at 2.0 s and subsequently decreased as shown in Figure 10. It should also be noted that it showed a lower value than that of TS270, even though a higher value was expected. Figure 11 shows the cross-sectional microstructures of the failed samples after the cross-tension test. Similar to the TSS results, both grades at 1.0 s and the TS270 joint at 2.0 s were fractured across the SZ. The TS980 joint at 2.0 s and the TS270 joint at 3.0 s were fractured along the "rising-up" shape of the interface toward the surface of the upper sheet on both sides. However, at 3.0 s, both sides of the TS980 joint were fractured along the "hook" shape of the interface.

Figure 12 shows the TSS plotted by the SZ area, and it reveals that the TSS values of both grades increase with the SZ area, whereas the TSS values of TS270 are lower than those of TS980 at the same SZ area.

The CTS of TS270 also increased with the SZ area, as shown in Figure 13. The CTS of TS980 also increased with the weld time for the SZ area. However, it showed a peak at 7.8 mm2 and subsequently decreased. In contrast to the TSS results, the CTS values for TS270 were higher than those of TS980 for the same SZ area.

4. Discussion

Both the TSS and CTS values were supposed to increase with the SZ area in both grades. The results shown in Figure 12 reveal that the TSS increased almost linearly with the SZ area in both grades, which agrees with previous reports [5]. It is supposed that if the fracture passes across the SZ of TS980, in which the microhardness is higher than that of TS270, the TSS value of TS980 will be higher than that of TS270, just as in the case of RSW [7].

To evaluate the influences of the SZ area and microhardness in the SZ on the resulting TSS, the prediction formula developed for RSW, as given in Eq.(2), was applied [12].

$$TSS = (\pi d^2 / 4) \cdot (H_{vn}) / 3\sqrt{3}, \tag{2}$$

where d is the diameter of a nugget, and H_{vn} is its Vickers hardness. In FSSW, the center of the joint is open because of the plunge of the probe. Therefore, the equation was modified using the SZ area, which represents S, as shown in Eq.(3).

$$TSS = S \cdot (H_{vn}) / 3\sqrt{3}, \tag{3}$$

The relationship between the experimental and predicted TSS values is shown in Figure 14. The experimental TSS values are in good agreement with the predicted values, especially in the case of

TS980, which corresponds to the previous report [5]. The experimental TSS values for TS270 tended to be higher than those of the predicted TSS, and a possible factor for this is discussed later. This result suggests that the most influential factors for the TSS of an FSSW joint are the SZ area and microhardness in the SZ.

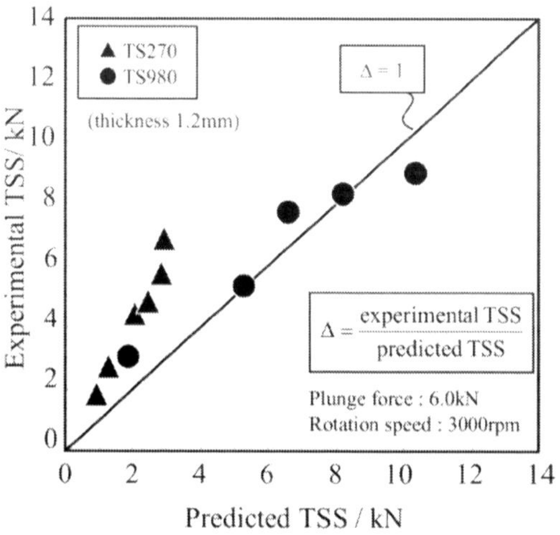

Figure 14. Correlation between experimental and predicted TSS values for TS270 and TS980.

It is also supposed that the SZ area basically increases the CTS in both cases. However, the lower CTS values for TS980 compared to those for TS270 and the decrease in the CTS with an SZ area larger than 7.8 mm^2, as shown in Figure 13, cannot be explained using only the SZ area and the microhardness in the SZ. Therefore, the shape and state of the interface should be considered to understand this difference between TS270 and TS980.

It is suggested that the "hook" shape of the interface may cause the decrease in the CTS in spite of the increase in the SZ area. This "hook" shape is formed around the SZ and passes throughout the upper sheet, as seen in Figure 2(b3), and the shape of the interface is supposed to cause the fracture propagation along the interface under low stress in the direction to open the lapped sheets. This corresponds to the fracture along the "hook" shape shown in Figure 11(b3), and results in a decrease in the CTS. The "rising-up" shape is also considered to have a detrimental effect on the CTS, because the extension of the "rising-up" interface can be the initial crack inside the upper sheet, which can be seen in the correspondence between the shape of the interface in Figure 2(b2) and the crack direction, as shown by the arrow in Figure 11(b). However, the effect of the "rising-up" shape is considered to be relatively small compared to that of the "hook" shape because the "rising-up" shape does not reach the surface of the upper sheet, whereas the "hook" shape does. Therefore, the "rising-up" does not cause a large decrease in the CTS.

It is also supposed that the joining across the "rising-up" interface has a large influence on the value of the CTS. In the case of TS270, the interface beside the SZ was joined as shown in Figure 6(a4). In addition, it can be seen that the grains grew across the interface at the root of the "rising-up" shape, as shown in Figure 6(a5). It is believed that the joining can prevent the crack propagation along the interface and increase the CTS. In contrast to TS270, the interface of TS980 beside the SZ inside the upper sheet can be seen as a sharp edge in Figure 7(b4). The complete separation of the interface in the TS980 weld is considered to cause crack initiation under a lower stress in the direction to open the lapped sheets, which results in a lower CTS than that of TS270. Further investigation is needed to reveal the reasons for the joining across the "rising-up" interface and the difference in the states of the joining between TS270 and TS980, although it is believed that diffusion joining affected the joining across the "rising-up" joining.

The crack initiation from the TS980 interface may also have been affected by the hardness distribution. The HAZ of TS980 was widely hardened, as shown in Figure 4(b2), compared to the case of TS270, in which the microhardness increase in the HAZ was relatively small, as shown in Figure 5. It is suggested that the harder plastic deformation in the HAZ of TS980 caused the higher stress concentration and fracture under a lower stress than the case of TS270, as mentioned in the case of RSW [9–10].

5. Conclusion

To ensure the mechanical properties during the FSSW of automotive steel, this study investigated the different factors influencing the joint strengths and failure modes of 270 N/mm^2 grade steel and 980 N/mm^2 grade steel, and the following conclusions were reached.

Both the TSS and CTS values of the FSSW joints of the 270 N/mm^2 and 980 N/mm^2 grade steels basically increased with the SZ area. A higher TSS corresponded to a higher microhardness value in the SZ in both grades.

The shape of the interface played a significant role in the CTS. A "hook" shape for the interface, which passed through the upper sheet, largely decreased the CTS because of the ease with which a fracture occurred along the interface, whereas a "rising-up" shape for the interface had a relatively small effect on the CTS.

The state of the interface had an influence on the CTS. When the "rising-up" interface was joined, as in the case of the 270-N/mm^2 grade steel, the crack propagation could be prevented and the CTS increased. However, when the joining across the "rising-up" interface was hard to achieve, as in the case of the 980 N/mm^2 grade steel, the CTS became relatively low.

References

[1] Mitlin D, Radmilovic V, Pan T, Chen J, Feng F, Santella M L 2006 *Mater. Sci. Eng. A* **441** 79
[2] Badarinarayan H, Yang Q, Hunt F 2008 *SAE International* 2008-01-0147
[3] Wang D A, Lee S C 2007 *J.Mater.Process.Technol.* **186** p291
[4] Taendl J, Nambu S, Inoue J, Enzinger N, Koseki T 2012 *Sci. Tech. Weld. Join.* **17(3)** 244
[5] Khan M I, Kuntz M L, Su P, Gerlich A, North T, Zhou Y 2007 *Sci. Tech. Weld. Join.* **12(2)** 175
[6] Back S, Choi D, Lee C, Ahn B, Mo Y, Song K, Jung S 2010 *Mater. Trans.* **51(5)** 1044
[7] Chuko L W, Gould J E 2002 *Weld J.* 1S
[8] Mimer M, Svensson L E, Johansson R 2004 *Welding in the World* **48** 14
[9] Pouranvari M, Marashi S P H, Safanama D S 2011 *Mater. Sci. Eng. A* **528** 8344
[10] Taniguchi K, Matsushita M, Ikeda R, Endo S 2011 *Pre-prints of the National meeting of JWS(in Japanese)* **89** 4
[11] Yang Q, Mironov S, Sato Y S, Okamoto K 2010 *Mater. Sci. Eng. A* **527** 4389
[12] Ono M 2005 *Welding technology of High strength and zinc coated steel sheets, NMS-ISIJ(in Japanese)* **184-185** 139

Transmittance properties and TEM observation of metal doped glass by field-assisted ion exchange

Souta Matsusaka, Taketsugu Nomura, Hirofumi Hidai, Akira Chiba and Noboru Morita
Department of Mechanical Engineering, Chiba University,
1-33, Yayoi-cho, Inage-ku, Chiba-shi, Chiba, 263-8522, Japan

E-mail: matusaka@faculty.chiba-u.jp

Abstract. Metal (silver or copper) ions were doped into borosilicate glass using an electric field-assisted ion exchange method. The optical transmittance of the metal doped glass was measured to determine why the doped glass exhibited an excellent laser micro-machinability. The doped metal ions were found to have enhanced the optical absorption of the glass, especially in the ultraviolet range. This in turn facilitated the efficient absorption of incident laser irradiation, and hence improved laser machinability of the glass. The metal doped glass also exhibited some absorption in the visible range, leading to a slight yellow-brown coloration. Transmission electron microscope (TEM) observations indicated that the metal ions had penetrated the glass and therein formed nanometer-sized (~6 nm) fine particles. In an attempt to control the optical characteristics in the ultraviolet-visible range, metal doped glass was heat-treated following the ion exchange doping step. In the case of silver-doped glass with heat treatment at 723 K, silver nanoparticles aggregated locally yielding an inhomogeneous structure. The heat-treated samples had a high optical absorption in the ultraviolet range.

1. Introduction

Glass materials have some unique characteristics such as high transparency, mechanical hardness, and thermal and chemical stability; and are therefore widely used as substrates for micro-machines, micro-electro-mechanical systems (MEMS) and devices in analytical chemistry. However, the laser micro-machining of glass materials has proven to be difficult. Some recent studies have considered the development of laser-machinable glass [1-4]. Typically, laser-machinable glass was prepared by metal (usually silver or copper) doping to glass surfaces using an ion exchange technique. The resulting glasses, showed excellent laser micro-machinability, without cracking or chipping around processed area [5]. In this study, the optical characteristics of metal doped glasses were examined in order understand their enhanced laser machinability. More fundamentally, we wish to understand how the penetrating dopant ions can modify both the optical characteristics and the laser machinability of glass. Accordingly, the structure of metal doped glass was studied using a transmission electron microscope (TEM). We will discuss the relationship between the microscopic structure of the metal dopant within the glass matrix, the optical characteristics and the laser machinability of metal doped glass. Finally, in order to better control the optical characteristics of doped glass, a heat treatment method was tested.

Content from this work may be used under the terms of the Creative Commons Attribution 3.0 licence. Any further distribution of this work must maintain attribution to the author(s) and the title of the work, journal citation and DOI.
Published under licence by IOP Publishing Ltd

2. Experimental procedure

Borosilicate glass slides (Schott, Borofloat 33, 25 mm × 25 mm × 1.1 mm) and pure silver or copper foils (> 99.9 mass%, 20 mm × 20 mm × 0.01 mm) were used for the ion exchange process. The glass slides and metal foils were rinsed in acetone, ethanol and distilled water before the procedure. A glass slide and a metal foil were placed between two copper electrodes in a high vacuum chamber (< 10^{-3} Pa). The experimental setup for ion exchange is illustrated in figure 1. After evacuation and temperature elevation to 623 K with a tungsten heater and a plate heater, a DC voltage was applied with the metal foil serving as the anode. The ion exchange conditions are listed in table 1. After ion exchange, the samples were heat-treated under two different sets of conditions. Some samples were heat-treated immediately after the ion exchange, i.e. the samples were held inside the vacuum chamber and heat-treated after the applied voltage was removed. The other samples were cooled in the vacuum chamber and then transferred to an electric furnace for heat treatment. The heat treatment temperatures applied in both methods were 623 K and 723 K.

Following the ion exchange and heat treatments, the glass samples were detached from the metal foil and rinsed in distilled water. The rinsed samples exhibited a slight frosted appearance on their underside due to the outflow of alkali ions. This was removed by mirror polishing of the underside only. The optical transmission spectra of the samples were then obtained using an ultraviolet-visible spectrophotometer (UV-VIS) (JASCO, V-550). The metal doped areas were examined using TEM (JEOL, JEM-2100F).

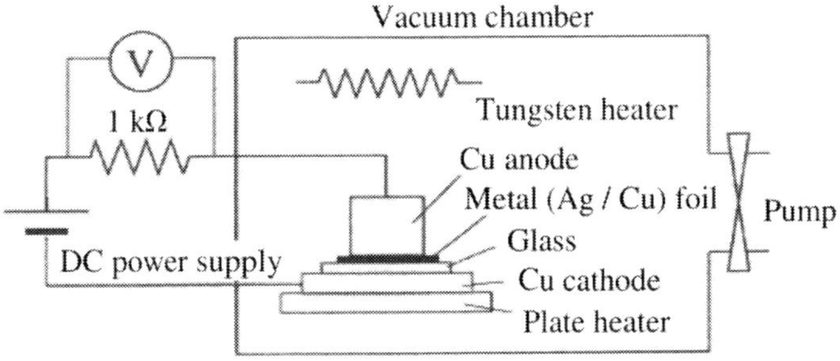

Figure 1. Schematic illustration of the experimental setup for ion exchange.

Table 1. Ion exchange conditions.

Doped element	Ag	Cu
Glass	Borofloat 33	
Temperature [K]	623	
Voltage [V]	100	
Experiment time [ks]	21.6	

3. Results and discussion

3.1. Optical characteristics and glass structure of the metal doped area

Figure 2 shows the optical transmittance for silver doped and copper doped glasses. The transmittance curve is clearly shifted towards the visible region with metal doping, i.e., the UV absorption of metal doped glass is greater than that of un-doped glass. The UV absorption is increased to a particularly large extent in the case of copper doped glass. Our previous report [5] on UV laser micro-machinability of metal doped glass showed that the diameter of a laser-ablated hole on copper doped glass was larger than that formed on silver doped glass under the same laser irradiation conditions. In other words, the energy threshold for laser processing in copper doped glass was lower than that for silver doped glass. The difference in UV absorption between copper doped and silver doped glass explains the higher laser-machinability of copper doped glass. Therefore, it was suggested that the efficient absorption of incident laser irradiation contributed to the much enhanced laser micro-machinability of the metal doped material. Less desirable, however, was the slight yellow-brown coloration of the doped glasses, arising from absorption in the visible range, also illustrated in figure 2. This reduced transparency may be unacceptable for many industrial applications. The development of laser-machinable glass with both high optical absorption in the UV range and high transparency in the visible range is necessary.

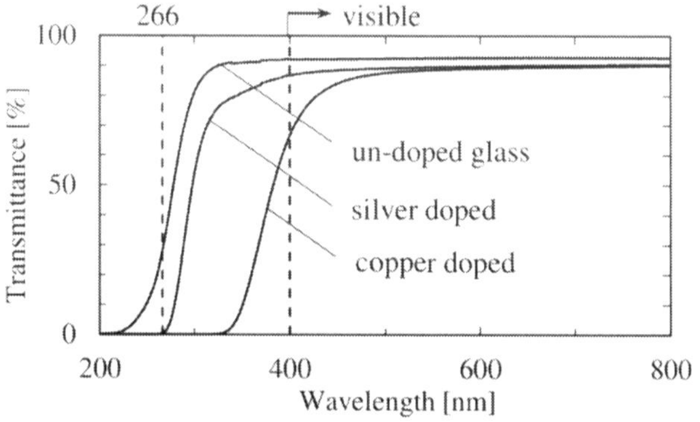

Figure 2. Transmittance curves of un-doped and metal doped glasses.

In order to investigate the relationship between the structure and the optical characteristics of metal doped glass, a cross-sectional TEM image of the glass surface was obtained. Figures 3a and b show HAADF-STEM images of silver and copper doped areas, respectively. In these figures, the bright granules are metal nanoparticles. The particle size was approximately 2-6 nm for both silver and copper, with no evident size difference between the two metals. It should be noted that the apparent difference in particle density between the two glasses is due to different thicknesses of the TEM samples. It is known that metal nanoparticles embedded in a dielectric matrix exhibit a characteristic light absorption due to localized surface plasmon resonance (LSPR) [6]. We therefore attribute the absorption edge shift from the UV into the visible range, shown in figure 2, to LSPR of metal nanoparticles in the glass matrix. The LSPR frequency and wavelength depend strongly on nanoparticle size, composition and structure [6]. To achieve specific intended optical characteristics, it will be necessary to control particle size and structure precisely.

At present, the growth mechanism from ions to nanometer-sized particles is not fully understood. In the ion exchange process, however, each metal ion moves from the glass surface to its interior by ionic self-diffusion and electrical drift [5]. This means that the transition from ions to particles occurs during a cooling process after ion exchange. We therefore expect that a heat treatment following the ion exchange would be effective for controlling the metal particle size.

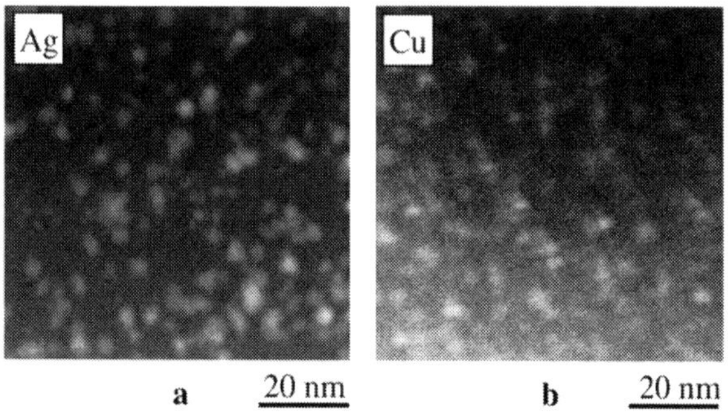

Figure 3. Cross-sectional HAADF-STEM images of a) silver and b) copper doped glasses.

3.2. Heat treatment of silver doped glasses

Two types of heat-treated silver doped glasses were prepared, as mentioned above. Heat treatment times were 3.6 ks for samples before cooling, and 43.2 ks for samples after cooling. Silver doped glass was selected for the heat treatment experiments because of the proximity of its UV absorption edge to the 266 nm line. Figure 2 shows the transmittance curve of silver doped glass approaching zero at around 270 nm. A 266 nm wavelength UV laser is widely used in micro-machining. A small shift of the silver doped glass transmittance curve towards the visible is expected to enhance the UV absorption by the material. Since this might be achieved by particle growth at elevated temperatures, heat treatment of silver doped glass was carried out. Correspondingly, a rapid cooling of copper doped glass may be expected to shift the transmittance curve towards the UV region. However, a rapid cooling in our vacuum chamber was not experimentally feasible because of protection of electrical heaters. Only the heat treatment results of silver doped glasses are reported here.

Figures 4a and b show transmittance curves of heat-treated silver doped glasses. Figure 4a is for the samples heat-treated before cooling and figure 4b is for heat-treated after cooling. Clearly UV absorption increases with increasing heat treatment temperature. Heat treatment at 723 K resulted in a particularly strong enhancement of UV absorption. It was also found that the transmittance curves of heat-treated samples at 723 K had a characteristic shoulder around 325 nm. These transformations in optical characteristics may be caused by a structural change in embedded nanoparticles.

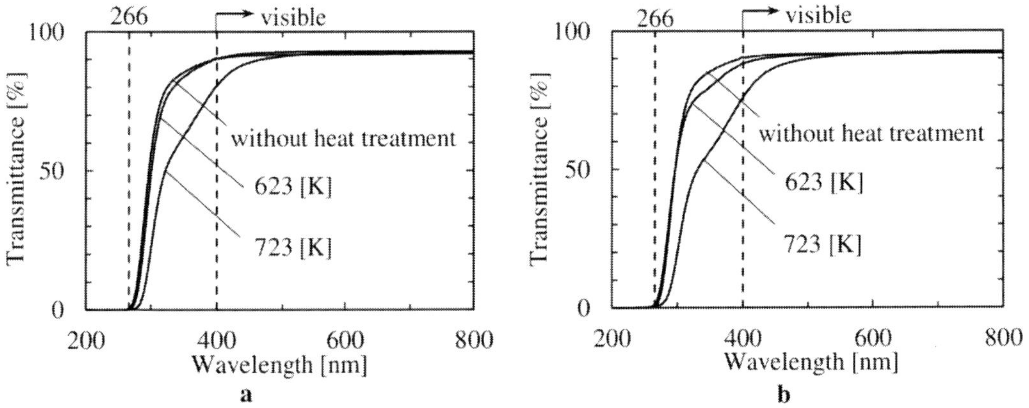

Figure 4. Transmittance curves of silver doped glasses after heat treatment. a) heat-treated before cooling and b) heat-treated after cooling.

Figure 5 shows HAADF-STEM images of heat-treated silver doped glasses. As shown in figure 5, the obvious growth of individual silver particles was not observed in samples heat-treated at either 623 K or 723 K. However, the local aggregation of particles did occur in samples heat-treated at 723 K. In the case of heat treatment at 723 K for 3.6 ks (figure 5b), a few particles aggregated. At 723 K for 43.2 ks (figure 5e) numerous particles aggregated and formed cloudlike structures. Figure 5f is a low magnification view of figure 5e, showing an inhomogeneous structure unevenly distributed in the glass. These aggregated structures will alter the optical characteristics of glass. We believe the 325 nm shoulder feature in the transmittance curves in figure 4 was caused by the aggregation of silver particles.

As shown in figure 4, an 'ideal' laser-machinable glass, which has high UV absorption and high transparency in the visible range, has not yet been developed. However, thermal treatment, i.e. heat treatment or rapid cooling, is one of the promising candidates for the improvement in optical characteristics of metal doped glass. For precise control of optical characteristics, it will be necessary to clarify the growth and aggregation mechanisms of metal nanoparticles in the glass substrate.

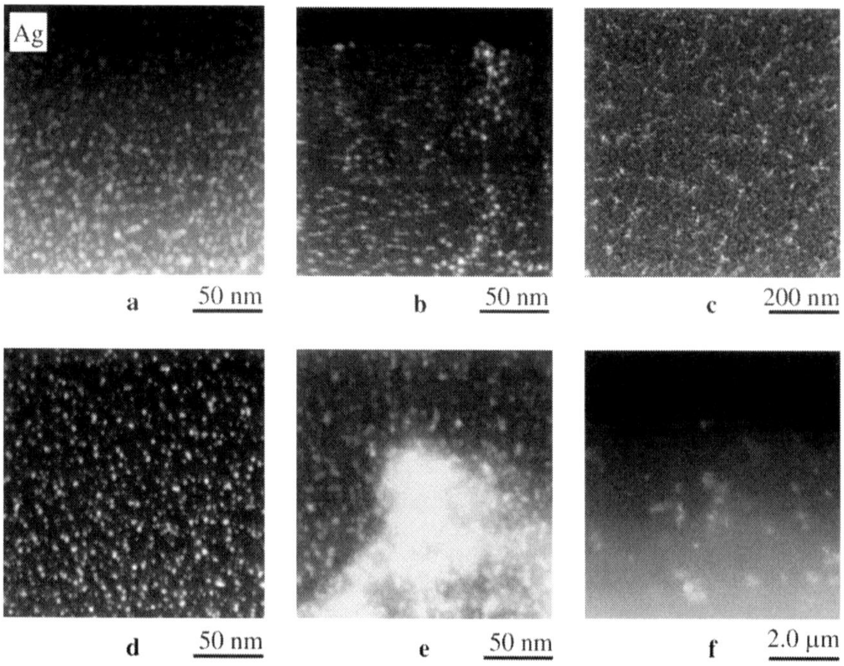

Figure 5. Cross-sectional HAADF-STEM images of heat-treated silver-doped glasses. a) before heat treatment, b) heat-treated at 723 K for 3.6 ks, c) low magnification view of photograph b), d) heat-treated at 623 K for 43.2 ks, e) at 723 K for 43.2 ks, f) low magnification view of photograph e).

4. Conclusions

Metal (silver or copper) ions were doped into borosilicate glass using an electric field-assisted ion exchange method. TEM imaging revealed that penetrated metal ions formed fine metal particles of diameter of 2-6 nm. These particles increased the UV absorption of the glass and led in turn to its improved laser micro-machinability. A heat treatment for controlling the optical characteristics was tested on silver doped glass. UV and visible absorption increased with increasing heat treatment temperature. TEM imaging suggested that the changed optical characteristics were not due to individual particle growth, but rather to local aggregation of particles.

Acknowledgement

This research was financially supported by A-STEP from Japan Science and Technology Agency and Grant-in-aid for scientific research from Japan Society for the Promotion of Science. The authors are deeply grateful to Prof. T. Itoi (Chiba University) for his cooperation in TEM observation.

References

[1] Tsunetomo K and Koyama T 1997 *Opt. Lett.* **22** 411
[2] Rosenbluh M, Antonov I, Ianets D, Kaganovskii Y and Lipovskii A 2003 *Opt. Mater.* **23** 401
[3] Matsusaka S and Watanabe T 2008 *J. Mat. Proc. Tech.* **202** 514
[4] Matsusaka S and Watanabe T 2010 *Scripta. Mater.* **62** 141
[5] Matsusaka S, T Kobayakawa, H Hidai and N Morita 2012 *J. Phys: Conf. Ser.* **379** 012035
[6] Link S and El-Sayed M 1999 *J. Phys. Chem.* B **103** 8410

IOP Conf. Series: Materials Science and Engineering 61 (2014) 012012 doi:10.1088/1757-899X/61/1/012012

Microstructure of interface between Si chip and tape-type Ag nanoparticle sheet

Hayato Mutoh[1*], Naoki Moriyama[1], Keiya Kaneko[1], Yasuyuki Miyazawa[1] and Hitoshi Kida[2]**

[1]Tokai University, Hiratsuka-shi, Kanagawa-ken, 259-1292, Japan
[2]Alent Japan Co., Alpha, Hiratsuka-shi, Kanagawa-ken, Japan

*E-mail: 3bazm011@mail.tokai-u.jp, **E-mail: hkida@alent.com

Abstract. A Ag nanoparticles-based technology for joining semiconductor materials through sintering has been developed. Currently, a paste consisting of Ag nanoparticles is used for die bonding in the electronics industry. A binder-free Ag nanojoining material has been developed for fabricating electronic devices. In this study, we investigated the joining ability of this binder-free tape-type Ag nanojoining material with respect to Si chips. A paste-type Ag nanojoining material that contained an organic binder was also investigated for comparison. The microstructures of the Si-joining material interfaces were observed using a focused ion beam system. The cleaved surfaces of the joints were observed using scanning electron microscopy. High-quality joints that contained only a few pores could be formed between a Au thin film electroplated on the Si chip and the tape-type material after sintering, owing to the simplicity of the joining process and the binder-free nature of the joining material.

1. Introduction

A Ag nanoparticles-based technology that involves sintering has been developed for joining semiconductor materials [1]. A unique feature of this technology is that joints can be formed quickly by applying small pressures. Currently, pastes consisting of Ag nanoparticle are used for die bonding in the electronics industry [2], and numerous studies have investigated the formation of joints using such pastes [3-6].

Generally, joint formation using a paste of Ag nanoparticles consists of a "drying" step, a "sintering and neck-growth" step, a "densification" step, and an "elimination of pores" step [3]. Organic shells are adsorbed on the surfaces of the Ag nanoparticles; these shells help disperse and stabilize the nanoparticles [4]. Further, a binder is also required for producing the paste of Ag nanoparticles. The removal of the organic shells is a prerequisite for sintering to occur. In addition, the removal of the binder is a prerequisite for joining to occur. However, this can result in relatively high process temperatures and complex joining steps [3].

A binder-free Ag nanojoining material has been developed recently [7]. It contains organic shells but not an organic binder. The sublimation and decomposition temperature of the organic shells present in this material is lower than that of the shells present in conventional joining materials. In this study, we investigated the joining ability of this binder-free Ag nanojoining material with respect to Si chips on the basis of observations of the interfacial microstructure.

Content from this work may be used under the terms of the Creative Commons Attribution 3.0 licence. Any further distribution of this work must maintain attribution to the author(s) and the title of the work, journal citation and DOI.
Published under licence by IOP Publishing Ltd

2. Experimental procedure

In addition to the binder-free nanojoining material, which was of the tape type, a paste-type Ag nanojoining material was also investigated for comparison. In both materials, the Ag nanoparticles were covered with organic shells, which prevented the nanoparticles from agglomerating at room temperature. However, the paste-type material consisted of a mixture of Ag nanoparticles and an organic binder. Thus, a drying step was required to remove the binder by evaporation. On the other hand, as mentioned previously, the tape-type material consisted of Ag nanoparticles only, and thus a drying step was not necessary; the joining process consisted only of the sintering and joining steps. Si chips electroplated with a Au thin film were used as substrates.

Figure 1 shows the typical device configuration employed in this study. Because it was noticed that the Ag nanoparticles increased in size during joint formation, we decided to examine the effects of a post-joining heat treatment as well.

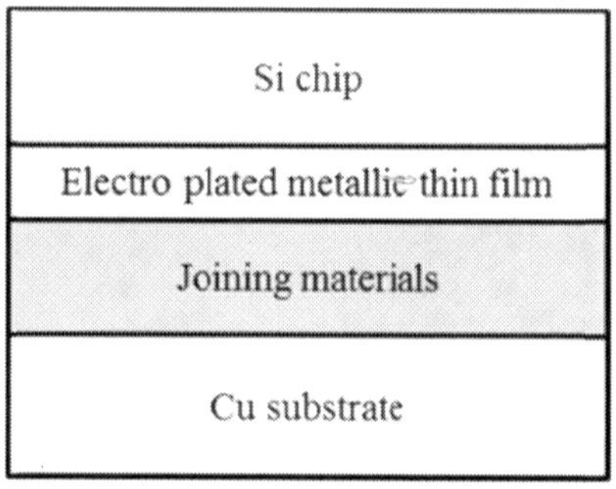

Figure 1. Schematic of a typical specimen configuration.

Table 1. Preparative conditions of specimens in this study.

Specimen No.	Substrate	Nano Ag	Drying Temp., Time	Sintering Temp., Time, Pressure	Heat treatment Temp., Time
No.1	Si chip with Au film	Paste Type	130°C, 30min	Not	Not
No.2	Si chip with Au film	Tape Type	Not	250°C, 90s, 10MPa	Not
No.3	Si chip with Au film	Tape Type	Not	250°C, 180s, 10MPa	Not
No.4	Si chip with Au film	Tape Type	Not	250°C, 180s, 10MPa	250°C, 500h
No.5	Si chip with Au film	Paste Type	130°C, 30min	250°C,180s, 10MPa	Not
No.6	Si chip with Au film	Paste Type	130°C, 30min	250°C, 180s, 10MPa	250°C, 500h

Table 1 shows the conditions for preparing the various specimens investigated. The paste-type joining material was patterned on the surface of a Cu substrate using a metal mask, and the Si chip was placed on the pattern. The specimen was then dried in an oven at 130°C for 30 min. The sintering of the specimen was performed using a hot-pressing machine. In the case of the tape-type joining material was used, the tape was placed on directly on the Cu substrate, and the Si chip was then placed on the tape. In this case too, the sintering of the specimen was performed using a hot-pressing machine. The heat treatments of the specimens were performed using a conventional constant-temperature oven.

The sintered and heat-treated specimens were then cut using a medium-speed cutter. The cut surfaces were polished using a focused ion beam (FIB) system so that the interfacial microstructures could be observed. The cross-sections were observed using scanning electron microscopy (SEM). The elemental distributions of the cross-sections were determined using an electron probe microanalyzer (EPMA). In addition, a set of the specimens was cooled in liquid nitrogen and broken by being subjected to an impact stress so that the cleaved surfaces could be observed using SEM.

3. Results and discussion

Figures 2 (a) and (b) show SEM images of the cleaved surfaces at the Si chip/paste-type material (specimen 1) interface and the Si chip/tape-type material interface (specimen 2), respectively.

Numerous pores were observed at the interface as well as among the Ag nanoparticles when the paste-type material was used, as can be seen from Figure 2 (a). It seems that the pores between the Ag nanoparticles were formed by the sublimation and decomposition of the organic shells and the binder when the specimen was at heated at 130°C for 30 min. Thus, it can be surmised that drying alone does not result in a sufficient degree of joining.

As can be seen from Figure 2 (b), a high-quality joint was obtained when the tape-type material was used. Furthermore, significant neck growth was also achieved; this was because the sample was heated and subjected to a pressure during the sintering process.

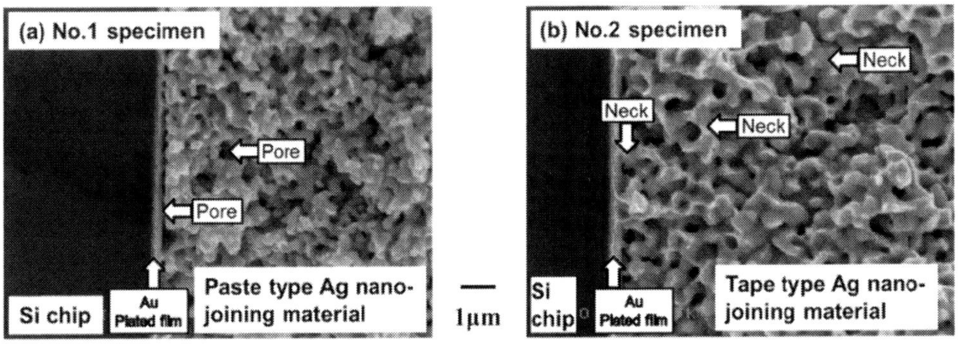

Figure 2. SEM images of the cleaved surfaces of the specimens.

(a) SEM image of the cleaved surface of specimen 1; paste-type Ag nanojoining material; drying at 130°C for 30 min; no sintering or heat treatment.

(b) SEM image of the cleaved surface of specimen 2; tape-type Ag nanojoining material; no drying; sintering at 250°C for 90s under 10 MPa; no heat treatment.

Figure 3 shows the effects of the heat treatment on the microstructure of the interfaces formed between the Si chip and the nanojoining materials. Images of the Si chip/tape-type material interface before the heat treatment (specimen 3) and after the heat treatment at 250°C for 500 h (specimen 4) are shown in Figures 3 (a) and (b), respectively.

A few pores could be observed among the Ag particles of specimen 3. Further, as can be seen from Figure 3 (a), the joint formed between the electroplated Au thin film and the tape-type Ag nanojoining material after sintering was of good quality. It was found that neck growth among the Ag nanoparticles as well as the growth of the Ag nanoparticles themselves occurred when the sample was heated 250°C for 180 s under a pressure of 10 MPa during the sintering step. According to previous studies [4], the strength of joints formed using a paste of Ag nanoparticles is also affected by the size of the bonding area. However, when the bonding area is large, it is difficult to decompose the organic binder because of the lack of contact with air. This results in insufficient sintering and a lower bonding strength. However, it was found that sound joints with a larger bonding area could be formed readily by using the binder-free tape-type Ag nanojoining material.

After the heat treatment, the pores that were noticed among the Ag nanoparticles disappeared, while the Ag nanoparticles grew in size, as shown in Figure 3 (b). It is likely that the growth of the Ag nanoparticles during the sintering process and the subsequent heat treatment was not attributable only to densification but also because of the recrystallization of Ag at 250°C. This is because it has been reported that the recrystallization temperature of rolled pure silver is approximately 100°C [8].

Figures 4 (a) and (b) show ion images, obtained using a FIB system, of the interface of the Si chip/paste-type material (specimen 5) before the heat treatment and after the heat treatment at (specimen 6) 250°C for 500 h, respectively.

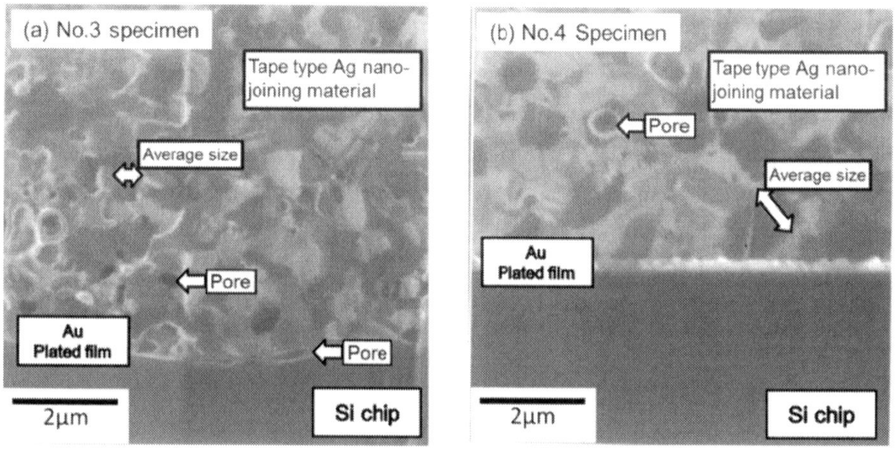

Figure 3. Effect of heat treatment on microstructure of the interface.

(a) Ion image of the joint interface of specimen 3 after FIB polishing; no drying; sintering at 250°C for 180 s under 10 MPa; no heat treatment.

(b) Ion image of the joint interface of specimen 4 after FIB polishing; no drying; sintering at 250°C for 180 s under 10 MPa; heat treatment at 250°C for 500 h.

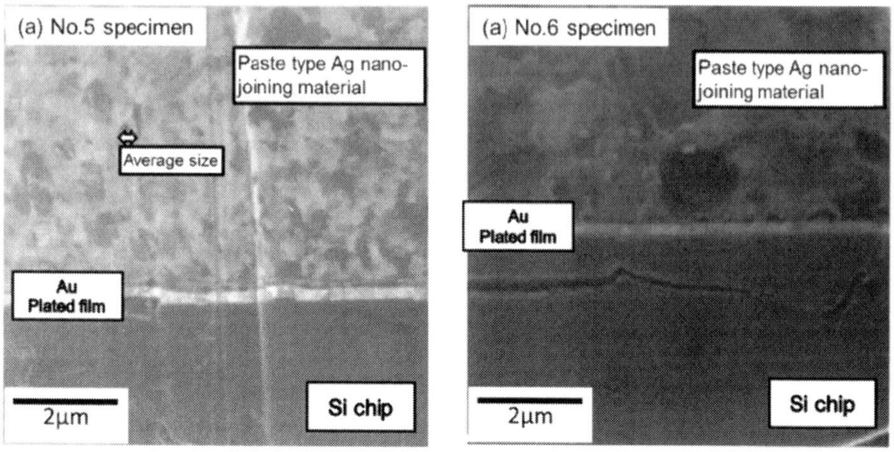

Figure 4. Effect of heat treatment on microstructure of the interface.

(a) Ion image of the joint interface of specimen 5 after FIB polishing; drying at 130°C for 30 min; sintering at 250°C for 180 s under 10 MPa; no heat treatment.

(b) Ion image of the joint interface of specimen 6 after FIB polishing; drying at 130°C for 30 min; sintering at 250°C for 180 s under 10 MPa; heat treatment at 250°C for 500 h.

It can be seen that that the joint between the electroplated Au thin film and the paste-type material was of good quality. The grain size of specimen 5 was smaller than that of specimen 3. It is possible that the residue formed after the sublimation and decomposition of the organic shells and binder during the drying and sintering steps retard both neck growth and the growth of the Ag nanoparticles. The joint formed using the tape-type material was of desirable quality because the joining process was simple and the material was did not contain an organic binder.

4. Conclusions

The ability of a recently developed binder-less Ag nanojoining material to form joints with Si chips was investigated on the basis of the microstructure of the interface formed between the two materials. The main results of the study were the following:

(1) Joints of good quality and containing few pores could be formed between the electroplated Au thin film on the Si substrate and the tape-type Ag nanojoining material after sintering.

(2) The high neck growth noticed between the Ag nanoparticles was caused by the heating of the joint and the application of a pressure on it during the sintering process.

(3) Sound joints having a wide bonding area could readily be obtained using the tape-type Ag nanojoining material, because of the organic binder-free nature of the material and the simplicity of the joining process.

(4) The growth of the Ag nanoparticles during the sintering and heat-treatment steps was caused by the densification and recrystallization of Ag at 250°C.

References

[1] Hirose A 2011 *Journal of the JWS* **80** 702
[2] Kim S. Siow 2012 *Journal of Alloys and Compounds* **514** 6
[3] Jianfeng Yan, Guisheng Zou, Ai-ping Wu, Jialie Ren, Jiuchun Yan, Anming Hu and Zhou Y 2012 *Scripta Materialia* **66** 582
[4] Shuai Wang, Mingyu Li, Hongjun Ji and Chunqing Wang 2013 *Scripta Materialia* **69** 789
[5] Rongwei Zhang, Kyoung-sik Moon, Wei Lin and C. P. Wong 2010 *Journal of Materials Chemistry* **20** 2018
[6] Shuai Wang, HongjunJi, MingyuLi, Chunqing Wang 2012 *MaterialsLetters* **85** 61
[7] Kim S. Siow 2014 *Journal of Electronic Materials* DOI: 10.1007/s11664-013-2967-3
[8] Hori S, Tai H and Kawaguchi S 1981 *Zairyou* **30** 556

Brazing of C/C composites and Ni-based alloy using interlayer

Toshi-Taka Ikeshoji[1], Tatsuya Tokunaga, Akio Suzumura and Takahisa Yamazaki

Mechanical & Aerospace Engineering, Tokyo Institute of Technology, O-okayama 2-12-1, Meguro-Ku, Tokyo, 152-8552, Japan

ikeshoji.t.aa@m.titech.ac.jp

Abstract. The brazing of C/C composites and Inconel-600 Ni-based heat resistant alloy was conducted using Fe-Ni-Cr-P-Si brazing filler alloy with inserting various thickness of Nb foil as stress relief interlayer. SEM observation of cross section of brazing interface revealed that Nb foil was resolved into the brazing filler layer on C/C composites side. Nb diffused to the surface of C/C composites and acted as the active metal element to enhance the wettability of molten metal on graphite matrix of C/C composites during the brazing process. The variation in shear strength values of the brazed joint with Nb layer thickness suggested that the Nb layer should be remained at least 100 μm.

1. Introduction

C/C composites have the high heat resistance. To utilize the C/C composites' heat resistance, the brazing techniques of C/C composites and the heat resistant alloys are expected. The brazing of C/C composites to the heat resistant alloys is, however, difficult due to C/C composites' poor wettability to the ordinary brazing filler alloys, and the relatively large mismatch in the coefficient of thermal expansion (CTE) [1]. In this research, the poor wettability is attempted to be overcome by using the Fe-Cr-Ni-(Si,P) brazing filler, which consists of the carbide forming elements, *e.g.* Fe and Cr. The residual thermal stress induced by the mismatch in CTE is attempted to be relieved by inserting the Nb foil. The Nb's CTE value is 7.3×10^{-6} K^{-1}, which is intermediate between the CTE value of C/C composites, $1\times10^{-6} - 8\times10^{-6}$ K^{-1}, and that of Inconel-600, 12×10^{-6} K^{-1}. By changing the thickness of Nb foil, the brazing test was conducted. And for the obtained joints, the shear strength was measured. Their microstructure at the brazing interface was made clear using the SEM and EPMA elements mapping.

2. Experimental details

The provided materials were a cross-ply laminated C/C composites plate in the dimensions of 10×10×3 mm (Across Inc.) and Inconel-600 (Ni-15.5Cr-8Fe0.25Si-0.2Mn-0.08C-0.001S in mass%) plate in the same dimensions. The interlayer was the Nb foil with the thickness of 100 – 500 μm. The brazing filler was Fe-42Ni-20Cr-(10– 12)(Si,P) alloy (in mass%, TB-2720, Tokyo Braze Co., Ltd.). The Ø100μm Mo wires were inserted as spacers in brazing filler layers (figure 1). The brazing of the specimen was conducted in the vacuum atmosphere. The brazing temperature was 1353 K (1080°C), and the keeping time was 300 s. The cross section of the brazed joints were observed by SEM and the thickness of residual Nb layer was measured. The elements distribution over the cross section of the brazed joint was obtained by EPMA. For the obtained brazed joints, the shear fracture test was

Content from this work may be used under the terms of the Creative Commons Attribution 3.0 licence. Any further distribution of this work must maintain attribution to the author(s) and the title of the work, journal citation and DOI.
Published under licence by IOP Publishing Ltd

conducted to measure the shear strength. For the shear test specimen, their fillet of the brazed joint was not removed.

3. Results

For all the thickness of Nb foil, the brazed joints were obtained. As shown in figure 2, by the SEM observation, the cracks along the interface of C/C composites and the brazing filler layer were found for the brazed joints with Nb foils with the initial thickness of 100 – 300 μm. On the case of Nb foil with the initial thickness of 500μm, the crack could not be observed. For all the case, the observed Nb layer was thinner than the initial Nb foil thickness. The Nb layer thickness was proportional to the initial Nb foil thickness (figure 3). The regression line was $y = 1.011\,x - 0.075$ mm, and its intercept with x-axis was 74 μm. The average of difference between the Nb layer thickness and the initial Nb foil thickness was 72 μm. So, the resolved thickness of Nb foil was approximately 72- 75 μm under the brazing conditions of this experiment.

The element distribution over the brazed joint interface on the case of Nb foil thickness of 100 μm are shown in figure 4. The Inconel-600 side of Nb layer was basically the solidified zone of Fe-Ni-Cr alloy. In the C/C composites side, (Fe,Ni)-(Cr,Nb) eutectic phase was observed. The Nb mapping shows it distributes mainly C/C composites side, and especially along the C/C composites surface. Cr also distributed along the C/C composites' surface. Therefore, Nb and Cr were considered to form the carbides on C/C composites' surface, and acted as the active elements during the brazing process. The crack was observed near the interface of the solidified brazing filler and C/C composites. In figure 4, the dashed lines shows the outline of the crack. Their shape did not agree because the crack shape changed by depth, perpendicular direction of the picture image.

The shear strength of the obtained brazed joints are plotted with variation in the residual Nb layer thickness in figure 5. The brazed joint with 32μm of the residual Nb layer was too weak to be subjected to the shear fracture test. Other joints' shear strength values varied from 2 – 16 MPa. The average value of shear strength was high for the residual Nb layer thickness of 127 μm. The average value once decreased for the Nb thickness of 227 μm, and recovered for thickest case.

4. Discussions

The change in the shear strength is considered to be affected by two factors, the thickness of Nb layer and the thickness of brazing filler layer. In this experiments, the solidified brazing filler was considered as Fe-Cr-Ni alloy. The 18-8 austenitic stainless steel, a typical Fe-Cr-Ni alloy, has the CTE value of 16×10^{-6} – 18×10^{-6} K^{-1}, which is larger than Inconel-600's CTE value, 12×10^{-6} K^{-1}. Adding to it, the thickness of the solidified brazing filler layer was thicker than or comparable with the residual Nb layer.

To examine the effect of thickness of Nb layer and the solidified brazing filler layer, the elastic-plastic FEM analysis was conducted for the same shape with the brazed joint in the experiments (figure 6). The length of sides was 10 mm. The thickness of Nb layer was changed from 0.1 to 5 mm. The solidified brazing filler layer was supposed to have the same physical properties with 18-8 austenitic stainless steel. The thickness of the solidified brazing filler layer was changed from 0 to 0.3 mm. The calculation areas was considered at 1353 K at initial, and cooled immediately to the room temperature, 300 K.

After cooling down of virtual specimen, the stress was induced by the thermal shrinking, *i.e.* the thermal residual stress. The von Mises' equivalent stress of the thermal residual stress on the C/C composites surface at the interface with the brazing filler layer was plotted (figure 7). The von Mises' equivalent stress value decreased with the thickness of Nb layer, and reached to a constant value (figure 7(a)). That value increased with thickness of brazing filler layer (figure 7(b)), though the amount of increase was not so larger compared with decrease with Nb layer thickness.

The difference between the bonding strength of the interface and the von Mises' equivalent stress of the thermal residual stress can be regarded as the apparent bonding strength of interface. When the value of von Mises' equivalent stress at the interface is large, the apparent bonding strength of interface becomes small. According to the FEM results, when Nb layer was thin, the value of von Mises' equivalent stress is large, and the apparent bonding strength of interface was small. With increase in Nb thickness, the value of von Mises' equivalent stress decrease, and the apparent bonding strength increased. Therefore, the thicker Nb layer is considered to enhance the shear strength. On the contrary, the thicker brazing filler layer is considered to weaken the brazed joint.

In this FEM analysis, the brazing filler layer was considered as 18-8 austenitic stainless steel, whose CTE value was 18×10^{-6} K^{-1}. The previous results suggests that the brazing filler layer on the Inconel-600 side was Fe-Ni-Cr alloy, but on the C/C composites side, it was (Fe,Ni)-(Cr,Nb) eutectic phase. The CTE value of brazing filler layer on the Inconel-600 side could have the similar value of

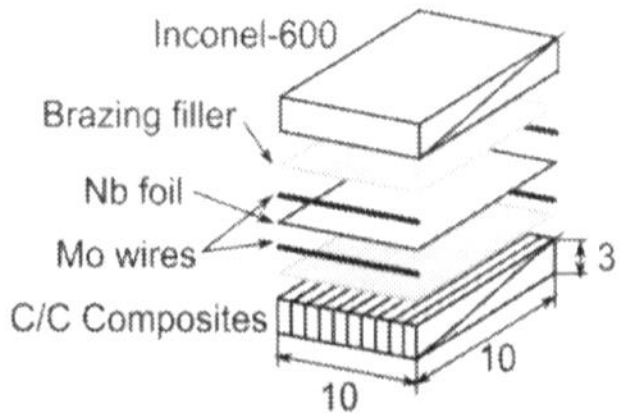

Figure 1. Brazing specimen assembly.

Figure 2. Micro-structure images of the brazed interface of C/C composites and Inconel 600 with various thickness Nb interlayers.

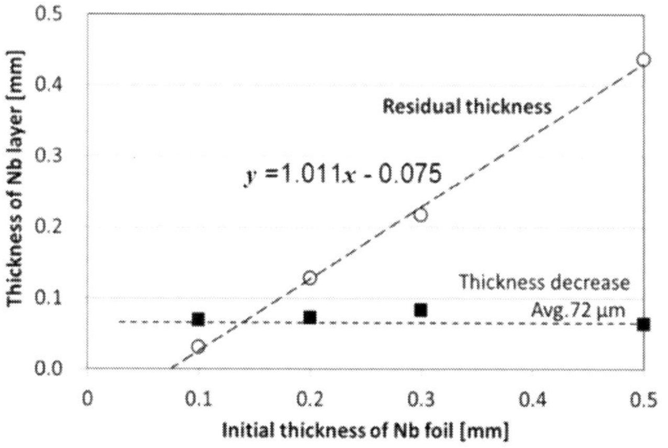

Figure 3. Thickness of residual Nb layer in the brazed joint interface.

18-8 austenitic stainless steel. That value on the C/C composites' side is unknown to the authors' knowledge. According to that CTE values of Fe-Ni alloy is 10×10^{-6} – 20×10^{-6} K^{-1}[3], and that Cr and Nb are 4.9×10^{-6} K^{-1} and 7.3×10^{-6} K^{-1}, respectively, CTE value of (Fe,Ni)-(Cr,Nb) eutectic phase might be around their average value, 10.7×10^{-6} K^{-1} in rough approximation. When the CTE value of the brazing filler layer on C/C composites' side might be lower than the 18-8 austenitic stainless steel's value, the residual stress would be estimated smaller than the estimated values shown in figure 7. But it will not change the tendency of von Mises' equivalent stress at the interface between C/C composites and the solidified brazing filler layer to the Nb foil thickness. For the further discussion, the CTE value should be measured actually.

In this experiment, the shear strength should have increased with the Nb layer thickness according to the above FEM results. For 227 μm Nb layer, the residual Nb layer thickness and the solidified

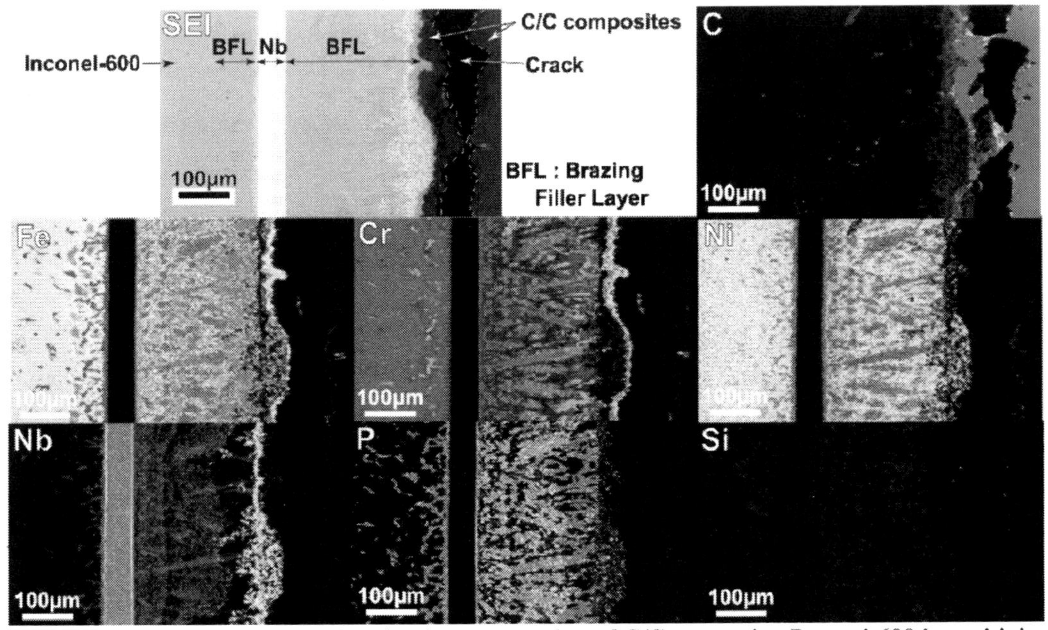

Figure 4. EPMA element mapping for the cross section of C/C composites/Inconel 600 brazed joint with 100 μm Nb foil interlayer.

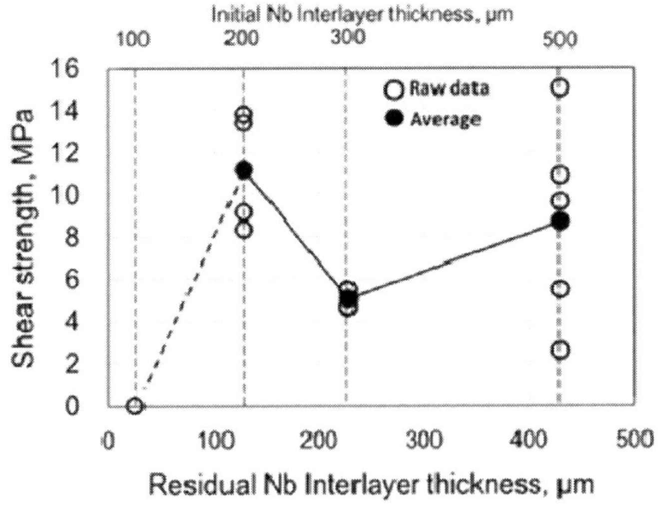

Figure 5. Shear strength of the brazed joints and the residual Nb layer thickness.

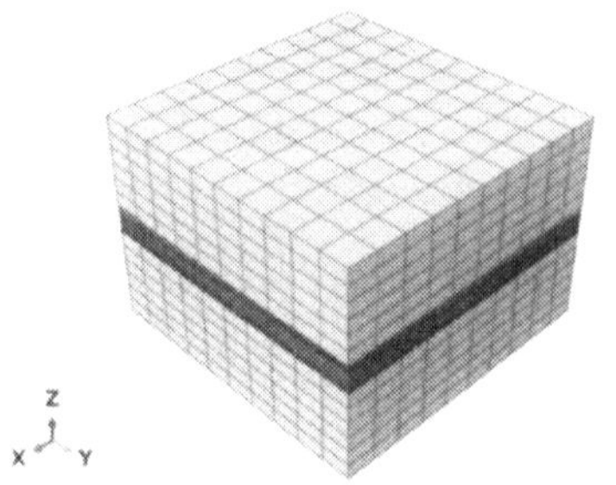

Figure 6. The calculation model of elastic-plastic FEM analysis for the brazing specimen.

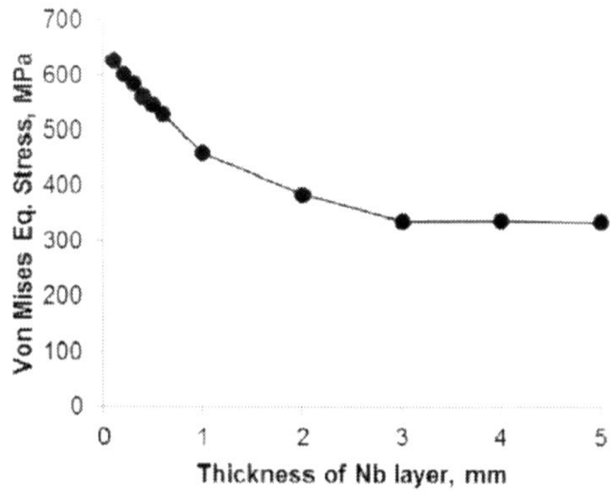

(a)

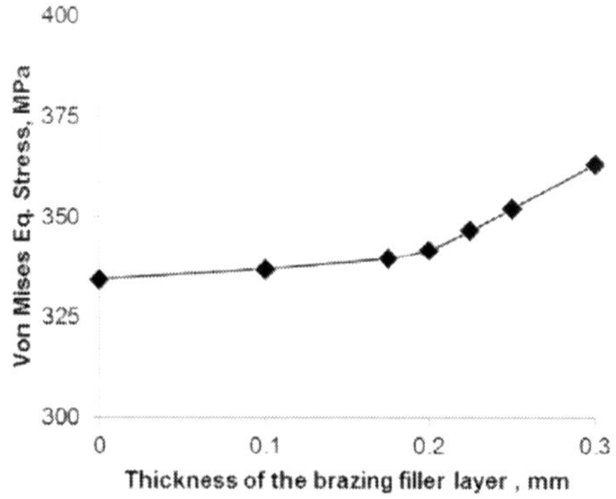

(b)

Figure 7. The results of elastic-plastic FEM analysis for the brazing specimen. (a) von Mises' equivalent stress vs Nb layer thickness on C/C composites surface at the interface with brazing filler layer. The brazing filler layer's thickness was 100 μm constant. (b) von Mises' equivalent stress vs thickness of the solidified brazing filler layer. The Nb layer thickness was 300 μm constant.

brazing filler layer thickness became comparable by chance, which lead to the small fluctuation in the shear strength values. But the comparable thickness of brazing filler layer seemed to lower the shear strength in average. The average value for the brazed joints with 427 μm Nb layer must be larger than the value of the brazed joints with 127 μm Nb layer. But the solidified brazing filler was not controlled properly to the constant thickness due to the larger dissolution of Nb into the brazing filler, the experimental results' value of shear strength was considered scattered in wide range.

5. Conclusions

The brazing of the C/C composites and Inconel-600 inserting Nb foil interlayer using Fe-Cr-Ni-(Si,P) brazing filler was conducted. The Nb interlayer resolved and segregated along the C/C composites surface. It was suggested that Nb could be active element in this brazing process. To obtain the firmly brazed joint, the Nb layer should be remained at least 100 μm.

References

[1] Ikeshoji T-T 2013 *Advances in Brazing: Sicence, technology and application*, ed. Sekulic D P Woodhead Publishing Ltd. 394.

[2] Ikeshoji T-T Amanuma T Suzumura A and Yamazaki T 2012 *Preprints of National Meeting of Japan Welding Society* **90** 76 *in Japanese.*

[3] Matsui M and Chikazumi S 1978, *Journal of the Physical Society of Japan* **45** 458.

Effect of stainless steel chemical composition on brazing ability of filler metal

Yasuyuki Miyazawa, Kei Ohta and Akira Nishiyama

Tokai University, 4-1-1, Kitakaname, Hiratsuka-city, Kanagawa 259-1292, Japan

E-mail: ymiyazawa@tokai-u.jp

Abstract. Many kinds of stainless steel have been used in the engineering field. So it is necessary to investigate the effect of SUS chemical compositions on the brazing ability of filler metal. In this study, SUS315J containing Cr, Ni, Si, Cu, and Mo was employed as a base metal. Excellent spreading ability of the molten nickel-based brazing filler on SUS315J was obtained as compared with that on SUS316. Copper and silicon influenced the significant spreading ability of the filler.

1. Introduction

Many kinds of stainless steel have been developed, and they have been used for many applications ranging from kitchenware and tableware to automobiles, construction, and electrical applications. They have been used for domestic hot water services such as electric hot water tanks and kerosene-burning water heaters [1]. A few alloying elements such as Cu or Si are added to improve the properties of stainless steel. However, the influence of the addition of a few alloying elements to the joining ability (brazing ability) has not been investigated until now. SUS315J2 stainless steel (JIS), which has good stress corrosion cracking (SCC) resistance and crevice corrosion resistance properties, has been developed [1]. SUS315J2 contains Cu and Si as alloying elements to improve SCC resistance properties [2].

The brazing ability of a paste-type Ni-based brazing filler metal as an all-purpose brazing filler metal to SUS315J2 was investigated in this study.

2. Experimental procedure

2.1. Base Metal and Brazing Filler Metal

SUS315J2 stainless steel (JIS) was employed as the base material. The chemical composition of this stainless steel is shown in Table 1; SUS316 stainless steel was used as the standard. Paste-type Ni–Cr–Si–P and BNi_2 brazing filler metals were used in this study. The Ni–Cr–Si–P brazing filler metal did not contain B and had a relatively low melting point with good corrosion resistance properties [3-4]. The chemical composition of both fillers is shown in Table 2.

2.2. Specimens

A T-joint specimen was used to investigate fillet formability and observe the cross-sectional microstructure at the brazed joint. A single-lap specimen was used to estimate the mechanical properties of the brazed joint. Spreading tests were employed to investigate the spreading ability of the

Content from this work may be used under the terms of the Creative Commons Attribution 3.0 licence. Any further distribution of this work must maintain attribution to the author(s) and the title of the work, journal citation and DOI.
Published under licence by IOP Publishing Ltd

molten Ni–Cr–Si–P and BNi_2 brazing filler metals on the SUS315J2 stainless steel (JIS). Schematic diagrams of each specimen are shown in Figures 1 and 2.

Table 1. Chemical compositions of SUS315J2 and SUS316 (JIS) used in this study.

SUS No.	Chemical composition (mass%) Fe	C	Si	Mn	P	S	Ni	Cr	Mo	Cu
315J2	Bal.	<0.08	2.5-4.0	<2.0	<0.045	<0.03	11.0-14.0	17.0-20.5	0.5-1.5	0.5-3.5
316	Bal.	<0.08	<1.0	<2.0	<0.045	<0.03	10.0-14.0	16.0-18.0	2.0-3.0	-

Table 2. Chemical compositions and temperature of brazing filler metals used in this study.

	Chemical compositions (mass%)						Temperature (°C)	
	Ni	Cr	Si	Fe	P	B	Solidus	Liquidus
BNi2	Bal.	6.0-8.0	4.0-5.0	2.5-3.5	-	2.75-3.50	970	1000
Ni-Cr-Si-P	Bal.	27.5-31.5	3.8-4.2	-	5.6-6.4	-	980	1030

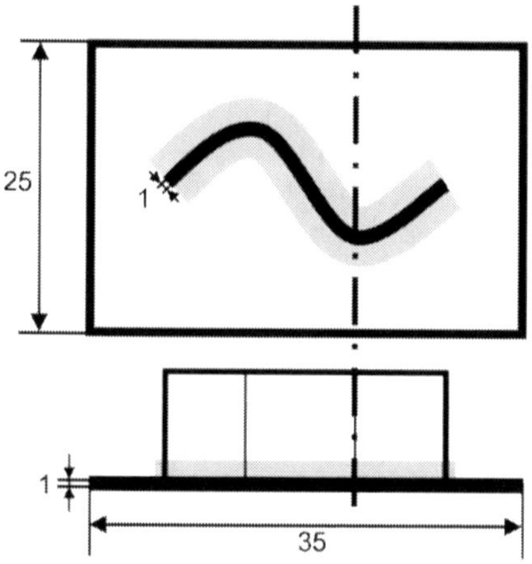

Figure 1. Schematic diagram of T-joint specimen.

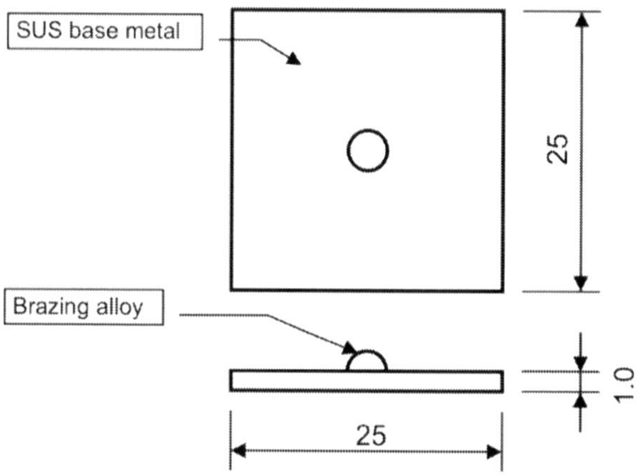

Figure 2. Schematic diagram of spread test specimen.

2.3. Brazing Process and Estimation Method

Before the spreading test, a constant amount of paste-type brazing filler metal was placed on the center of the SUS315J2 plate using a dispenser. After the test or after solidification of the molten filler on the stainless steel plate, the shape of the solidified brazing filler metal region was observed and its area was measured. The spreading area was defined as the measured area. The spreading ability of the molten brazing filler metal was investigated by using the measurement results of the spreading area. Furthermore, the edge of the solidified brazing filler metal region was observed and analyzed with an electron probe microanalysis (EPMA) system in order to investigate the mechanism behind the spreading of the molten brazing filler metal.

Brazing was done by using a vacuum-type (5×10^{-3} Pa) conventional electric heated furnace under Ar gas (purity: 99.995%) atmosphere or vacuum atmosphere. The brazing temperatures were 1080 and 1130 °C with consideration of the liquidus temperature of the brazing filler metal. The brazing times were 5 or 30 min.

The top view of the specimens and the cross-sectional microstructures at the brazed joints of the T-joint specimens were observed with an optical microscope (OM) and scanning electron microscope

(SEM). Also, elemental distributions were analyzed with using an EPMA system. The mechanical properties of the brazed joint were estimated with a single-lap specimen and shear tests.

3. Results and Discussion

3.1. Outside appearance and cross-sectional microstructure of the T-joint specimen

Figure 3 shows the outside appearance of the T-joint specimens brazed with the Ni–Cr–Si–P filler alloy. Excellent wettability and spreading ability of the Ni–Cr–Si–P filler on SUS315J2 as compared with that on SUS316 were obtained.

The EPMA analysis of the interfacial microstructure of the SUS315J2 joint brazed with Ni–Cr–Si–P was similar to that of the SUS316 brazed joint, as shown in Figure 4. We did not observe an alloyed layer or alloy phase containing Cu, Si, or Mo by interfacial reaction during wetting and spreading at the brazed interface between the base metal and brazed layer through either SEM observations or EPMA analytical results.

3.2. Spreading test

Figure 5 shows the spreading test results and Figure 6 shows the spreading area measurement results as a function of brazing atmosphere. According to these results, good joinability of SUS315J2 was obtained. In the case of brazing under vacuum atmosphere, excellent wettability of the Ni–Cr–Si–P filler to SUS315J2 was obtained as compared with SUS316. On the other hand, in the case of brazing under Ar gas atmosphere, the wettability of the Ni–Cr–Si–P filler to SUS315J2 was similar to that to SUS316.

Figure 7 shows typical top view SEM images of the edge of the solidified brazing filler metal region and typical EPMA elemental distribution in the case of vacuum atmosphere. It was found that similar surface microstructures of the solidified brazing filler metal regions formed. Surface conditions such as grain boundary grooves of the base materials hardly influenced the wetting and spreading phenomenon of the molten filler in both cases. It is recognized that the molten filler solidified from the edge of the spreading filler metal region and that the influence of solidification shrinking was minimal. Consequently, it appears that the driving force for the spreading phenomenon of the molten brazing filler metal can be estimated from elemental distribution results.

According to the elemental distribution, these results, in that case of using SUS315J2, it is recognized that Cu is contained in molten brazing filler metal. And in that case of using SUS316 as standard, it is recognized that Cu is hardly contained in molten brazing filler metal. It is considered that existence of Cu and Si element in molten brazing filler is caused by dissolution phenomenon of the SUS315J2 during brazing. Consequently, it appears that the presence of Cu and Si changed the specific characteristics of the molten brazing filler metal and that spreading progressed as a result of these changes.

Similar elemental distributions results were obtained in the case of brazing under Ar gas atmosphere. An Ar gas atmosphere is not a reducing atmosphere, and thus the formation of a Cr oxide thin film on the stainless steel surface decreased the effect of changes by Cu or Si.

The effect of the addition of Cu to the brazing filler metal on spreading ability was investigated. Figure 8 shows top views of T-joint specimens brazed with the Ni–Cr–Si–P filler alloy with added Cu. Figure 9 shows the effect of Cu addition on the spreading area. The addition of Cu to the brazing filler metal and increasing the amount of added Cu facilitated spreading of the brazing filler metal slightly in both stainless steels. Figure 10 shows the effect of adding Si to the brazing filler metal on the spreading ability. According to the results, the addition of Si to the Ni–Cr–Si–P filler affected the spreading ability of SUS315J2 slightly.

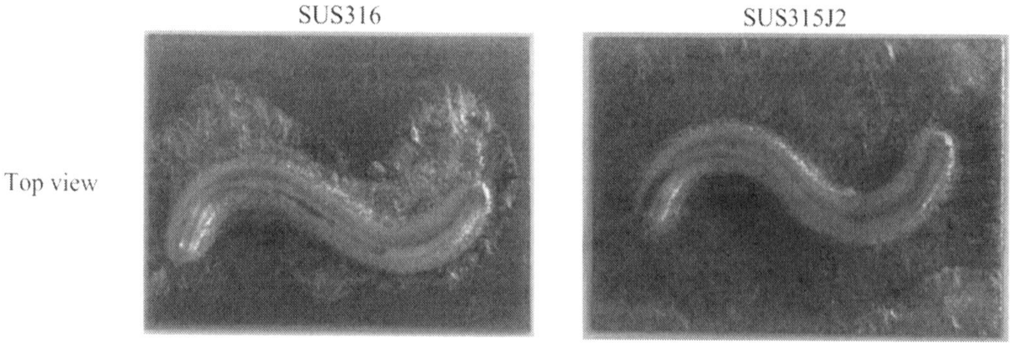

Figure 3. Typical outside appearance of the T-joint specimens brazed with Ni–Cr–Si–P filler alloy. Brazing temperature was 1080 °C. Brazing time was 5 min. Brazing atmosphere was vacuum.

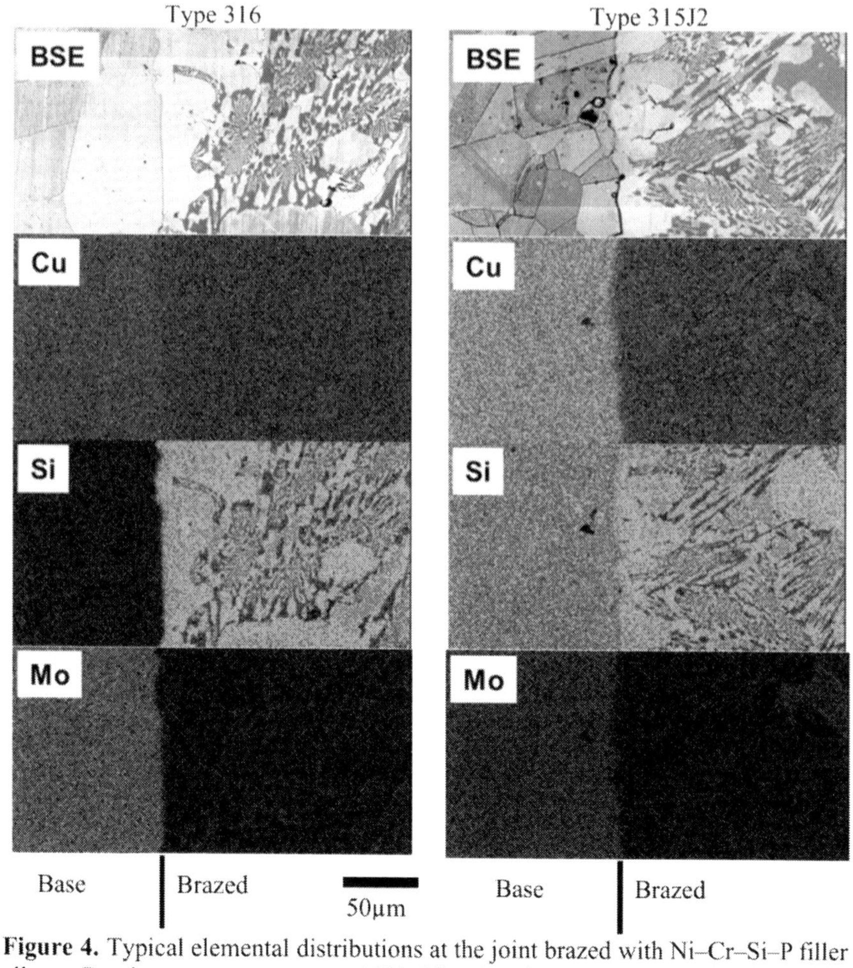

Figure 4. Typical elemental distributions at the joint brazed with Ni–Cr–Si–P filler alloy. Brazing temperature was 1080 °C. Brazing time was 5 min. Brazing atmosphere was vacuum.

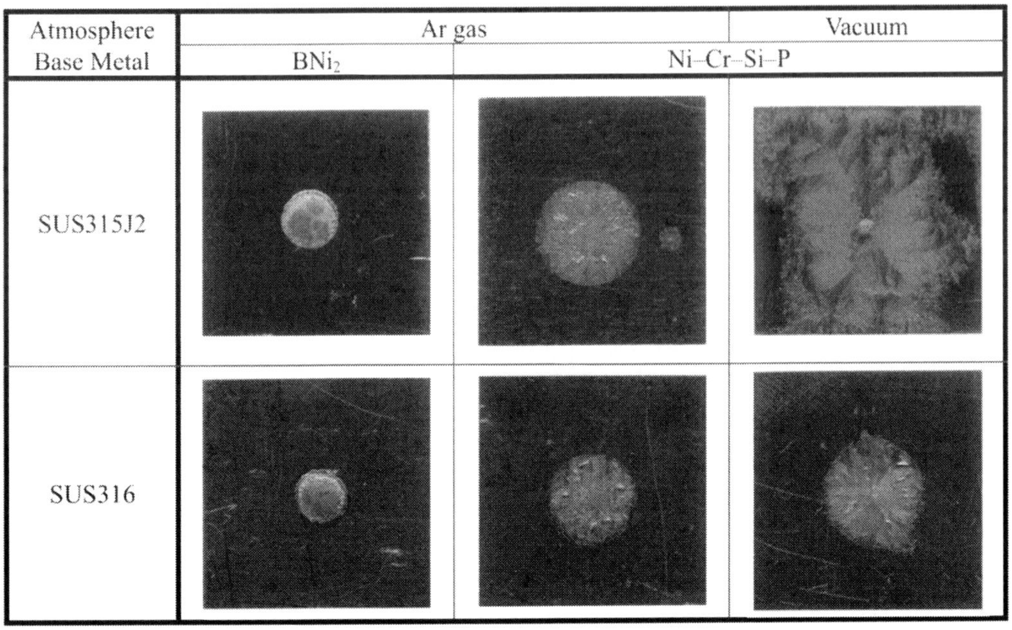

Figure 5. Typical outside appearance of the spread specimen after test. Brazing temperature was 1080 °C. Brazing time was 5 min.

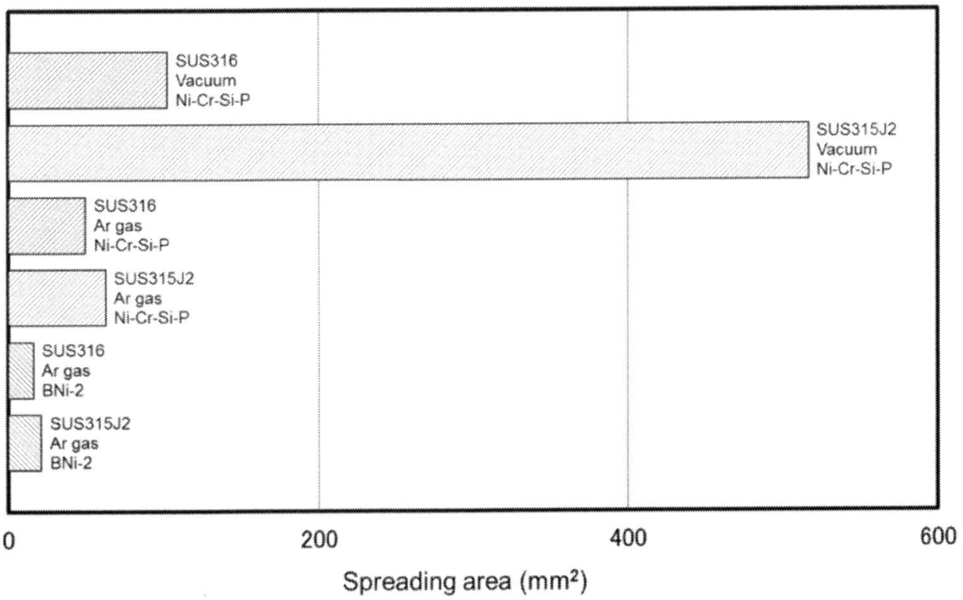

Figure 6. Spreading area measurement results. Brazing temperature was 1080 °C. Brazing time was 5 min.

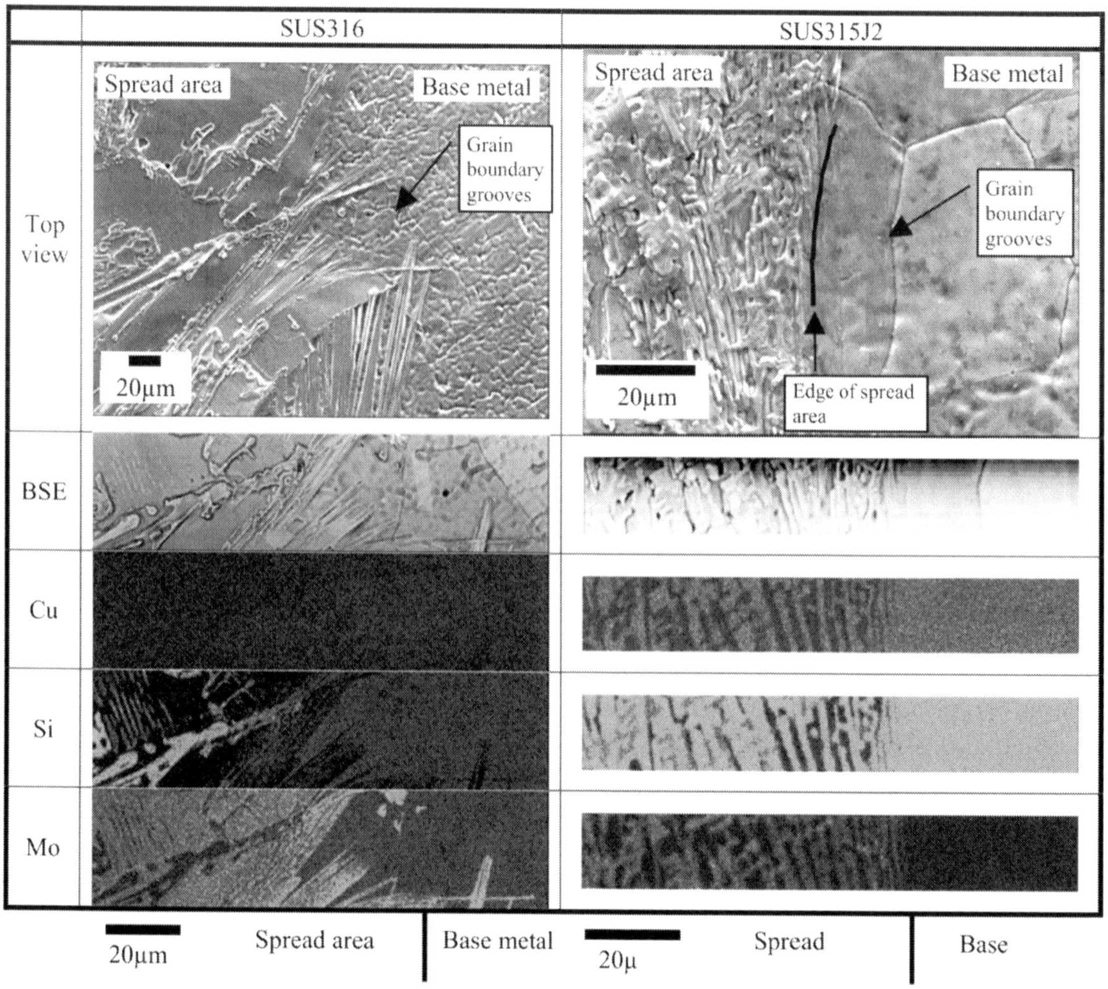

Figure 7. Typical top view outside appearance of the spread specimen and elemental distributions at the edge of spread area with EPMA. Brazing temperature was 1080 °C. Brazing time was 5 min. Brazing atmosphere was vacuum.

Base metal	Vacuum Ni–Cr–Si–P	Vacuum Ni–Cr–Si–P + 1 mass% Cu	Vacuum Ni–Cr–Si–P + 2 mass% Cu
SUS315J2		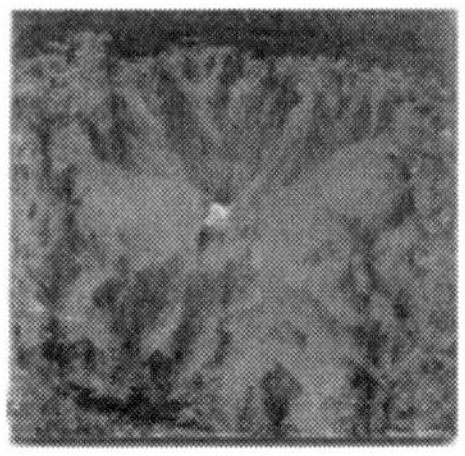	
SUS316		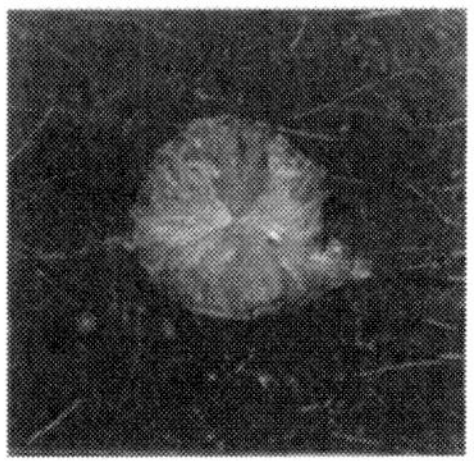	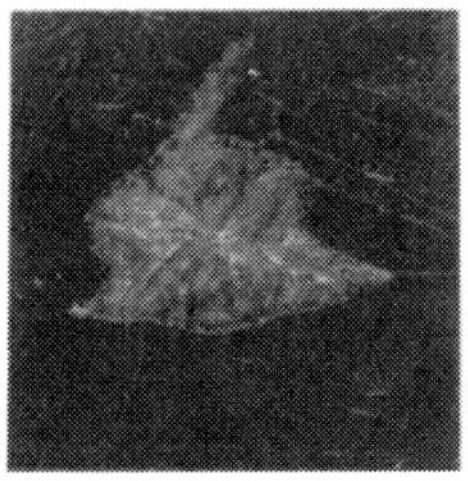

Figure 8. Effect of Cu addition to Ni–Cr–Si–P filler on spread area. Brazing temperature was 1080 °C. Brazing time was 5 min.

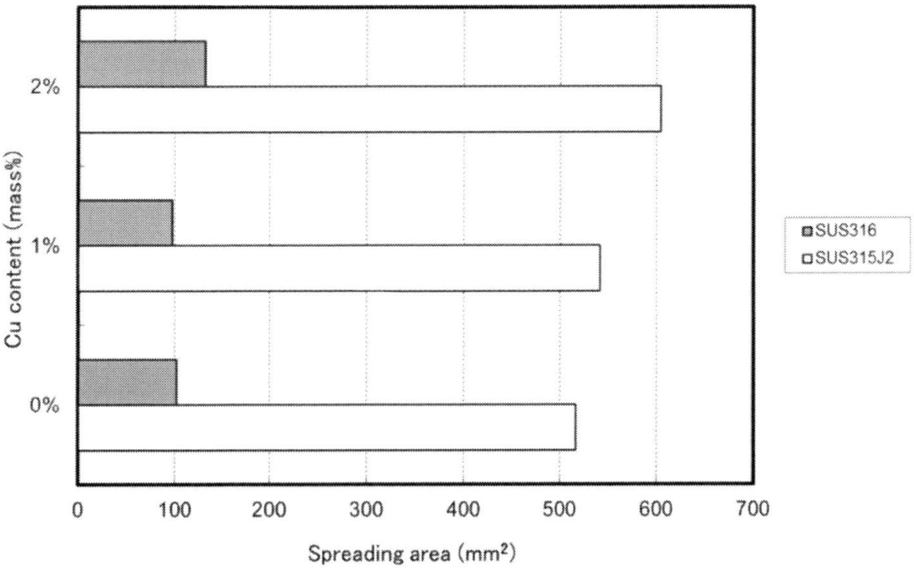

Figure 9. Spreading area measurement results. Brazing temperature was 1080 °C. Brazing time was 5 min.

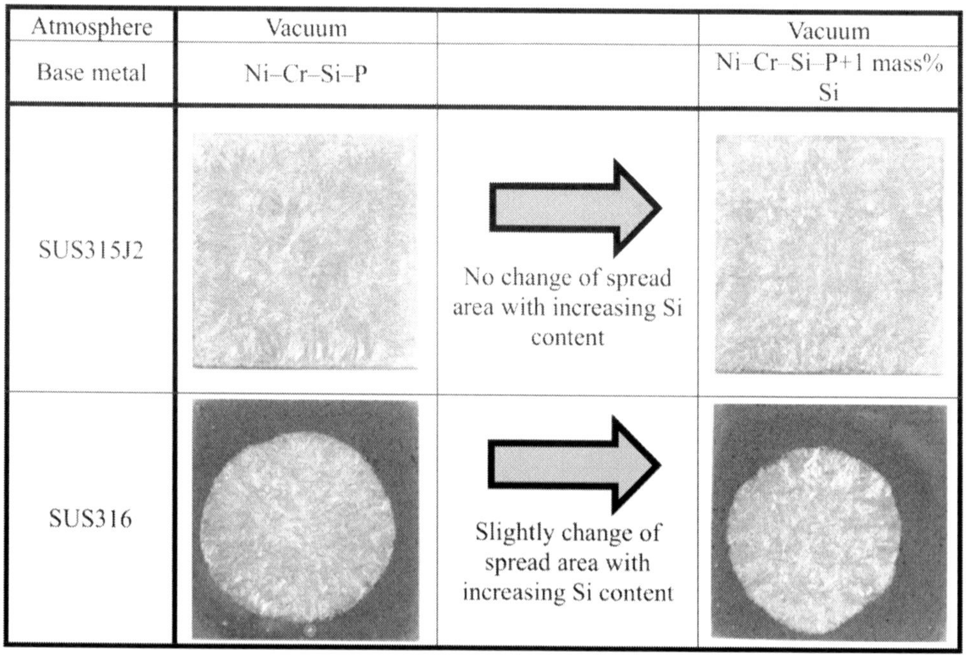

Figure 10. Effect of Si addition to Ni–Cr–Si–P filler on spread area.

The normal thickness of a Cr_2O_3 thin film as a passivation film on stainless steel is 1–3 nm [5]. On the other hand, Si oxidizes on the stainless steel with increasing temperature when the stainless steel contains Si [6]. The self-diffusion coefficient of metallic ions in a passivation film on stainless steel is extreme low at 1000 °C [7]. Therefore, it was difficult to imagine that Si was oxidized by oxygen from the outer atmosphere.

The formation energy of Cr_2O_3 was higher than that of Si oxide [8], and so it appeared that Cr_2O_3 was reduced by the oxidation of Si on the stainless steel. Also, Cr_2O_3 was reduced by the extremely low oxygen partial pressure caused by the clean vacuum atmosphere. The passivation film on stainless steel was thus broken by the reduction of the Cr_2O_3 thin film. A molten Ni-based brazing filler metal was in direct contact with the metallic surface of the stainless steel, as shown in Figures 5 and 8. Extreme wetting and spreading took place in the case of using SUS315J2 stainless steel.

4. Conclusions

The brazing ability of a paste-type Ni-base brazing filler metal as an all-purpose brazing filler metal for SUS315J2 stainless steel (JIS) was investigated in this study, and the following results were obtained:

(1) A sound SUS315J2 brazed joint with a Ni–Cr–Si–P brazing filler metal was obtained.
(2) In the case of brazing under vacuum atmosphere, excellent wettability of the Ni–Cr–Si–P filler to SUS315J2 was obtained as compared with that to SUS316.
(3) It appears that the presence of Cu and Si changed the specific characteristics of the molten brazing filler metal and that spreading was facilitated by these changes.
(4) It appears that a passivation film on the stainless steel was broken by the reduction of the Cr_2O_3 thin film. The molten Ni-based brazing filler metal was in direct contact with the metallic surface of the stainless steel. Extreme wetting and spreading took place in the case of using SUS315J2 stainless steel.

References

[1] Adachi T, Nishikawa M and Yoshii T 1989 *Nisshinseikogiho* **60** 56
[2] Katoh T, Kusaka K and Hisada T 1980 *Denkiseiko* **51 4** 252
[3] Nagai S, Tanaka K and Hidaka K 2000 *Proc. of 2000 Powder Metallurgy World Congress (Kyoto)* p 875
[4] Miyazawa Y, Denda K and Ariga T 2003 *International Brazing and Soldering Conference Proc.* (*San Diego*) p6.6
[5] Ohmi T, Nakagawa Y, Nakamura M, Ohki A and Koyama T 1996 *J. Vac. Sci. Technol.* A. **14 4** 2505
[6] Fukaya M, Maeda S and Yamanaka M 2002 *Current advances in materials and processes report of the ISIJ meeting* **15 6** 1299
[7] Hashimoto M 2007 *Stainless Steel in japanese* (Tokyo: Kogyo Chosakai Publishing Co., Ltd.)
[8] The Japan Institute of Metals 1993 *Metals data book 3rd ed. in japanese* (Tokyo: Maruzen Co., Ltd.)

Spot brazing of aluminum to copper with a cover plate

Junya Hayashi and Yasuyuki Miyazawa

Metallurgical Engineering (Materials Science), Tokai University,
4-1-1, Kitakaname, Hiratsuka-shi, Kanagawa-ken, 259-1292, Japan

E-mail: ymiyazawa@tokai-u.jp

Abstract. It is difficult to join dissimilar metals when an intermetallic compound is formed at the joining interface. Spot brazing can be accomplished in a short time by resistance heating. Therefore, it is said that the formation of a intermetallic compound can be prevented. In this study, aluminum and copper were joined by spot brazing with a cover plate. The cover plate was used to supply heat to base metals and prevent heat dissipation from the base metals. The ability to braze Al and Cu was investigated by observation and analysis. Pure aluminum (A1050) plate and oxygen-free copper (C1020) plate were used as base metals. Cu–Ni–Sn–P brazing filler was used as the brazing filler metal. SPCC was employed as cover plate. Brazing was done with a micro spot welder under an argon gas atmosphere. Brazing ability was estimated by tensile shear strength and cross sectional microstructure observation. Al and Cu can be joined by spot brazing with Cu-Ni-Sn-P brazing filler and cover plate.

1. Introduction

Joining of dissimilar metals is a very important technology for the manufacturing field [1]. However, it is difficult to join dissimilar metals when intermetallic compounds are formed at the joining interface. However, brazing is effective as a joining method for dissimilar metals. Therefore, many research results have been reported [2–8]. Spot brazing can be accomplished in a short time by resistance heating. Therefore, it is said that the formation of intermetallic compounds can be prevented.

Combinations of dissimilar metals largely depend on the intended use. Joining of Al and Cu is expected to be applied to cooling products for electronic devices.

Specific resistance and heat conductivity are very important because spot brazing or spot welding is accomplished by a heat source using resistant heating. Al and Cu have low specific resistance and high heat conductivity. Therefore, it is difficult to ensure the heat amount required for brazing (welding). Studies using spot welding with a cover plate have been carried out to solve this problem [9–11]. Figure 1 shows a schematic diagram of spot brazing with a cover plate. The cover plate is used to supply heat to the base metals and prevent heat dissipation from the base metals. No such research on spot brazing with a cover plate has yet been reported.

In this study, Al and Cu were brazed by spot brazing using a cover plate. The ability to braze Al and Cu was investigated by observation and analysis.

Content from this work may be used under the terms of the Creative Commons Attribution 3.0 licence. Any further distribution of this work must maintain attribution to the author(s) and the title of the work, journal citation and DOI.
Published under licence by IOP Publishing Ltd

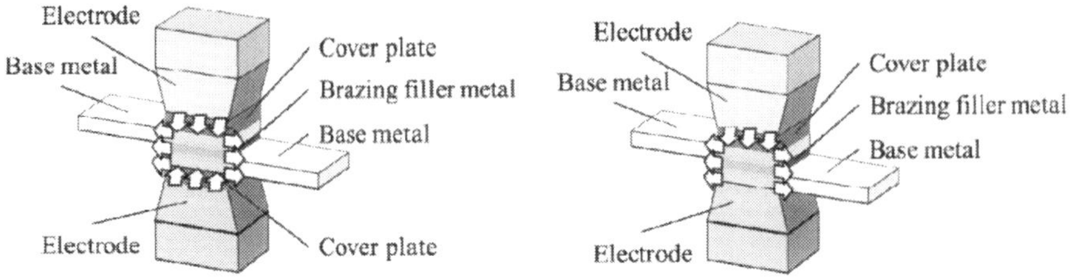

Figure 1. Schematic diagram of spot brazing with cover plate.

2. Experimental procedure

2.1. Base metals, brazing filler metal and cover plate

Pure aluminum (A1050) plate (width 7 mm, length 50 mm, and thickness 0.8 mm) and oxygen-free copper (C1020) plate (width 7 mm, length 50 mm, and thickness 0.8 mm) were used as the base metals. Cu–Ni–Sn–P brazing filler ($Cu_{Bal.}Ni_{5.7}Sn_{9.7}P_7$: width 5 mm, length 5 mm, and thickness 20 μm) was employed as the brazing filler metal. The Cu–Ni–Sn–P brazing filler (MBF2005® Metglas®, Inc.) was produced by liquid quenching solidification. As a preliminary experimental result, Al–Cu joints brazed with Cu–Ni–Sn–P brazing filler have higher tensile shear strength than Al–Cu joints brazed with other filler metals. Therefore, Cu–Ni–Sn–P brazing filler was used in this study.

The cover plate was of suitable material with a high specific resistance and low heat conductivity. In this study, SPCC (cold-rolled steel sheet: width 7 mm, length 12 mm, and thickness 0.5 mm) was used as the cover plate.

Figure 2 shows a schematic diagram of the specimen. Figure 3 shows the outside appearance of the specimen.

2.2. Brazing condition

Brazing was done with a micro spot welder (Figure 4) in an argon gas atmosphere. Figure 5 shows a schematic diagram of the spot brazing process. Table 1 shows brazing conditions. Brazing currents were 700, 800, 900, 1000, and 1100 A at 3.5 V for 2 s under a 4-kgf as brazing load.

2.3. Microstructure observation and EPMA

After brazing, the outside appearance was examined, The specimen was then cut along the center line, and cross sections were polished using emery paper and a buffing cloth. Sodium hydroxide was used to etch the Al base metal side, and ferric chloride was used to etch the Cu side. After etching, the cross-sectional microstructure was examined by optical microscope. Furthermore elemental distributions at the brazed joint were analyzed by EPMA to investigate the interfacial reaction between the base metal and the molten brazing filler metal.

2.4. Tensile shear test

Tensile shear strength (fracture load) of the Al–Cu brazed joint was measured to estimate its mechanical properties. Tensile shear strength was measured by a tensile test machine (Instron® 3367) with a cross-head moving speed of 1 mm/min.

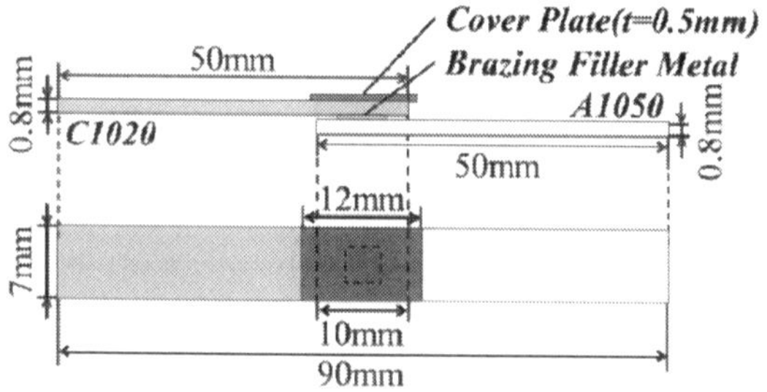

Figure 2. Schematic diagram of specimen when the cover plate was set Cu side.

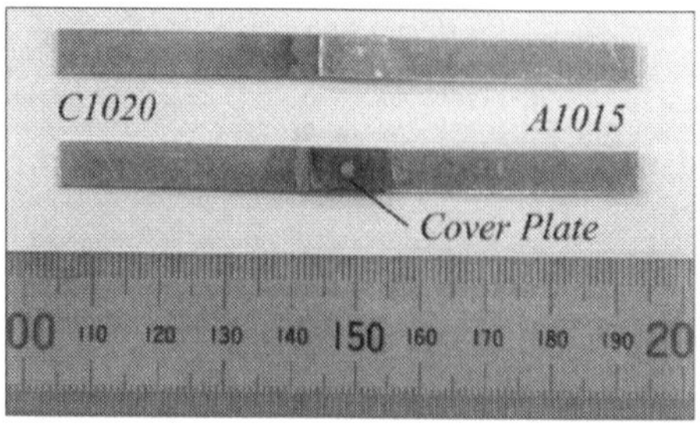

Figure 3. Outside appearance of specimen.

Table 1. Brazing conditions.

Brazing current value	700～1100A
Brazing voltage value	3.5V
Brazing time	2sec
Brazing load	4kgf
Electrode	Cu-W
Atmosphere gas	Ar

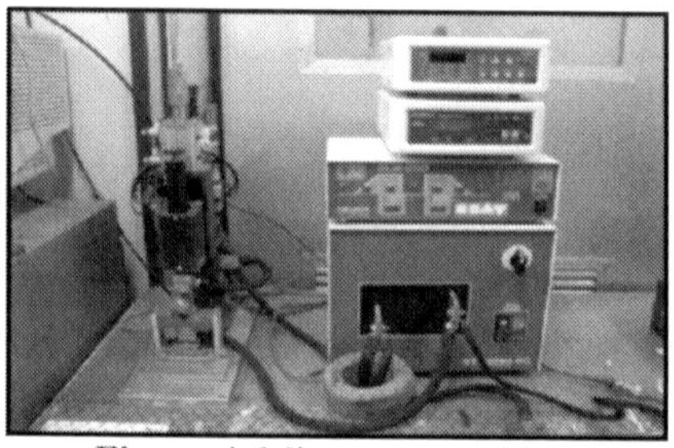

Figure 4. Micro-spot welder.

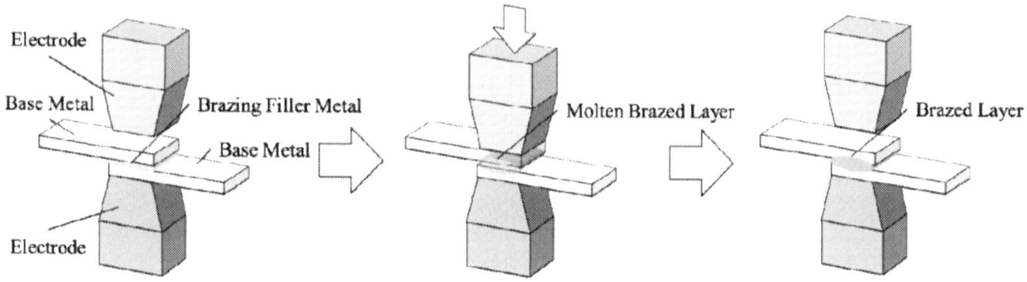

Figure 5. Schematic diagram of spot brazing process.

3. Results and Discussion

3.1. Effect of cover plate

Figure 6 shows tensile shear strength of the specimens brazed using and not using a cover plate. According to Figure 6, when the cover plate was used, the tensile shear strength was increased by approximately 40%.

Figure 7 shows typical cross-sectional microstructure of specimens brazed with and without using a cover plate. According to Figure 7, when the cover plate was used, the volume of the brazed layer was larger. The cover plate prevents heat dissipation from the base metals, thereby generating more heat. Furthermore, the cover plate became a heat source due to its low heat conductivity, and it supplied heat to the base metals. Therefore, it seems that the amount of heat at the brazed layer is increased, and an interface reaction occurred drastically. When the cover plate was set on the Al side or on both the Al and Cu sides, the Al side fully melted, and brazing was impossible. Therefore, the cover plate was set Cu side only.

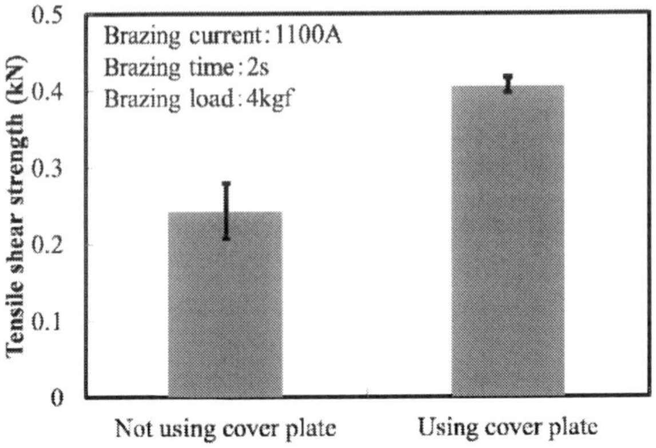

Figure 6. Tensile shear strength of the specimen using and not using a cover plate.

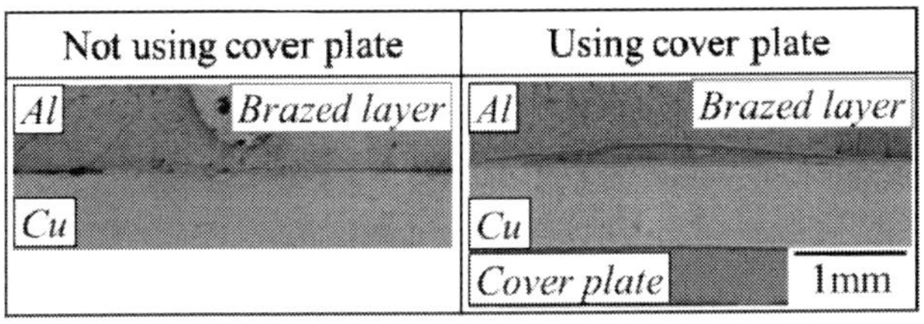

Figure 7. Typical cross-sectional microstructures of the specimen using and not using a cover plate. Brazing current: 1100A, brazing time: 2 s, brazing load: 4 kgf.

3.2. Microstructure observation

Figure 8 shows the relation between brazing current and the cross-sectional microstructure of the Al–Cu brazed joint with Cu–Ni–Sn–P brazing filler and a cover plate. According to Figure 8, the volume of the brazed layer increases with increased brazing current. It seems that the generated amount of heat increases with increasing brazing current. Also, the cover plate and Cu were joined above 800 A. The cover plate surface was sprayed with boron nitride solution to prevent joining of the cover plate to the Cu. However, the cover plate and Cu were still join. Furthermore, brazing filler metal did not melt at 700–1000 A. causing brazing filler metal foil to remain. However, brazing filler metal was fully melted at 1100 A.

Figure 9 shows typical cross sectional microstructure at the Al–Cu brazed joint. According to Figure 9, an interface reaction on the Al side occurred drastically. This seems to have been due to the difference between the melting points of Al and Cu. The brazed layer contained voids, but this was caused by polishing. Before polishing, a good void-free brazed layer was formed.

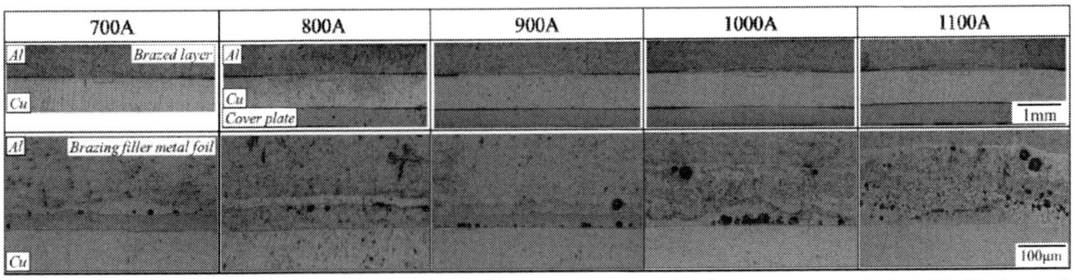

Figure 8. Relation between brazing current and cross-sectional microstructure of the Al–Cu brazed joint with Cu–Ni–Sn–P brazing filler and a cover plate. Brazing time: 2 s, brazing load: 4 kgf.

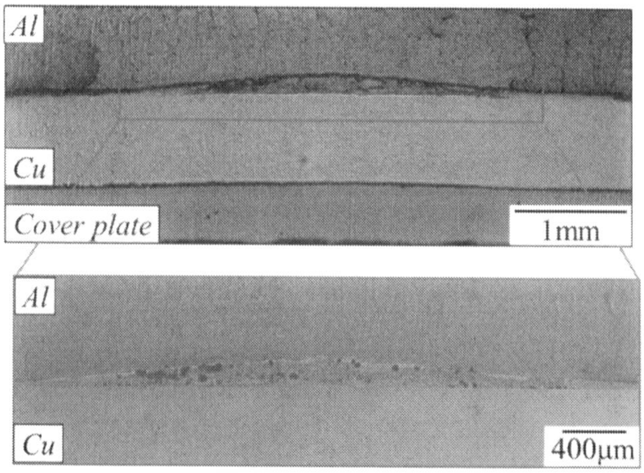

Figure 9. Typical cross sectional microstructure at the Al–Cu brazed joint. Brazing current: 1100 A, brazing time: 2 s, Brazing load: 4 kgf.

3.3. EPMA analysis

Figure 10 shows the elemental distributions by EPMA at the Al–Cu brazed joint with Cu–Ni–Sn–P brazing filler and a cover plate. As seen in Figure 10, the brazed layer mainly consists of an Al-rich layer and a Cu-rich layer. The Al-rich layer consists of Al solid solution, Al–P compound, and Cu_3P or CuP_2 compound by Al–P and Cu–P phase diagrams. The Cu-rich layer consists of Cu solid solution, Cu–Ni alloy phase, Cu_3P or CuP_2 compound, and Al–Cu compound by Cu–Ni and Cu–P phase diagrams. Nickel is detected in the Cu-rich layer. From phase diagrams, Ni–Al alloy phase is precipitated at the Cu-rich layer. Sn concentrates at the interface between the brazed layer and Cu base metal where Cu–Sn compound is formed. It seems that this compound has little effect on tensile shear strength. Phosphorus exists in the overall brazed layer.

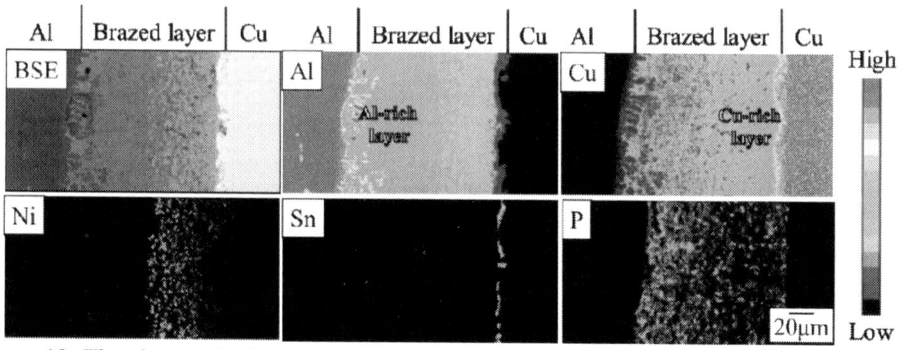

Figure 10. The elemental distributions by EPMA at the Al–Cu brazed joint with Cu–Ni–Sn–P brazing filler and cover plate. Brazing current: 1100 A, brazing time: 2 s, brazing load: 4 kgf.

3.4. Tensile shear test

Figure 11 shows the relation between brazing current and tensile shear strength of the Al–Cu joints brazed with Cu–Ni–Sn–P brazing filler and cover plate. As seen in Figure 11, tensile shear strength increases with increasing brazing current. It seemed that the brazed joint area increases with increasing brazing current (Figure 8). Also, the volume of brazing filler metal foil decreases with increasing brazing current. When the brazing filler metal foil did not melted, the Al–Cu brazed joints have low tensile shear strength. However, when the brazing filler metal foil is melted, the Al–Cu brazed joints have high tensile shear strength.

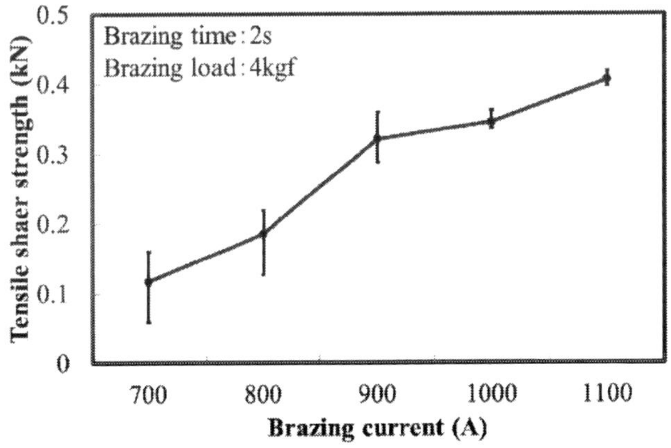

Figure 11. Relation between brazing current and tensile shear strength of the Al–Cu joints brazed with Cu-Ni-Sn-P brazing filler and cover plate.

Figure 12 shows the outside appearances of specimens after the tensile shear test. The figure shows fracture positions on the brazed layers of the specimens brazed at 700–1000 A. Also, a fracture position is seen on the Al base metal in the specimen brazed at 1100 A.

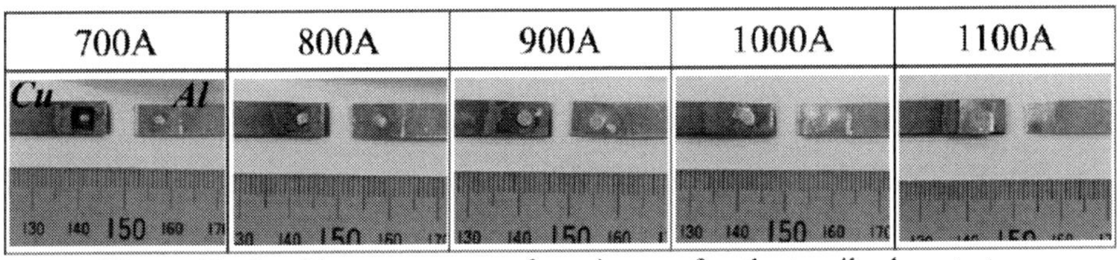

Figure 12. Outside appearances of specimens after the tensile shear test.

4. Conclusions

In this study, Al and Cu were joined by spot brazing using a cover plate. The ability to braze Al and Cu was investigated by observation and analysis. The results are as follows;

(1) In when a cover plate was used, tensile shear strength was increased by approximately 40 percent.

(2) The volume of the brazed layer was increased with increasing brazing current. The generated amount of heat was increased with increasing brazing current.

(3) An interface reaction on the Al side occurred drastically. This was due to the difference between the melting points of Al and Cu.

(4) Tensile shear strength was increased with increasing brazing current. The brazed joint area increased with increasing brazing current. When the brazing filler metal foil is not melted, the Al–Cu brazed joints have low tensile shear strength. However, when the brazing filler metal foil is melted, the Al–Cu brazed joints have high tensile shear strength.

(5) The brazed layer mainly consists of an Al-rich layer and a Cu-rich layer. The Al-rich layer consists of Al solid solution, Al–P compound, and Cu_3P or CuP_2 compound. The Cu-rich layer consists of Cu solid solution, Cu–Ni alloy phase, Cu_3P or CuP_2 compound, and Al–Cu compound. Nickel is detected in the Cu-rich layer. Ni–Al alloy phase is precipitated at the Cu–rich layer. Sn concentrates at the interface between the brazed layer and Cu base metal. Cu–Sn compound is formed. Phosphorus exists in the overall brazed layer.

(6) Al and Cu can be joined by spot brazing with Cu–Ni–Sn–P brazing filler and a cover plate.

References

[1] Nakata N and Ushio M 2002 *J. JWS.* **71 6** 418
[2] Hayashi J, Miyazawa Y and Ariga T 2012 *Proc. Sch. Eng. Tokai Univ., Ser. E.* **52 2** 165
[3] Hayashi J, Miyazawa Y and Ariga T 2013 *Proc. 10th Int. Conf. Brazing, High Temperature Brazing and Diffusion Bonding (LOT2013).* 287
[4] Hayashi J and Miyazawa Y 2013 *Proc. Int. Symp. Interfacial Joining and Surface Technology (IJST2013).* 113
[5] Koyama K, Shinozaki K, Ikeda K, Miki K and Kuroki H 2002 *J. JILM.* **40 9** 403
[6] Koyama K, Miki K, Masumoto S, Mochida A, Ikeda K, Yoshida M and Shinozaki K 2003 *J. JILM.* **41 2** 75
[7] Koyama K, Miki K, Masumoto S, Mochida A, Dejun L, Yoshida M and Shinozaki K 2003 *J. JILM.* **41 9** 409
[8] Laik A, Tewari R, Dey G K, Shirzadi A A, Kumar A and Jayakumar T 2013 *Met. Mat. Trans. A.* **44 5** 2212
[9] Qiu R, Iwamoto C and Satonaka S 2009 *J. Mat. Pro. Tec.* **209 9** 4186
[10] Shi H, Qiu R, Zhu J, Zhang K, Yu H and Ding G 2010 *Mat. Des.* **31 10** 4853
[11] Satonaka S, Iwamoto C, Murakami G and Matsumoto Y 2012 *Wel. Wor.* **56 7** 44

IOP Conf. Series: Materials Science and Engineering **61** (2014) 012016 doi:10.1088/1757-899X/61/1/012016

Brazing of copper to stainless steel with a low-silver-content brazing filler metal

Tatsuya Fukikoshi*, Yūki Watanabe*, Yasuyuki Miyazawa*, and Fumio Kanasaki**
* Metallurgical Engineering (Materials Science), Tokai University, 4-1-1, Kitakaname, Hiratsuka-shi, Kanagawa-ken, 259-1292, Japan
** K. K. Saginomiya Mfg. 535, Sasai, Sayama-city, Saitama-ken, Japan

E-mail: ymiyazawa@tokai-u.jp

Abstract. The brazing of copper to stainless steel (SUS304 JIS) was performed using a low-silver-content brazing filler metal, Ag-50Cu, under an Ar gas atmosphere with a conventional furnace, owing to the potential economic benefits of using low-silver-content filler metals. The brazeability of the low-silver-content brazing filler metal to copper and SUS304 was investigated. A good joint was obtained, and a drastic dissolution reaction occurred at the copper side. Molten BAg8 penetrated along the crystal grain boundary of the copper base metal when BAg8 was used as the filler metal. This was caused by the dissolution of Ni from the stainless steel into the molten filler metal. Ag-50Cu, which was investigated in this work, can be used instead of BAg8 filler metal.

1. Introduction

Cu/SUS304 (JIS) joints are used extensively in refrigerators [1, 2]. The brazing of a copper pipe to a stainless steel pipe has been traditionally performed using Ag-based brazing filler metals. The eutectic Ag-based filler metal JIS BAg8 exhibits good brazing ability; however, it is expensive because of its high Ag content. A low-silver-content brazing filler metal is less expensive; however, it has a higher liquidus temperature compared to eutectic Ag-based filler metals, as indicated in Figure 1. High brazing temperatures inhibit the joining to the base metal, and the fluidity of brazing filler metals diminishes with increasing Cu content. Therefore, in this study, the brazing of copper to stainless steel was performed using a low-silver-content brazing filler metal. The brazeability of the low-silver-content brazing filler metal is investigated and discussed.

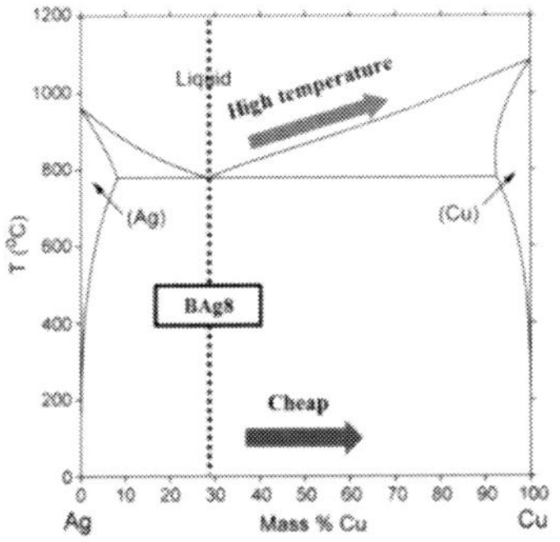

Figure 1. Ag–Cu phase diagram.

Content from this work may be used under the terms of the Creative Commons Attribution 3.0 licence. Any further distribution of this work must maintain attribution to the author(s) and the title of the work, journal citation and DOI.
Published under licence by IOP Publishing Ltd

2. Experimental procedure

2.1 Base metal and brazing filler metal

Copper (phosphorous-deoxidized copper), austenitic stainless steel (SUS304 JIS), and ferritic stainless steel (SUS444 JIS) were used as microstructure observation specimens. A schematic diagram of the pipe-joint specimen used in this work is presented in Figure 2. The chemical compositions of the base metals are listed in Table 1. Ag-based (BAg8 JIS) and low-silver-content brazing filler metal wires were used as the brazing filler metals. The chemical compositions and melting temperatures of the brazing filler metals are listed in Table 2. Brazing was performed under an Ar gas atmosphere using a conventional furnace. The brazing temperatures were 840 °C and 940 °C, and the brazing time was 2 min.

2.2 Microstructure observation and EPMA analysis

The microstructures of cross-sections were observed using an optical microscope, and the elemental distributions were analyzed by electron probe micro-analysis (EPMA).

2.3 Tensile shear tests

The tensile shear strength of a Cu/SUS304 brazed joint was measured to estimate the mechanical properties. A schematic diagram of the single-lap specimen is presented in Figure 3. The tensile shear strength was measured using an Instron® 3367. The cross-head moving speed was 1 mm/min.

2.4 Micro-Vickers hardness tests

The hardness at the brazed joint was measured using a micro-Vickers hardness tester. The applied load was 4.0 N, and the holding time was 10 s.

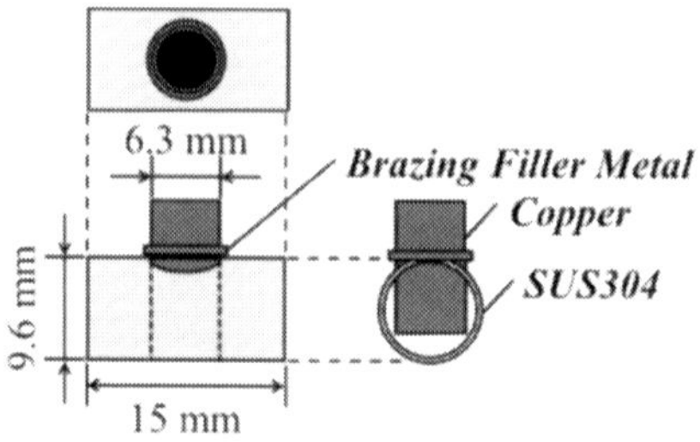

Figure 2. Schematic diagram of the pipe joint.

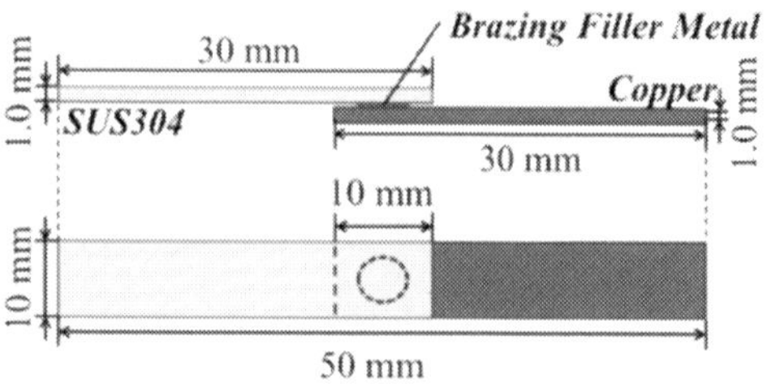

Figure 3. Schematic diagram of the single lap.

Table 1. Chemical compositions of the base metals.

Type	Chemical compositions, mass%								
	Cu	P	Fe	C	Si	Mn	S	Ni	Cr
Cu	99.96	0.025	-	-	-	-	-	-	-
SUS304	-	-	Bal.	0.04	0.59	0.28	0.003	9.13	18.20
SUS444	0.2	0.035	Bal.	0.01	0.29	1.01	0.002	0.17	18.37

Table 2. Chemical compositions and solidus and liquidus temperatures of the brazing filler metals.

Type	Chemical compositions, mass%		Temperature, °C	
	Ag	Cu	Solidus	Liquidus
BAg8	71.0–73.0	Bal.	780	780
Low-silver-content	50	50	780	880

3. Results and discussion

3.1 Microstructure observation and EPMA analysis

Figure 4 shows typical cross-sections of the Cu/SUS304 brazed joint. BAg8 exhibits excellent fluidity on the basis of its good fillet shape; this good fluidity is attributed to the eutectic composition of the filler metal. However, the fluidity of the low-silver-content brazing filler metal was inferior to that of BAg8.

According to the results in Figure 4, dissolution of the Cu base metal occurred preferentially. In the case where the low-silver-content filler was used, the dissolution of the Cu base metal was more extensive compared with the case of BAg8 because of the relative high liquidus temperature of the low-silver-content filler.

When BAg8 was used, molten BAg8 penetrated along the crystal grain boundaries of the copper base metal. Gabbay et al. reported that copper brazed with BAg8 does not result in grain-boundary damage [3, 4]. Figure 5 presents the distributions of elements at the Cu/SUS304 joint brazed with BAg8 and with the low-silver-content filler, obtained by EPMA. These results show that molten BAg8 containing Ni, which originates from the stainless steel base metal during brazing, penetrates along the crystal grain boundary of the copper base metal. This is caused by dissolution of Ni from the stainless steel to the molten filler [5, 6].

Copper base metal was brazed with BAg8 and low-silver-content filler to ferrite stainless steel that did not contain Ni. Figure 6 shows typical cross-sectional microstructures at the Cu/ferrite stainless steel brazed joint. These results show that when the Cu/ferrite stainless steel joint was brazed with BAg8, no grain boundary penetration occurred. When a Ni-bearing base metal such as austenitic stainless steel was brazed to copper with BAg8, grain boundary penetration into the Cu base metal was observed.

3.2 Tensile shear tests

Figure 7 presents the results of the tensile shear tests of the Cu/SUS304 brazed joint and the tensile strength of the base metals. In the single-lap shear test, a fracture occurred at the copper base metal. Good joint shear strength was achieved in this study, and the tensile shear strength of the joints brazed with low-silver-content brazing filler metal was approximately equal to that of the joints brazed with BAg8. Numerous primary crystals of Cu solid solution formed at the brazed layer with low-silver-content brazing filler metal, as compared to that with BAg8; these phases did not adversely affect the joint strength.

3.3 Micro-Vickers hardness tests

Figure 8 presents the results of the micro-Vickers tests of the Cu/SUS304 brazed joint. The micro-Vickers hardness was measured at the base metals and the brazed layer. These results showed that the hardness at the Cu/SUS304 joint brazed using BAg8 was similar to that at the joint brazed using the low-silver-content filler metal. Thus, the low-silver-content brazing filler metal can be used as an alternative to BAg8.

IOP Conf. Series: Materials Science and Engineering **61** (2014) 012016 doi:10.1088/1757-899X/61/1/012016

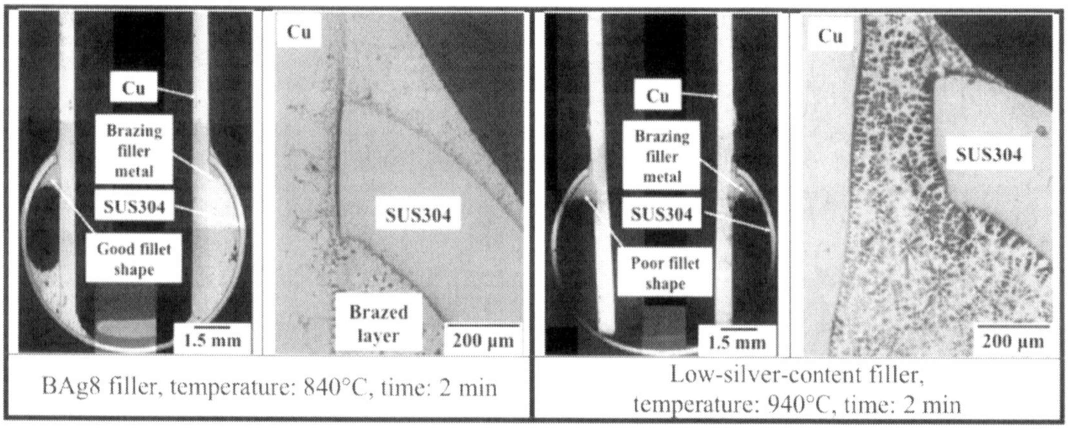

Figure 4. Typical cross-sections of the Cu/SUS304 brazed pipe joint.

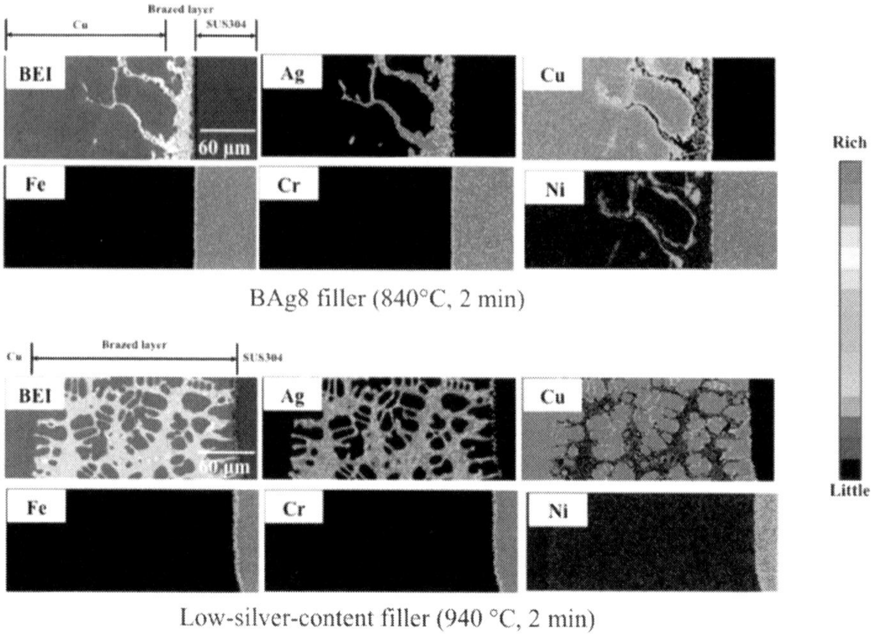

Figure 5. Elemental distributions determined by EPMA at the Cu/SUS304 pipe joint brazed using BAg8 and low-silver-content brazing filler metals.

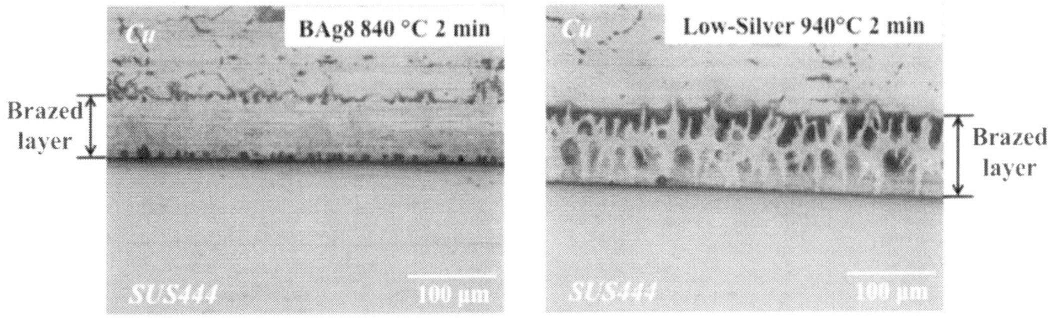

Figure 6. Typical cross-sections at the Cu/SUS444 plate joint brazed using BAg8 and low-silver-content brazing filler metals.

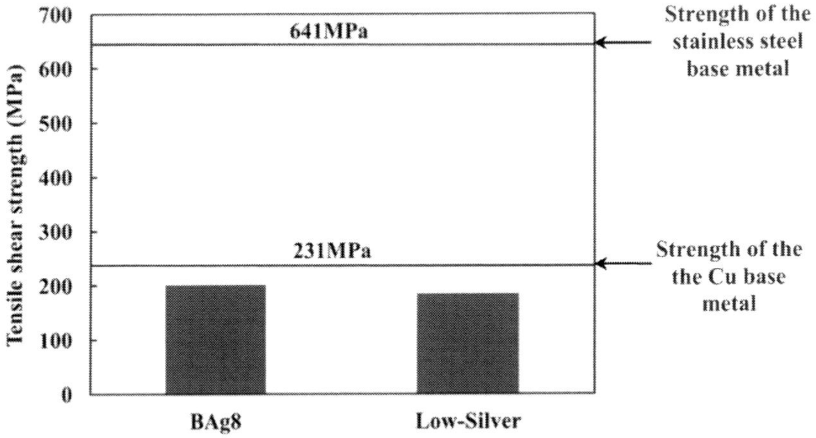

Figure 7. Tensile shear test results of the Cu/SUS304 joint brazed with BAg8 filler metal (840 °C, 2 min) and with the low-silver-content filler metal (940 °C, 2 min).

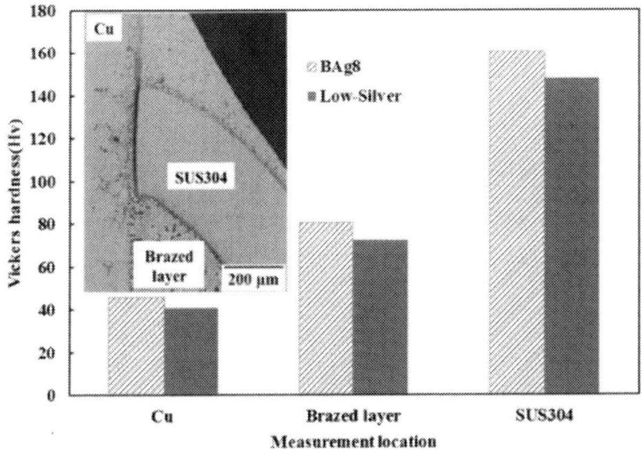

Figure 8. Micro-Vickers hardness test results for the Cu/SUS304 joint brazed with the BAg8 filler metal (840 °C, 2 min) and the low-silver-content filler metal (940 °C, 2 min).

4. Conclusions

Copper and SUS304 were brazed using a low-silver-content brazing filler metal. The brazeability was investigated based on the microstructural observations and EPMA analyses. The following results were obtained:

(1) A low-silver-content brazing filler metal exhibits good brazing ability.
(2) The fluidity of the low-silver-content brazing filler metal was approximately equal to that of BAg8.
(3) A dissolution reaction occurred at the copper side when the low-silver-content brazing filler metal was used. In contrast, molten BAg8 penetrated along the crystal grain boundary of the copper base metal when BAg8 was used. Penetration of the molten BAg8 filler along the crystal grain boundary was caused by the dissolution of Ni from the stainless steel into the molten filler.
(4) The Ag-50Cu low-silver-content brazing filler metal can be substituted for BAg8.

References

[1] Fukikoshi T, Watanabe Y, and Miyazawa Y 2013 *Pros. JWS.* **93** 10
[2] Fukikoshi T, Watanabe Y, and Miyazawa Y 2013 *Proc. Int. Symp. Interfacial Joining and Surface Technology(IJST2013)* 147
[3] Gabbay R, Wagner H. Z, Dirnfeld S. F, and Ramon J. J 1990 *Weld. J.* **69** **10** 378
[4] Dirnfeld S F, Ramon J J, Gabbay R, and Wagner H J 1991 *Mater. Charact.* **26** **1** 17
[5] Siewert T A, Heine R W 1977 *Metall Trans A* **8** **3** 515
[6] Luo H-T, Chen S-W 1996 *J Mater Sci* **31** **19** 5059
[7] Miller V. R, Schwanecke A. E 1978 *Weld J.* **57** **10** 303

Wetting and spreading behavior of molten brazing filler metallic alloys on metallic substrate

Satoshi Kogi, Tetsurou Kajiura, Yukiakira Hanada and Yasuyuki Miyazawa

Metallurgical Engineering (Materials Science), Tokai University,
4-1-1, Kitakaname, Hiratsuka-shi, Kanagawa-ken 259-1292, Japan

E-mail: 3bazm007@mail.tokai-u.jp, ymiyazawa@tokai-u.jp

Abstract. Wetting and spreading of molten brazing filler material are important factors that influence the brazing ability of a joint to be brazed. Several investigations into the wetting ability of a brazing filler alloy and its surface tension in molten state, in addition to effects of brazing time and temperature on the contact angle, have been carried out. In general, dissimilar-metals brazing technology and high-performance brazed joint are necessities for the manufacturing field in the near future. Therefore, to address this requirement, more such studies on wetting and spreading of filler material are required for a deeper understanding. Generally, surface roughness and surface conditions affect spreading of molten brazing filler material during brazing. Wetting by and interfacial reactions of the molten brazing filler material with the metallic substrate, especially, affect strongly the spreading of the filler material. In this study, the effects of surface roughness and surface conditions on the spreading of molten brazing filler metallic alloys were investigated. Ag-(40-x)Cu-xIn and Ag-(40-x)Cu-xSn (x=5, 10, 15, 20, 25) alloys were used as brazing filler materials. A mild-steel square plate (S45C (JIS); side: 30 mm; thickness: 3mm) was employed as the substrate. A few surfaces with varying roughness were prepared using emery paper. Brazing filler material and metallic base plate were first washed with acetone, and then a flux was applied to them. The filler, 50 mg, was placed on the center of the metallic base with the flux. A spreading test was performed under Ar gas using an electrically heated furnace, after which, the original spreading area, defined as the sessile drop area, and the apparent spreading area, produced by the capillary grooves, were both evaluated. It was observed that the spreading area decreased with increasing In and Sn content.

1. Introduction

Joining technologies are important fabrication processes, and joining-ability affects product quality. Brazing is a special joining technology and offers several advantages to the manufacturing field. Applications of brazing include heat exchanger, EGR cooler, and dissimilar-metal joining. Filler material properties, wetting and spreading by molten filler material, and interfacial reactions between molten filler and base materials are all important factors for getting a sound joint by brazing [1]. Wetting and spreading ability of molten brazing filler metallic alloy influence the brazing ability of a joint. Several investigations into wetting by and surface tension of molten brazing filler alloy have been carried out [2]. Effects of brazing time and temperature on the contact angle too have been investigated. In general, brazing of dissimilar metals and high-performance brazed joints are expected to be essential for the field of manufacturing in the near future. Consequently, wetting and spreading abilities of filler materials must be investigated in depth. Moreover, spreading of molten brazing filler

Content from this work may be used under the terms of the Creative Commons Attribution 3.0 licence. Any further distribution of this work must maintain attribution to the author(s) and the title of the work, journal citation and DOI.
Published under licence by IOP Publishing Ltd

material is affected by the surface roughness and surface conditions, and especially by the interfacial reactions between the filler and base materials. In this study, the effects of the surface roughness and surface conditions on the spreading of molten brazing filler metallic alloys are investigated.

2. Experimental procedure

2.1. Base and brazing filler materials

Ag-(40-x)Cu-xIn and Ag-(40-x)Cu-xSn (x = 5, 10, 15, 20, 25; x is mass%) alloy were used as brazing filler metallic alloys [3]. Chemical compositions and temperatures were shown in Table 1. Mild steel plate, S45C, and pure nickel plate were employed as the base. Specimens with values of maximum surface roughness, R_{max}, of 0.15 μm, 0.55 μm, and 3.2 μm were prepared by polishing with emery paper.

Table 1. Chemical compositions and temperature [3].

	Chemical compositions, mass%				Temperature, °C	
No.	Ag	Cu	In	Sn	Solidus	Liquidus
1	Bal.	35	5	-	728	776
2	Bal.	30	10	-	673	748
3	Bal.	25	15	-	610	704
4	Bal.	20	20	-	612	685
5	Bal.	15	25	-	605	677
6	Bal.	35	-	5	680	767
7	Bal.	30	-	10	602	718
8	Bal.	25	-	15	600	680
9	Bal.	20	-	20	521	640
10	Bal.	15	-	25	514	586

2.2. Spreading test

Brazing filler alloy and base metallic plate were washed with acetone, and had a flux applied to them. The filler alloy, 50 mg, was placed on the center of the base plate with the flux. Spreading test was performed under an Ar gas atmosphere using an electrically heated furnace. Test temperatures were 640 °C, 730 °C, and 860 °C, and test duration was 20 min. After the test, the original spreading area, defined as the sessile drop area; and the apparent (halo) spreading area, which includes the original spreading area and is produced by the capillary grooves characterizing surface roughness; were evaluated, as shown in Figure 1.

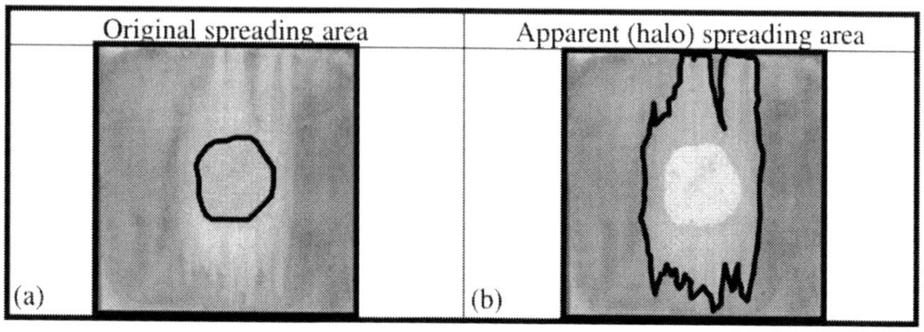

Figure 1. (a) Original spreading area (red region) and (b) apparent (halo) spreading area (including the original spreading area; blue region).

2.3. Estimation methods
To investigate the effect of surface roughness and surface condition on the spreading of the molten brazing filler alloy, microstructures at the cross-section along an edge of the original spreading area were observed. Elemental distributions were analyzed using an electron probe micro-analyzer (EPMA) along an edge of the original spreading from directly above the said area.

3. Results and discussion

3.1. Wetting and spreading on mild steel plate, S45C
Typical spreading test results are shown in Figure 2. According to these results, filler material composition and surface roughness of the base metal influence the original and halo spreading areas, the latter being produced by the capillary grooves that cause the surface roughness (blue region).

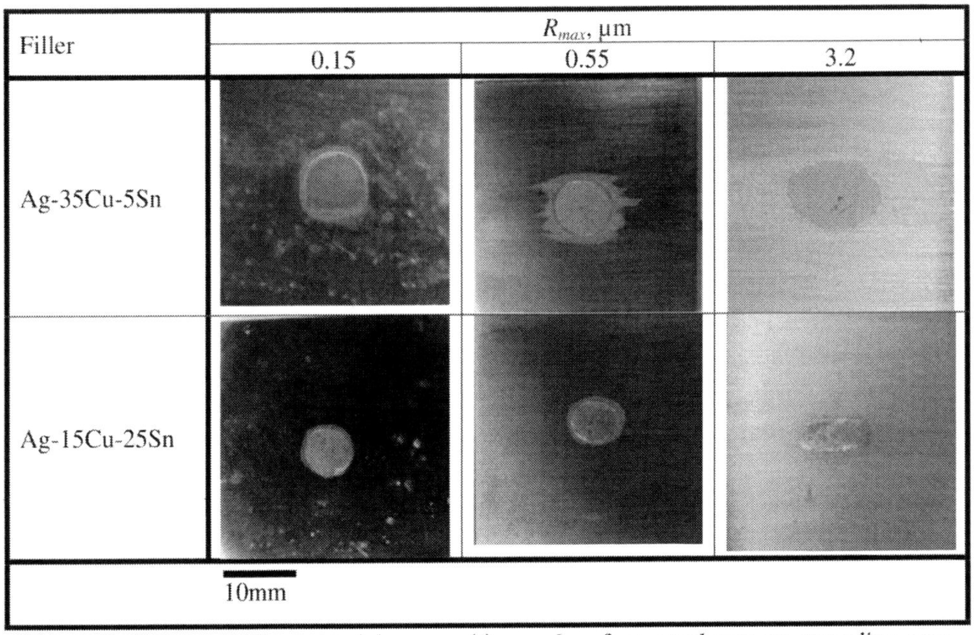

Figure 2. Effects of filler material composition and surface roughness on spreading area. Blue region shows the halo spreading area, produced only by the capillary grooves, that does not include original spreading area.

Formation of an alloyed layer, which can be caused by an unrestrained interfacial reaction between the molten filler and the base metal during spreading, was not observed in EPMA results at the cross-section along an edge of the original spreading area. Therefore, it can be inferred that interfacial reaction hardly influenced the spreading of the molten filler in this study. Figures 3, 4, 5, and 6 show the effects of surface roughness on original and halo spreading areas. In the case of fillers 1 and 6 from Table 1, the halo spreading area increased dramatically with increasing surface roughness. Figure 7 shows the scanning electron microscopy (SEM) image of the halo spreading area as viewed from directly above. In conjunction with this image, the elemental distribution results by EPMA show that a white phase, corresponding to a Ag solid solution containing In as the solid solute, and a grey phase, corresponding to a Cu solid solution, are present in fillers 1 and 6, and that consequently liquation did not occur. Figure 7 also shows that the molten brazing filler solidified along open microgrooves at the halo spreading area. It is appeared that a halo spreading area is caused by spreading of molten filler through open microgrooves [4]. Figure 8 shows the surface microstructure of the base metal after heat treatment. According to Figures 7 and 8, a small white phase at the halo spreading area was formed along the austenite grain boundaries at test temperature. The halo spreading area decreased slightly with increasing surface roughness in that case of filler 5 (Ag-15Cu-25In) and increased slightly with increasing surface roughness in that case of filler 10 (Ag-15Cu-25%Sn). Detailed explanations for this phenomenon are as yet unknown. In that case of Ag-Cu-In fillers, the original spreading area decreased slightly with increasing surface roughness. However, the original spreading area was not influenced by surface roughness in that case of Ag-Cu-Sn fillers.

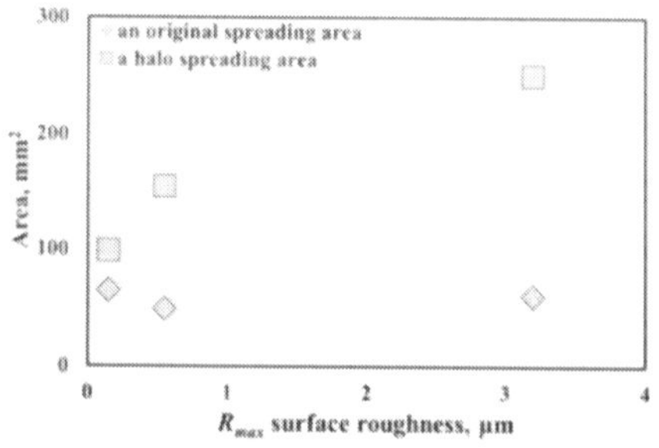

Figure 3. Effect of R_{max} surface roughness on original and halo spreading areas for filler 1 (Ag-35Cu-5In). Temperature was 860°C. Duration was 20 min.

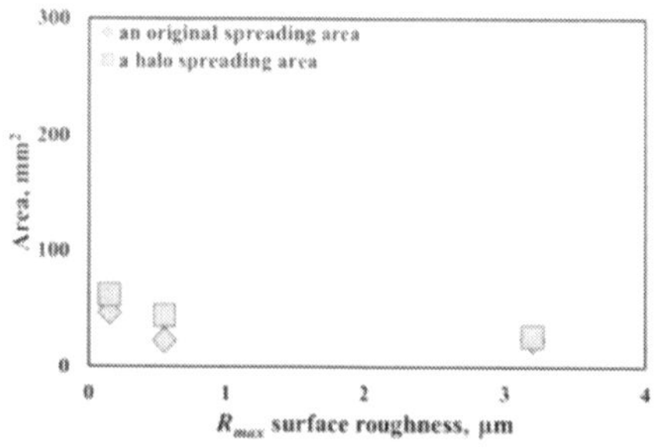

Figure 4. Effect of R_{max} surface roughness on original and halo spreading areas in for filler 5 (Ag-15Cu-25In). Temperature was 860 °C. Duration was 20 min.

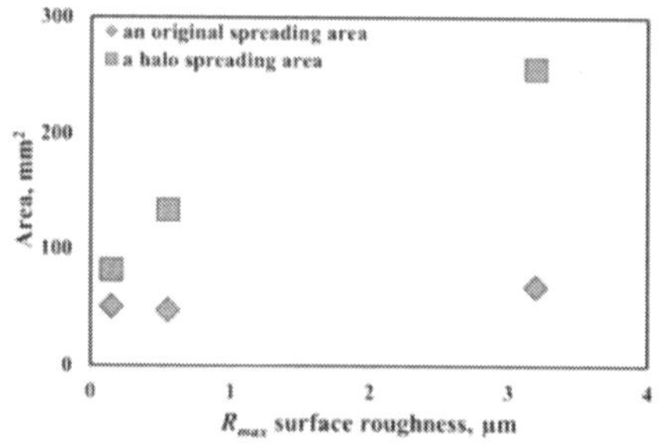

Figure 5. Effect of R_{max} surface roughness on original and halo spreading areas for filler 6 (Ag-35Cu-5Sn). Temperature was 860 °C. Duration was 20 min.

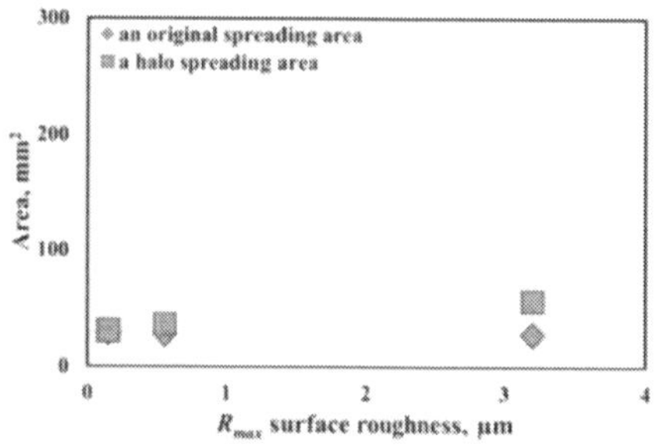

Figure 6. Effect of R_{max} surface roughness on original and halo spreading areas for filler 10 (Ag-15Cu-25Sn). Temperature was 860 °C Duration was 20 min.

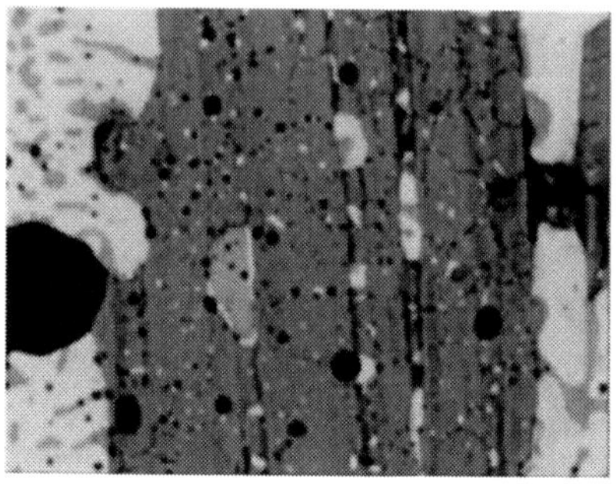

10μm

Figure 7. Surface microstructure at the halo spreading area.

Figure 8. Surface microstructure of the base metal after only heat treatment.

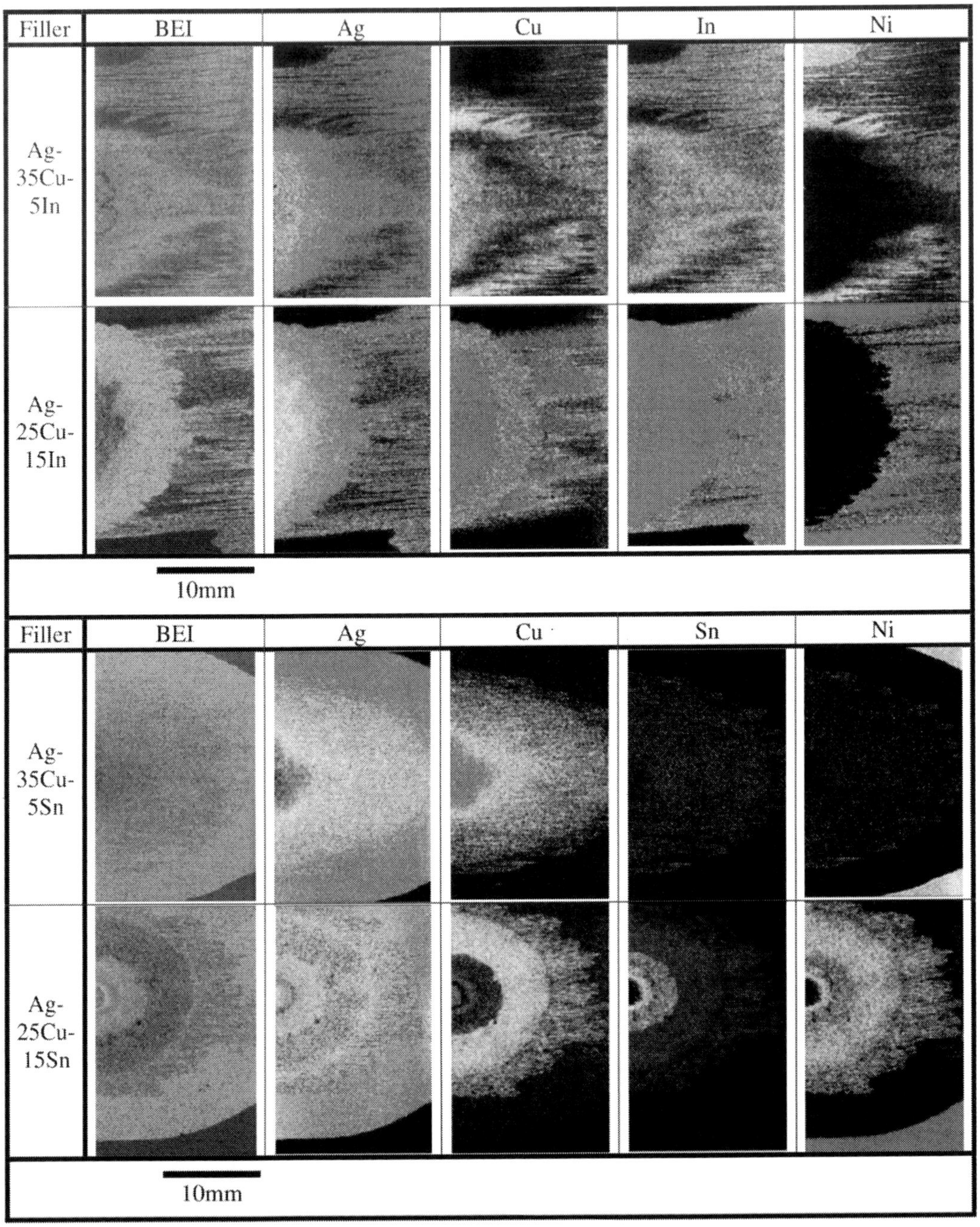

Figure 9. Elemental distributions at the halo spreading area on pure Ni plate.
Test temperature: 860°C. Test duration: 20 min. Surface roughness: 3.2 μm.

3.2. Wet and spread on pure Ni plate

Figure 9 shows the elemental distributions at a halo spreading area on pure Ni plate, according to which, liquation occurred during brazing. Liquation in brazing is defined as the tendency of the lower-melting constituents of the brazing filler alloy to separate out and flow away by capillary action from the higher-melting constituents upon heating. [5] When liquation occurs, low-melting constituents spread preferentially. However, Ag, despite being high-melting constituent, too spreads preferentially. It thus appears that different phenomena occur simultaneously. In this study, Ag and In spread on the base metal along the front edges of the spreading area during wetting and spreading in the case of the Ag-Cu-In filler. Solidified In, especially, was detected spread across the entire halo spreading area after wetting and spreading. On a Ni plate, the Ag and In components of the molten alloy spread preferentially. Therefore, Ag and In constituents of Ag-Cu-In fillers affect the wetting-ability and the halo spreading area. According to the EPMA results, in that case of Sn-based filler alloys, the Ag component of the molten filler spreads preferentially on a pure Ni plate. A molten brazing filler composed mainly of Ag has relatively high liquidus temperature, as can be seen from the Ag-Cu-Sn ternary phase diagram [6]. It is thus inferred that liquation does not occur in the filler alloys studies.

4. Conclusions

In this study, the effects of surface roughness and surface condition on the spreading of molten brazing filler materials were investigated. The results are as follows.

(1) Original spreading area, which is defined as the sessile drop area, and the halo spreading area, produced by capillary grooves, were both observed.

(2) The halo spreading area formation is caused by the spreading of molten filler through open microgrooves.

(3) The interfacial reaction between molten brazing filler and base metal hardly influences spreading of the molten filler.

(4) Molten alloys containing mainly Ag and In spread preferentially on a Ni base plate. Therefore, Ag and In, which are both constituents of Ag-Cu-In fillers, affect the wetting-ability of the filler.

References

[1] American Welding Society C3 Committee on Brazing Soldering 2007 *Brazing Handbook Fifth Edition, (AWS2007)*, Chapter1,2.

[2] Kozlova O, Voytovych R, Eustathopoulos N, Protsenko P *J Mater Sci* **45 8**

[3] Samouth L, Kaneko K, Kawai Y, Miyazawa Y, Ariga T *2010 Proc. 9th Int. Conf. Brazing, High Temperature Brazing and Diffusion Bonding, (LOT 2010)*196

[4] Dusan P. Sekulic 2011 *Heat Transfer Eng,* **32** 648.

[5] D Kay 2010 *Weld J.* **89 9** 36

[6] Villars P, Prince A, Okamoto H 1995 *Handbook of ternary alloy phase diagrams* (ASM International

Development of plasma MIG brazing process for dissimilar metal joining of aluminum to steel

Shinichi Tashiro and Manabu Tanaka
Joining and Welding Research Institute, Osaka university, Osaka, Japan

E-mail: tashiro@jwri.osaka-u.ac.jp

Abstract. This study aims to develop a new brazing process employing plasma MIG. Because the energy density of the plasma produced by the plasma electrode is low, the base metal can be heated extensively without melting of the base metal, consequently improving the wettability of bead. This paper discussed the dissimilar metal joining of aluminum to steel by plasma MIG brazing process. Fracture occurred at the HAZ in the aluminum plate at 80 MPa.

1.Introduction

Dissimilar metal joining of aluminum and steel is significantly important from the viewpoint of the weight reduction of transport systems such as cars, trains, and airplanes. However, since it is difficult to conduct direct fusion joining owing to the formation of a brittle intermetallic compound layer [1], mechanical joining processes using bolts are also still employed. The mechanical joining processes considerably diminish the advantages weight reduction.

Recently, laser welding, which provides excellent energy control characteristics, and FSW (Friction Stir Welding), which enables solid–phase joining, have been adopted for dissimilar metal joining of aluminum and steel [2,3]; however, these processes involve high costs. Therefore, it is necessary to develop simple and inexpensive methods for dissimilar metal joining of aluminum and steel, such as arc process.

Although MIG blazing is one of the arc processes for dissimilar metal joining processes of aluminum and steel, its practical use is hindered by technical problems that need to be solved. Murakami et al reported that although the dissimilar metal joining of aluminum and steel using flux-cored wire is possible by controlling the heat input appropriately, the use of solid wire increases difficulty owing to low wettability [4].

This study aims to develop a new brazing process employing the plasma MIG process for realizing the dissimilar metal joining of aluminum and steel using solid wire, in order to reduce cost. Figure 1 shows a schematic of the plasma MIG process. Shielding gas is ionized with the plasma electrode before introduction to the arc gap by employing another power source than that for MIG arc, and hence, the shape of the arc is easily controlled by applying electromagnetic force, as shown in Figure 2 [5]. Therefore, the controllability of the arc and the stability of metal transfer in the plasma MIG process are remarkably enhanced compared to the conventional MIG process. Because of this advantage, plasma MIG can be employed for MIG welding in pure inert gas atmosphere. In addition, because the energy density of the plasma produced by the plasma electrode is low, the base metal can be heated extensively without melting of the base metal, thus improving the wettability of the bead.

Content from this work may be used under the terms of the Creative Commons Attribution 3.0 licence. Any further distribution of this work must maintain attribution to the author(s) and the title of the work, journal citation and DOI.
Published under licence by IOP Publishing Ltd

Furthermore, since the oxide layer is cleaned automatically by cathode spots produced on the base metal during the brazing process, a high quality weld joint can be obtained.

This paper discusses the dissimilar metal joining of aluminum and steel plates using solid wire by plasma MIG brazing process conducted for realizing fracture at the HAZ of the aluminum plate in transverse tensile tests of the welded joint.

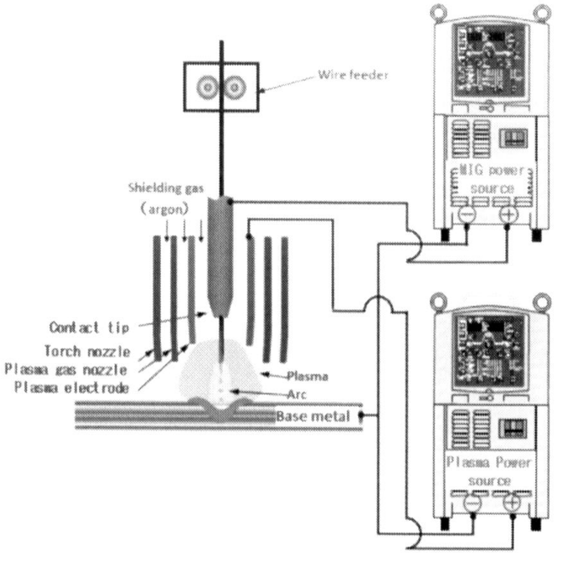

Figure 1. Schematic illustration of plasma MIG process.

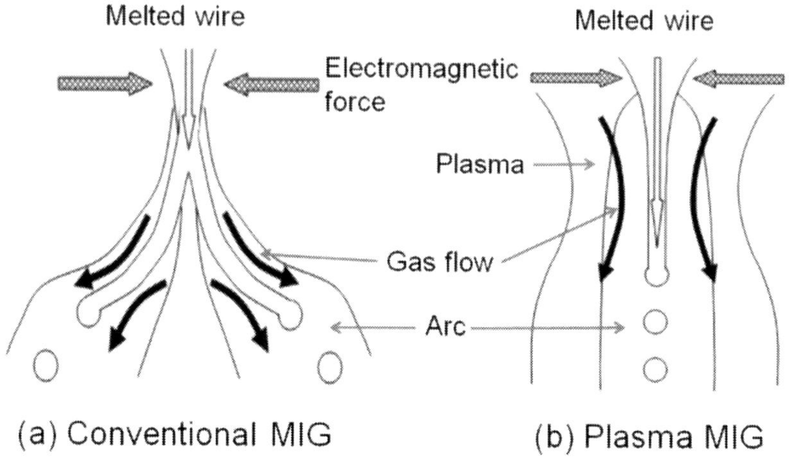

Figure 2. Schematic illustration of metal transfer.

2.Experimental procedure

The experimental apparatus consists of a plasma MIG torch, MIG power source with constant voltage characteristics, plasma power source with constant current characteristics, and base metal. In addition, metal transfer was imaged using a high speed video camera for evaluating the condition under which the metal transfer was stabilized.

The MIG power source generated the arc through the contact tip (conventional type) between the wire tip and the base metal surface, whereas the plasma power source generated the plasma through the plasma electrode with a diameter of 3 mm. An industrial pure aluminum plate (A1050) and cold rolled plain carbon steel plate (SPCC), with thicknesses of 2 mm each, were used as the base metal. Figure 3 shows a schematic diagram for setting the lap joint between the aluminum and steel plates. The aluminum plate was lapped on the steel plate without gap between both plates. The torch was aimed at the edge of the upper aluminum plate. Solid wire A4047WY with a diameter of 1.2 mm was employed as the electrode wire. A4047WY had a composition of Al-12.0%Si, which is suitable for suppressing growth of the intermetallic compound layer [4].

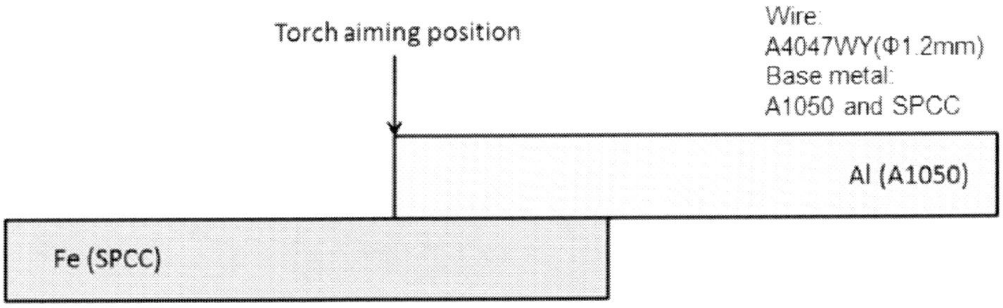

Figure 3. Schematic diagram for setting lap joint between aluminum top plate and steel bottom plate.

3.Results and discussion

Figure 4 shows metal transfer imaged using the high speed video camera. As seen in the figure, weak plasma was generated around the bright MIG arc under the wire tip, and the stability of the metal transfer was remarkably improved by the influence of electromagnetic force owing to the plasma current in pure argon atmosphere. Furthermore, many cathode spots were produced on the base metal surface and cleaned the oxide layer on the surface.

In order to realize the fracture at HAZ of the aluminum plate in transverse tensile tests for evaluating the shear strength of the welded joint, brazing tests were conducted by varying experimental parameters such as the wire feed rate, welding speed, voltage of the MIG power source, and plasma current.

This paper also presents an example of experimental results in which fracture at the HAZ of the aluminum plate occurred. The distance between the torch and the base metal was 10 mm, the wire feed rate was 2 m/min, the welding speed was 60 cm/min, the setting voltage of the MIG power source was 20 V, the plasma current was set to be 20 A, the shielding gas composition was pure argon, and the gas flow rate was 25 L/min.

Figures 5 and 6 show the bead appearance and the bead cross section observed using an optical microscope. The bead with high wettability was found to be obtained because of heating of the steel plate by the plasma.

Figure 7 shows the microstructures observed using an SEM (Scanning Electron Microscope) and the composition observed using an EDX (Energy Dispersive X-ray) spectroscope near the interface between the weld metal and the steel plate. The average thickness of the intermetallic compound layer was approximately 2 μm, which was less than the upper limit thickness of 2.5 μm required for fracture at HAZ in the aluminum plate previously reported [4] and was in Al-Fe-Si ternary phase.

Figures 8 and 9 show the results of the transverse tensile test of the welded joint and a fractured joint, respectively. It was confirmed that the fracture of the welded joint occurred at the HAZ in the aluminum plate at approximately 80 MPa.

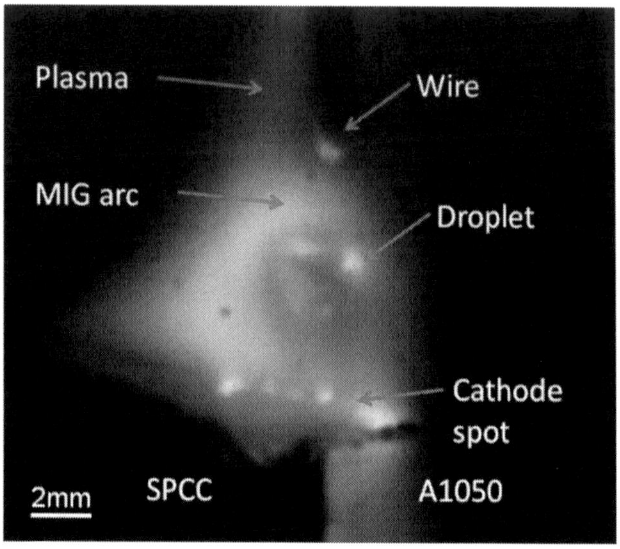

Figure 4. Metal transfer imaged using high speed video camera.

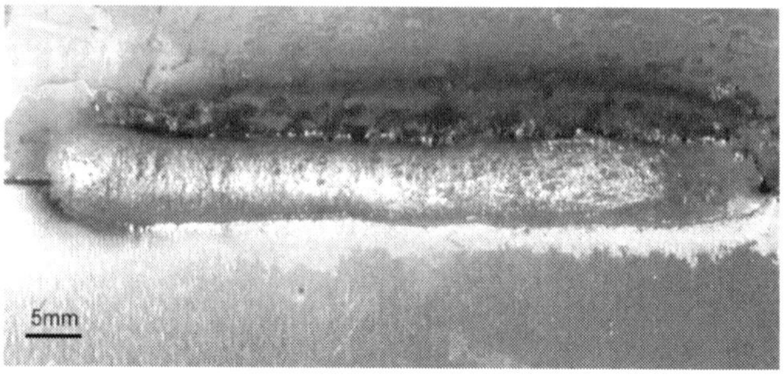

Figure 5. Bead appearance.

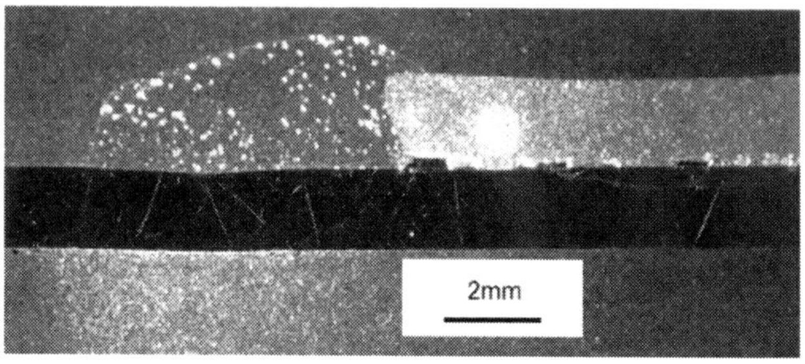

Figure 6. Bead cross section.

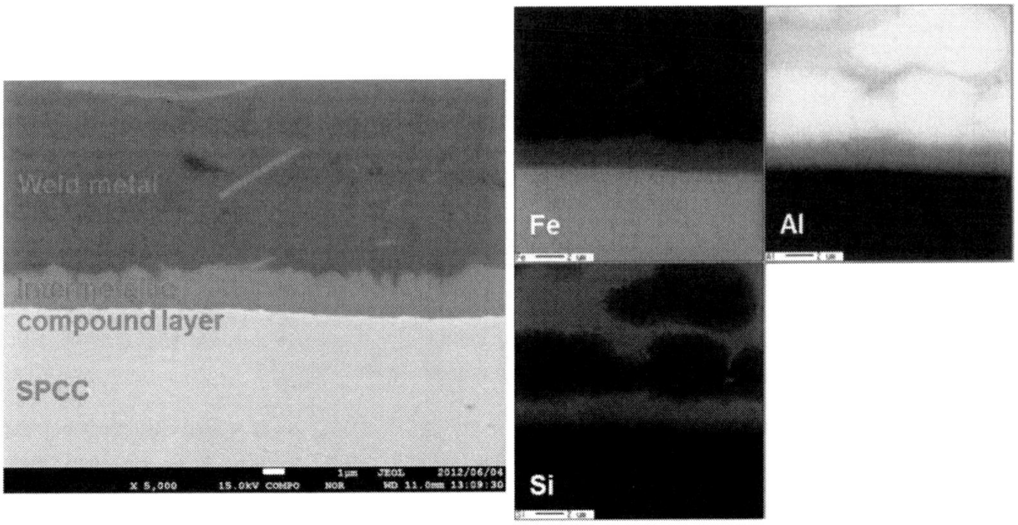

Figure 7. Microstructures and composition near interface between weld metal and SPCC.

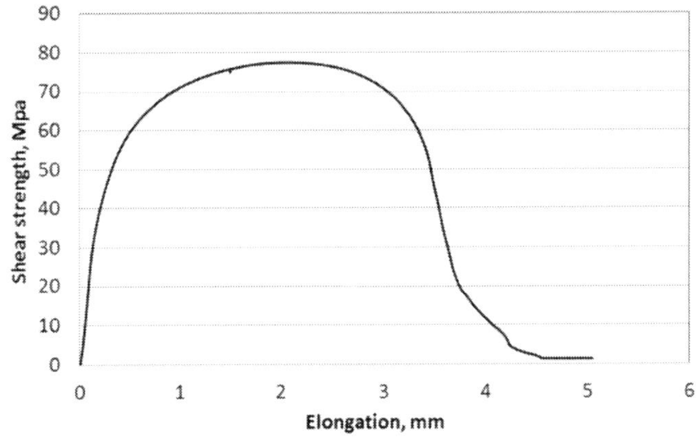

Figure 8. Dependence of shear strength on elongation.

Figure 9. Fractured joint.

4.Conclusions

This paper discussed the dissimilar metal joining of aluminum and steel plates using solid wire by plasma MIG brazing process conducted for realizing fracture at the HAZ of the aluminum plate in transverse tensile tests of the welded joint. The main conclusions are summarized as follows.

1) A bead with high wettability was obtained because of heating of the steel plate by the plasma.

2) Average thickness of the intermetallic compound layer was approximately 2 μm, which was less than the upper limit thickness of 2.5 μm required for fracture at the HAZ in the aluminum plate previously reported.

3) As a result of the transverse tensile test of the welded joint, the fracture of the welded joint occurred at the HAZ in the aluminum plate at 80 MPa.

References

[1] Oikawa H,Saito T, Yoshimura T, Nagase T and Kiriyama T 1996 *Q. J. Jpn. Weld. Soc.* **14** 69
[2] Schubert E, Klassen M, Zerner I, WalzC and Sepold G 2001 *J. Mater. Process. Technol.* **115** 2
[3] Kinya A, TAKAHASHI M and IKEUCHI K 2008 *Q. J. Jpn. Weld. Soc.* **26** 227
[4] Murakami T, Nakata K, Tong H and Ushio M 2003 *ISIJ International* **43** 1596
[5] Katayama T, Tashiro S and Tanaka M 2011 *Q. J. Jpn. Weld. Soc.* **29** 39s

Effect of composition of titanium in silver–copper–titanium braze alloy on dissimilar laser brazing of binder-less cubic boron nitride and tungsten carbide

Yoshihisa Sechi[1], Kimiaki Nagatsuka[2] and Kazuhiro Nakata[2]

[1]Kagoshima Prefectural Institute of Industrial Technology, 1445-1 Oda, Hayato-cho, Kirishima-city, Kagoshima 899-5105, Japan

[2]Joining and Welding Research Institute, Osaka University, 11-1 Mihogaoka, Ibaraki-city, Osaka 567-0047, Japan

E-mail: sechi@kagoshima-it.go.jp

Abstract. Laser brazing with Ti as an active element in silver–copper alloy braze metal has been carried out for binder-less cubic boron nitride and tungsten carbide, using silver–copper–titanium braze alloys with titanium content that varied between 0.28 mass% and 1.68 mass%. Observations of the interface using electron probe microanalysis and scanning acoustic microscopy show that efficient interface adhesion between binder-less cubic boron nitride and the silver–copper–titanium braze alloy was achieved for the braze with a titanium content of 0.28 mass%.

1. Introduction

The brazing process is used in many industrial fields for developing engineering structures and electronic devices [1-13]. This process has several advantages, making it suitable for use in joining dissimilar materials and materials that are difficult to connect precisely; furthermore, this process can be efficiently applied to the mass production of these structures and devices. It is also required applying to ceramic materials, which are frequently bonded to metals. Moreover, the development of high-functionality products in recent years has led to the demand for a new joining process for different materials.

Boron nitride ceramics is one of the products with various functional characteristics. In particular, cubic boron nitride (c-BN) has high hardness, thermal conductivity, and thermal and chemical stability that is second to that of diamond [14]; c-BN is suitable as a material for cutting tools, especially those for cutting hardened steel. Conventional sintered c-BN is manufactured commercially in a liquid-phase sintering process using binder material such as Al, and the volume fraction of the binder phase varies from 10 to 50%. Because the binder phase exists between c-BN crystal grains, it is difficult to produce effects expected from the ceramics' original excellent properties. Therefore, specifications of tools are partly determined by the types and amount of binder phase in the material.

To solve this problem, high-purity and binder-less c-BN was developed [14]. This binder-less, polycrystalline c-BN consists of fine particles with good wear resistivity owing to the direct contact of crystal grains in the absence of a binder [14]. Thus, this binder-less c-BN is expected to be applied in high-performance cutting tools. Binder-less c-BN is very expensive, however, and it is difficult to

Content from this work may be used under the terms of the Creative Commons Attribution 3.0 licence. Any further distribution of this work must maintain attribution to the author(s) and the title of the work, journal citation and DOI.
Published under licence by IOP Publishing Ltd

shape monolithically fabricated cutting tools made of binder-less c-BN because of its hardness. Therefore, dissimilar joining between binder-less c-BN and metals is necessary to bring out and utilize the former's outstanding properties. Certain problems can arise, however, while meeting these requirements, such as the formation of joint defects resulting from thermal stress in the joint field and material deterioration resulting from heating in the brazing process.

In general, the thermal expansion coefficient of metal is higher than that of ceramics; this may result in the generation of a large thermal stress at the brazing joint, which leads to the formation of defects at the interface [15]. The tungsten carbide–cobalt (WC–Co) alloy made by powder metallurgy has a low thermal expansion coefficient and high rigidity, which makes it a good complementary structural material to ceramics. However, only a few studies have focused on the joining of boron nitride and tungsten carbide alloy or other metals [5,8,10] by brazing in a furnace, which requires a long heating time.

Among brazing processes [16], laser brazing [10,11,13,17-30] has good characteristics for dissimilar joining: when compared with conventional furnace brazing, short heating time can be used only a small heating area is required, and damage to base materials is suppressed in the absence of a furnace. Recently, laser brazing with the addition of Ti as an active element in the Ag–Cu alloy as braze metal has been successfully applied to dissimilar joining of hexagonal BN (h-BN), single crystal diamond, and WC–Co alloy, after evacuation of the atmosphere to a vacuum of the order of 10^{-1} Pa followed by Ar (99.999% purity) substitution to avoid oxidation of the active element Ti [12,31].

This study examines the effect of titanium content in the silver–copper–titanium braze alloy on dissimilar laser brazing of binder-less c-BN and WC–Co. In order to investigate the characteristics of the dissimilar joint, cross-sectional observation as well as elemental and structural analyses of the joint interface were performed.

2. Experimental procedure

This study used commercially available WC–Co alloy (ISO K10 grade; WC: 94 mass%, Co: 6 mass%) and binder-less polycrystalline cubic boron nitride (binder-less c-BN; c-BN > 99.9 mass%) made by the direct-conversion sintering method without sintering additives. Binder-less c-BN disks were divided into quarters, and blocks of the quadrisections were used. A braze sheet of the Ag–Cu alloy was used; it had a eutectic composition that includes Ti as the major active ingredient for direct ceramic brazing and it had a thickness of 0.1 mm. The nominal composition of the alloy is summarized in Table 1. Before brazing, the binder-less c-BN block, braze alloy, and WC–Co plate were degreased by ultrasonic agitation for 10 min in acetone and dried in air.

The sample has the shape of a top hat. A braze alloy sheet was sandwiched with a binder-less c-BN block from the top and a WC–Co plate from the bottom in a vacuum chamber. A low pressure of 2.4 MPa was applied to prevent the workpiece from moving when the braze alloy sheet was melted, and no gap adjustments for the joint were performed. A chamber with a diameter of 100 mm and inner volume of 145 mL was used in the brazing experiments. To avoid the oxidation of Ti, this chamber was evacuated to less than 10^{-1} Pa after the samples were loaded; it was then filled with Ar gas after the evacuation. The purity of Ar gas was 99.999%. These evacuation and substitution cycles were carried out at least three times before brazing. During brazing, Ar gas continued to flow at a rate of about 5 L/min.

The laser brazing setup is shown in a perspective illustration in figure 1(a) and a cross-sectional schematic illustration in figure 1(b). The sample was located in a small vacuum chamber. The topside of the specimen was covered with a transparent quartz glass plate, which also served to fix the specimen in place. The generated YAG and LD laser beams were transferred with an optical fiber to the laser head unit and radiated through the transparent quartz glass plate to the topside of the WC–Co plate, with a radiation angle of 85°. Laser scanning of the WC–Co substrate was performed around the binder-less c-BN block. The laser brazing condition is summarized in table 2. The scanning speed of the laser was 0.6 mm/s during the first run and 1.0 mm/s during the second to fourth runs in order to heat the specimen efficiently in a circular manner. The temperature during laser brazing was monitored at the bottom of the WC–Co plate using a thermocouple.

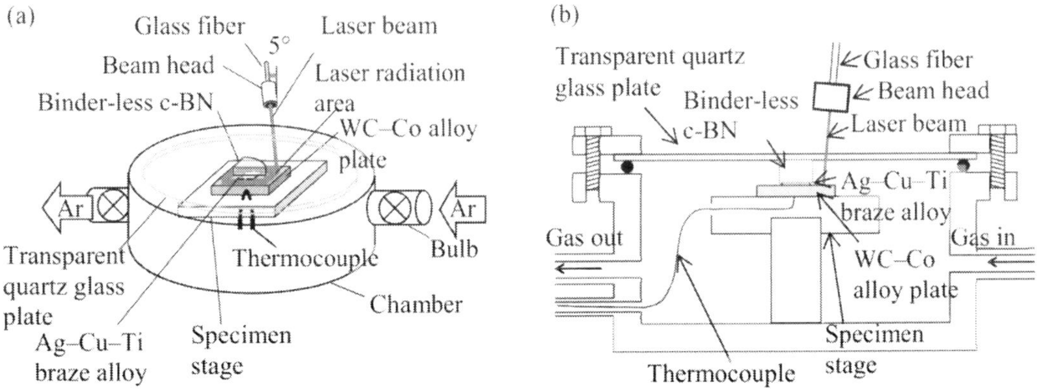

(a) Perspective illustration (b) Cross-sectional illustration

Figure 1. Schematic diagram of laser brazing.

Table 1. Chemical composition of Ag–Cu–Ti braze alloy.

No.	Elements (mass%)		
1	Ag	Cu	Ti
2	71.49	bal.	0.28
3	71.34	bal.	0.63
4	71.21	bal.	0.85
5	70.90	bal.	1.28
6	70.26	bal.	1.68

Table 2. Laser heating condition.

Pulsed YAG Average Output (kW)		0.134
Pulsed YAG wave length (nm)		1064
CW LD Output (kW)		0.02
CW LD wave length (nm)		808
Pulse frequency (Hz)		100
Scanning speed (mm/s)	(1st run)	0.6
	(2nd run)	1.0
	(3rd run)	1.0
	(4th run)	1.0
Laser beam diameter (mm)		0.5

Some of the samples were subsequently cross-sectioned by a diamond saw with water cooling. The cross-sectioned specimen was then mounted by epoxy resin, cured at room temperature for about 8–10 h, ground using SiC paper with grit size #120–#1200, and polished by 3 μm and 1 μm polycrystalline diamond to provide microstructural information. Cross-sectional observations and elemental analysis of the interface were performed using an electron probe micro-analyzer (JXA-8230, JEOL Co. Ltd.) and an X-ray diffractometer (D8, Bruker AXS Co. Ltd.) with Co K_α as the radiation source. A collimator with a diameter of 50 μm was used to determine the crystallographic phase of the micro-region. Interfacial observations and estimation of the interface area were performed using a scanning acoustic microscope (HSAM220, Hitachi Kenki FineTech Co., Ltd.) that can observe interfacial structures using ultrasonic echoes radiated from the probe in a water tank.

3. Results

In order to investigate the effect of titanium content of the silver–copper–titanium braze alloy on dissimilar laser brazing of binder-less c-BN and WC–Co, we performed cross-sectional observations, elemental and structural analyses, and adhesion evaluation of joint interfaces using several kinds of brazes containing Ti in concentrations varying from 0.28 mass% to 1.68 mass%.

Figure 2 shows the dependence of the brazing behavior on Ti composition in Ag–Cu–Ti braze alloy. The critical Ti content was determined by the shape of the braze fillet and whether the contact angle between the braze and binder-less c-BN was an acute angle. The addition of Ti seemed to be sufficient to induce good wettability between binder-less c-BN and WC–Co, even when the Ti content was just over 0.28 mass%, which is lower than that in h-BN [32].

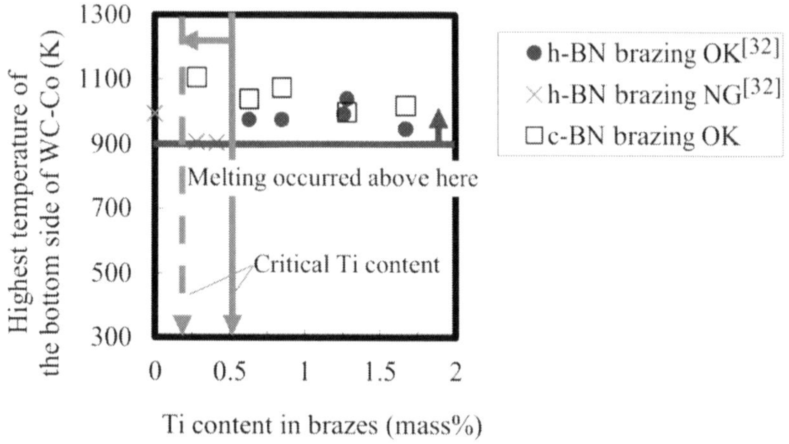

Figure 2. Dependence of brazing behavior on Ti content of braze alloy.

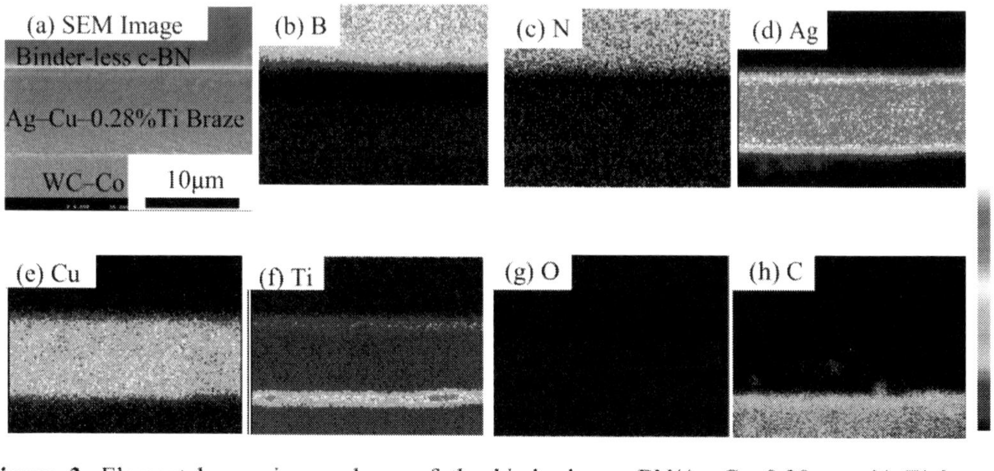

Figure 3. Elemental mapping analyses of the binder-less c-BN/Ag–Cu–0.28 mass% Ti braze alloy/WC–Co interfaces.

Figure 3 shows the results of elemental mapping analysis of the interface between binder-less c-BN and the 0.28 mass% Ti braze. Figure 3(a) shows the enlargement of the interface between binder-less c-BN and the Ag–Cu–0.28 mass%Ti braze alloy. Figures 3(b)–(h) show the distributions of B, N, Ag, Cu, Ti, O, and C, respectively, near the interface in the same area of (a). From these distributions of the elements, an accumulation and continuous layer of Ti near the interface between binder-less c-BN and the Ag–Cu–0.28 mass%Ti braze alloy were observed. Besides, O was not detected at the interface. Moreover, an accumulation and continuous layer of Ti near the interface between the Ag–Cu–0.28 mass%Ti braze alloy and WC–Co were also observed.

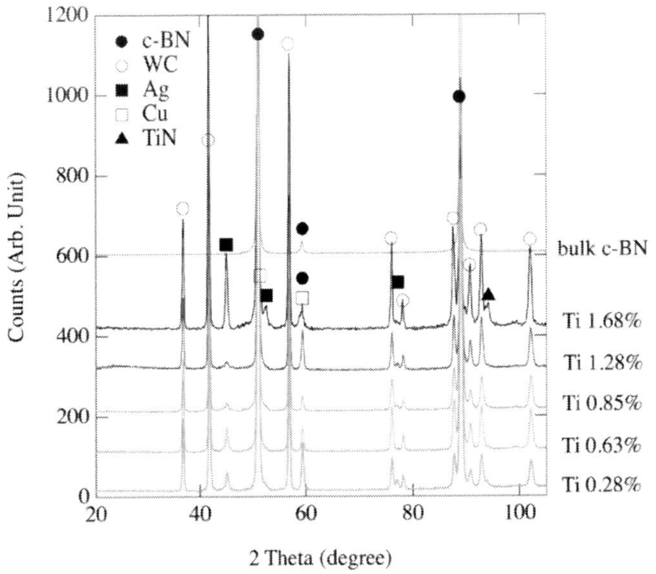

Figure 4. Dependence of XRD profiles of binder-less c-BN/Ag–Cu–Ti braze alloy/WC–Co interfaces on the composition of Ti in the Ag–Cu–Ti braze alloy. Owing to the thickness of the Ag–Cu–Ti braze alloy, which was lower than the diameter of the collimator (50 μm), diffraction data from the interface between the Ag–Cu–Ti braze alloy and WC–Co were also recorded simultaneously.

Figure 4 shows the dependence of XRD profiles of binder-less c-BN/Ag–Cu–Ti braze alloy/WC–Co interfaces on the composition of Ti in the Ag–Cu–Ti braze alloy. All of the measurements were focused on the interface between the binder-less c-BN and the Ag–Cu–Ti braze alloy, but because the thickness of the Ag–Cu–Ti braze alloy was thinner than the 50 μm diameter of the collimator, as shown in figure 3(a), diffraction data from the interface between the Ag–Cu–Ti braze alloy and WC–Co were also recorded simultaneously. In the XRD pattern of interfaces of the braze with 1.68 mass% of Ti, which produced a reaction layer, a peak from TiN was observed, suggesting that the reaction phase included TiN.

Figures 5(a) and (b) show the appearance of the specimen and the scanning acoustic microscopy images at the interfaces of binder-less c-BN/Ag–Cu–1.28 mass% Ti braze alloy/WC–Co, respectively. Figures 5(c) and (d) show the specimen containing the Ag–Cu–0.28 mass% Ti braze alloy. The braze alloy filled the interface between binder-less c-BN and WC–Co. Besides, a fillet existed around the binder-less c-BN block. As shown in the left side of Figure 5(a), there was a Ag–Cu–Ti braze alloy lump, which occasionally flowed out from the interface between binder-less c-BN and WC–Co, and its color maintained a metallic luster, which showed that there was not a large amount of titanium oxide at the surface of the braze alloy. There were no big voids where melted braze existed at the interface. Therefore, the wettability between the Ag–Cu–Ti braze alloy and binder-less c-BN, and that between the braze alloy and WC–Co appeared to be good even in the specimen containing only 0.28 mass% Ti in the braze alloy. As shown in the left center area of figure 5(d), there was a dense contrast region. To investigate this area, cross-sectional observations were performed. As seen in figure 3(a), significant gaps at the interface between binder-less c-BN and the Ag–Cu–0.28 mass%Ti braze were not observed, so the dense contrast region was presumed to be a ghost reflection.

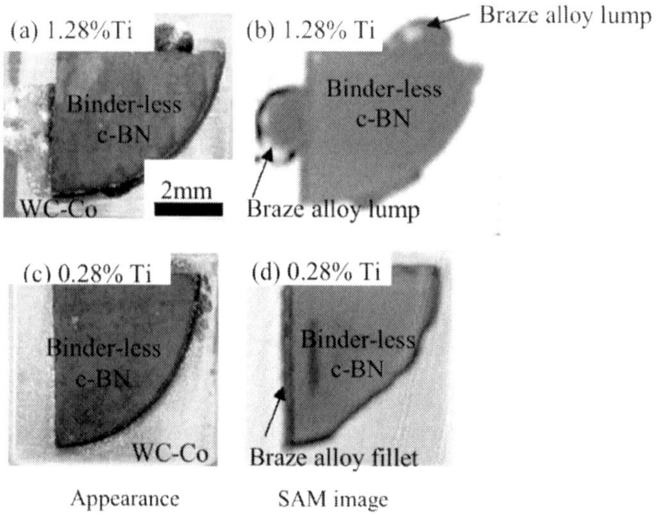

Figure 5. Images of the interfaces between binder-less c-BN and WC–Co of dissimilar joining specimens, obtained using scanning acoustic microscopy.

4. Discussion

As shown in figure 2, the wettability between binder-less c-BN and WC–Co was relatively satisfactory compared to that between h-BN and WC–Co, even with a small Ti content of 0.28 mass% in the braze alloy. Similar to the case of h-BN [32], the contact angle between c-BN and the Ag–Cu braze alloy without Ti was presumed to indicate non-wetting [6]. Judging from the fact that the addition of as little as 0.28 mass% Ti appeared to be sufficient to induce good wettability between binder-less c-BN and WC–Co, the critical Ti content must exist between 0 and 0.28 mass%.

Figure 3 shows that the distribution of N was decreased sharply at the interface. Meanwhile, an accumulation of Ti near the interface was observed for the specimen containing the braze alloy with a small Ti content of 0.28 mass%. As shown in figures 3(f) and (g), the distribution of oxygen at the interface was lower than the detection limit of the electron probe micro-analyzer. Thus, it is concluded that oxidation of Ti at the interface was not observed at least mesoscopically despite the easily oxidizable property of Ti [33]. Also, at the interface between the braze alloy and WC–Co, accumulations of Ti and C near the interface were observed, as shown in figures 3(f) and (h). Therefore, it is believed that Ti reacted with C in WC–Co [34], and a continuous reaction layer was formed at the interface between the braze alloy and WC–Co. These results agree with a similar tendency shown in an earlier study [12]. These results suggest the formation of a reacted layer such as that of Ti-N [6,12] or Ti-C [34] even in the case when the braze contained 0.28 mass% Ti. In the fabrication process of the Ag–Cu–Ti braze alloy, an intermetallic compound phase such as Cu_3Ti [35] is produced during melting of the raw materials, and it increases the hardness of the braze with increasing Ti content. From the point of view of cutting and forming of sheet-type braze alloys, it is preferable that the Ti content remains small because this intermetallic phase is hard. The results of this study show our capability to reduce Ti content below 0.3 mass% in Ag–Cu–Ti for brazing c-BN.

Compared to the case using h-BN [12], the configuration of the interface between binder-less c-BN and WC–Co was flat and its thickness was uniform, as shown in figure 3. These differences appeared to originate from the crystal grain shape and the sinterability. In case of c-BN, its grains are fine and granular, and it has a dense body [14]. On the contrary, the grains of h-BN are plate-like and the relative density of h-BN is low because of its low sinterability. The wettability of the Ag–Cu–Ti braze alloy against ceramics was good and it could fill up open pores at the surface, even in the case of h-BN [12,35]. Therefore, when binder-less c-BN was used, which has a low specific surface area, the amount of material in the reaction layer could be reduced compared to the case of h-BN [12].

From figure 4, a peak from TiN [2,6,10] was observed only for the specimen containing 1.68 mass% of Ti in the braze. This XRD peak is believed to have originated from the reacted phase at the interface. Besides, the standard free energies of the formation of nitrides [33], namely TiN and BN at 1173 K, are about -220 and -150 kJ/g mol N_2, respectively, which suggest that TiN is more stable than BN in all temperature ranges. In our study, TiN was observed only in the specimen containing 1.68 mass% of Ti in the braze because the amount of TiN was relatively small owing to the decrease in Ti content in the braze containing 1.28 mass% of Ti. Meanwhile, the formation of the Ti–N reaction layer was observed, as shown in figure 3(f), even in the specimen using the braze containing 0.28 mass% Ti. Therefore, a small Ti–N layer, which did not show metallic compound peaks in the XRD profile, could improve the wettability between binder-less c-BN and WC–Co in the Ag–Cu braze.

In both figure 5(b) and (d), there were no big voids at the interface, which indicate that the wettabilities between the Ag–Cu–Ti braze alloy containing 0.28 mass% Ti and binder-less c-BN, and also between the braze alloy and WC–Co were the same as the wettability shown by the braze alloy containing 0.28 mass% Ti. Furthermore, the configuration of the fillet in figure 5(d) indicates good wettability.

5. Conclusions

Dissimilar laser brazing of binder-less cubic boron nitride and tungsten carbide was accomplished. The dependence of the brazing behavior on the composition of Ti in the braze alloy can be summarized as follows:

(1) At the interface between binder-less c-BN and the Ag–Cu–Ti braze alloy, an accumulation of Ti near the 1 μm wide interface was observed for the braze containing 0.28 mass% Ti. A distribution of N was also observed in the same area. It was presumed that the reacted phase existed near the interface between binder-less c-BN and the braze.

(2) The wettability of these brazes by binder-less c-BN was also considered to be good. No large voids or cracks in the braze area were detected in non-destructive observation of the interface using scanning acoustic microscopy, even when the Ti content in the braze was 0.28 mass%.

Acknowledgement

This study was performed under the Cooperative Research Program of Institute for JWRI, Osaka University. The authors would like to thank Tanaka Kikinzoku Kogyo K.K. and Sumitomo Electric Industries, Ltd. for assistance and for providing brazes and binder-less c-BN materials.

References

[1] Nakao Y, Nishimoto K and Saida K 1989 *Trans of J.W.S.* **20** 66-76
[2] Nicholas M G, Mortimer D A, Jones L M and Crispin R M 1990 *J. Mat. Sci.* **25** 2679-89
[3] Nakao Y, Nishimoto K and Saida K 1990 *ISIJ Int.* **30** 1142-50
[4] Murakawa M and Takeuchi S 1991 *Mater. Sci. Eng.* A **140** 759-63
[5] Chattopadhyay A K and Hintermann H E 1993 *J. Mat. Sci.* **28** 5887-93
[6] Peteves S D 1996 *Ceram. Int.* **22** 527-33
[7] Fernandes A J S, Fonseca M J, Costa F M, Silva R F and Nazare M H 1999 *Dia. Rel. Mat.* **8** 855-8
[8] Felba J, Friedel K P, Krull P, Pobol I L and Wohlfahrt H 2001 *Vac.* **62** 171-80
[9] Huang S F, Tsai H L and Lin S T 2004 *Mater. Chem. Phys.* **84** 251-8
[10] Sechi Y, Takezaki A, Tsumura T and Nakata K 2008 *Smart Process Tech.* Vol 2 pp 27-30
[11] Rohde M, Suedmeyer I, Urbanek A and Torge M 2009 *Ceram. Int.* **35** 333-7
[12] Sechi Y, Tsumura T and Nakata K 2010 *Mat. Des.* **31** 2071-7
[13] Suedmeyer I, Hettesheimer T and Rohde M 2010 *Ceram. Int.* **36** 1083-90
[14] Sumiya H and Uesaka S 1999 *New Diamond* **15** 14-9
[15] Wlosinski W 1985 *Proc. 1st Euro. Symp. Eng. Ceram.* pp 45-61
[16] 2001 *Welding Handbook* Vol I ed C L Jenney and A O'Brien (American Welding Society)
[17] Witherell C E and Ramos T J 1980 *Weld J.* **59** 267S-77S
[18] Hanebuth H, Hoffmann P and Geiger M 1994 *Proc. SPIE-Int. Soc. Opt. Eng.* 2207 pp 146-153
[19] Whitehead D G and Foster R J 1995 *Assem. Auto.* **15** 17-9
[20] Dave V R, Carpenter R W, Milewski J O, and Christensen D T 2001 *Weld. J.* **80** 142.S-7.S
[21] Markovits T, Takacs J, Lovas A and Belt J 2003 *J. Mater. Process. Technol.* **143/144** 651-5
[22] Vollertsen F and Grupp M 2005 *Steel Res. Int.* **76** 240-4
[23] Saida K, Song W and Nishimoto K 2005 *Sci. Tech. Weld. Join.* **10** 227-35
[24] Mathieu A, Cicala E, Matteie S, Gewvey D, Pontevicci S and Viala J C 2006 *Mater. Sci. Eng.* A **435/436** 19-28
[25] Li M G, Sun D Q, Qiu X M, Sun D X and Yin S Q 2006 *Mater. Sci. Eng.* A **424** 17-22
[26] Mathieu A, Matteie S, Grevey D, Deschamps A and Martin B 2006 *NDT E Int.* **39** 272-6
[27] Saida K, Song W and Nishimoto K 2006 *Sci. Tech. Weld. Join.* **11** 694-700
[28] Saida K, Song W and Nishimoto K 2007 *Design Int. Str. Adv. Mat. Join. Sol. Sta. Phen.* (Part B) **127** 301-6
[29] Takemoto T, Kawahito Y, Nishikawa H, Katayama S and Kimura S 2008 *J. Lig. Met. Weld. Cons.* **46** 300-8
[30] Saida K, Song W and Nishimoto K 2008 *Mat. Sci. For.* **580/582** 271-4
[31] Sechi Y and Nakata K 2010 *Trans JWRI* **39** 340-1
[32] Sechi Y, Takezaki A, Matsumoto T, Tsumura T and Nakata K 2008 *Proc. IUMRS-ICA 2008* p JP-29
[33] Elliott J F, Gleiser M 1960 *Thermochemistry for Steelmaking* (Massachusetts: Addison-Wesley) vol 1
[34] Palavra A, Fernandes A J S, Serra C, Costa F M, Rocha L A and Silva R F 2001 *Dia. Rel. Mat.* **10** 775-80
[35] Sechi Y, Takezaki A, Matsumoto T, Tsumura T and Nakata K 2009 *Mat. Trans.* **50** 1294-9

IOP Conf. Series: Materials Science and Engineering **61** (2014) 012020 doi:10.1088/1757-899X/61/1/012020

Formation behaviour of reaction layer in Sn-3.0Ag-0.5Cu solder joint with addition of porous Cu interlayer

Nashrah Hani Jamadon[1]*, Yukio Miyashita[2], Farazila Yusof[1,4], Mohd Hamdi[1,4], Yuichi Otsuka[2], Tadashi Ariga[3,4]

[1]Department of Mechanical Engineering, Faculty of Engineering, University of Malaya, 50603 Kuala Lumpur, Malaysia

[2]Department of Mechanical Engineering, Nagaoka University of Technology, 1503-1 Kamitomioka, 940-2188, Nagaoka, Niigata, Japan

[3]Department of Metallurgical Engineering, School of Engineering, Tokai University, 1117 Kitakaname, Hiratsuka, 259-12, Kanagawa, Japan

[4] Centre of Advanced Manufacturing and Material Processing (AMMP Centre), Level 8, Engineering Tower, Faculty of Engineering, University of Malaya, 50603 Kuala Lumpur, Malaysia

*Email : nashrahhani@siswa.um.edu.my

Abstract. The morphology and growth of interfacial intermetallic compound (IMC) between Sn-3.0Ag-0.5Cu solder alloy and Cu substrate metal of solder joint is reported. The IMC morphology and IMC thickness layer were observed at three different porosities of porous Cu interlayer. The results revealed that during soldering process, Cu_6Sn_5 compound with scallop like morphology was formed at the interface of both the solder alloy and Cu substrate and at solder alloy and porous Cu interlayer. By adding porous Cu interlayer at the solder joint, the IMC thickness increased with increasing soldering temperature and the number of pores in porous Cu interlayer. The effect of porosity on increasing the IMC layer was also due to the slower cooling rate during solidification of molten solder.

1. Introduction

It is reported that Sn-Pb eutectic solder has been widely used for a long period of time [1]. However, due to the environmental and health concerns over the toxicity of lead, research activities to find suitable lead-free solders has intensified [2]. Various investigations have been made with regard to characterization of lead-free solder including solder joint microstructure and its interfacial reaction [3,4]. For instance, the interfacial reactions between Sn-Ag alloys and Cu substrate have been investigated in some details in order to obtain an understanding of the reactions at the micro level [5]. It has been acknowledged that in Sn-based joint soldered with Cu substrate, the solder will bind

Content from this work may be used under the terms of the Creative Commons Attribution 3.0 licence. Any further distribution of this work must maintain attribution to the author(s) and the title of the work, journal citation and DOI.
Published under licence by IOP Publishing Ltd

through the formation of a Cu-Sn intermetallic compound (IMC) layer. The IMC layer basically consists of Cu_6Sn_5 which will also be altered significantly by increasing the cooling time of the molten solder [6,7]. The rate at which a solder joint solidifies depends on how fast the molten solder solidified. The solidification of molten solder will subsequently affect the growth of IMC layer.

Recent developments in automotive electronic industries demand electronic components that can withstand high temperature. Therefore researchers have been searching for potential lead-free solder for use in the assembly of heat tolerance electronic components. There are also studies which showed that cellular structure like metal foam or porous metal can be applied as the heat transfer medium in semiconductor application where in actual practice, high thermal resistance of the system is very crucial [8].

Thus, in the present studies, the porous metal which is known for its high thermal conductivity was used as the heat dissipation agent in the solder joint. The main purpose of this research is to investigate the formation behavior of reaction layer especially on their microstructure features in Sn-3.0Ag-0.5Cu solder joint with addition of porous Cu interlayer.

2. Experimental procedures

Sn-3.0Ag-0.5Cu (wt%) of lead free solder paste with the constant amount of 0.1g was used to join two Cu rods (99.99%) with diameter and length of 8 mm and 30 mm respectively as illustrated in Figure 1. The porous Cu interlayer was inserted in the middle of joining part as shown in the figure. Three different porosities namely at 15 ppi – pores per inch (P15), 25 ppi (P25) and 35 ppi (P35) were used where the pore size was smaller at higher ppi. Prior to be inserted in the middle of joining part, the porous Cu interlayer was rolled to obtain a uniform thickness of 0.1 mm. Figure 2 shows the microstructure of the porous Cu interlayer after rolling process. The soldering process was conducted in the furnace with an Argon gas flow. The holding temperature was controlled at 267°C and 287°C with a holding time of 60 s. Solder joint without adding porous Cu interlayer was also prepared as the control sample. The *in-situ* temperature measurement was conducted during the soldering process by positioning the end of thermocouple at the spot where the porous Cu interlayer was located or directly at the solder layer for the control. After the sample was soldered, it was cut perpendicular to the joint and the cut surface was polished to observe the interface of joining part. Scanning electron micrograph (SEM) was used to observe the IMC thickness at the interfaces between the solder and Cu rod as well as between solder and porous Cu. Energy dispersive X-ray spectroscopy (EDS) was utilized to determine the elemental composition of IMC layer.

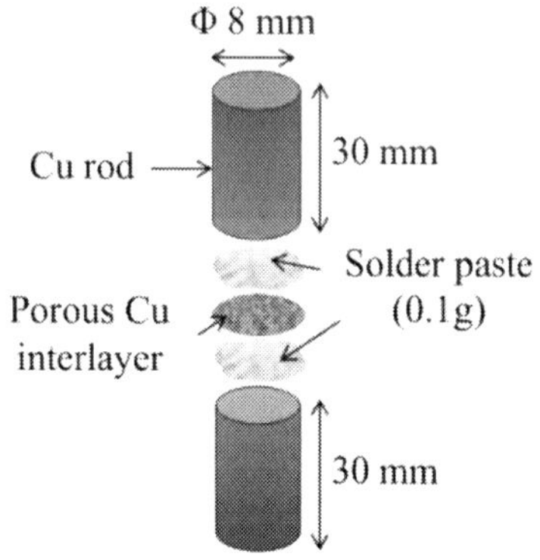

Figure 1. Configuration of solder joint.

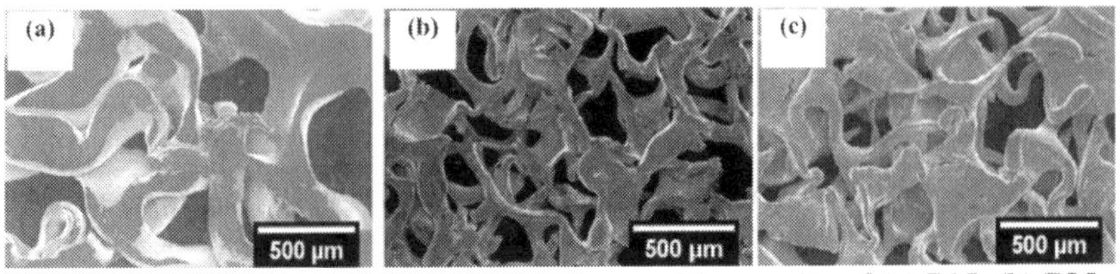

Figure 2. SEM image of porous Cu interlayer with porosity of (a) P15, (b) P25 and (c) P35 after rolled.

3. Results and Discussion

3.1. Cross-sectional morphology

Micrographs from typical cross-sections of the sample soldered with or without porous Cu interlayer are shown in Figure 3. It is observed that the thickness of soldered sample without porous Cu interlayer (Figure 3a) is thicker compared to the soldered sample with porous Cu interlayer (Figures 3b, c and d); a constant amount of solder alloy being used in both conditions. A thinner solder layer was further observed at the soldered sample with P15 of porous Cu interlayer when compared to those of the soldered sample with P25 and P35 of porous Cu interlayer. The bigger pores at P15 of porous Cu interlayer caused the molten solder alloy to penetrate easily inside the interlayer. In contrast the solder layer for soldered sample with P25 and P35 were thicker. This is because the molten solder was unable to penetrate into small pores. It was observed that the solder layer with P25 and P35 porous Cu interlayer have almost similar thickness of approximately 80 µm.

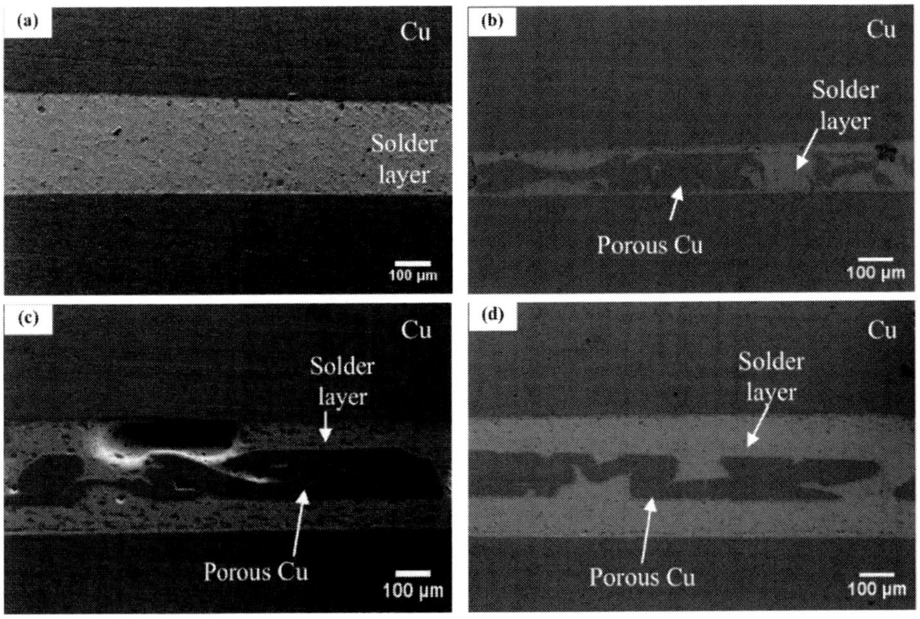

Figure 3. Cross-section of solder joint of Sn-3.0Ag-0.5Cu solder alloy and Cu rod at 287°C, 60 s. (a) no porous (b) P15, (c) P25 (d) P35.

3.2. Formation of IMC layer

Figure 4 shows IMC layer for solder joint without porous Cu added and soldered at temperature and time of 287ºC and 60 s respectively. Elemental analysis of the IMC layer at Table 1 showed the main element that exists is Cu_6Sn_5. It was observed the IMC layer for solder joint without porous Cu appeared as rough and scallop-like structures. Similar result was obtained for the soldered sample at the same soldering time of 60 s and a lower temperature of 267 ºC except that the IMC/ solder layer appeared slightly thinner.

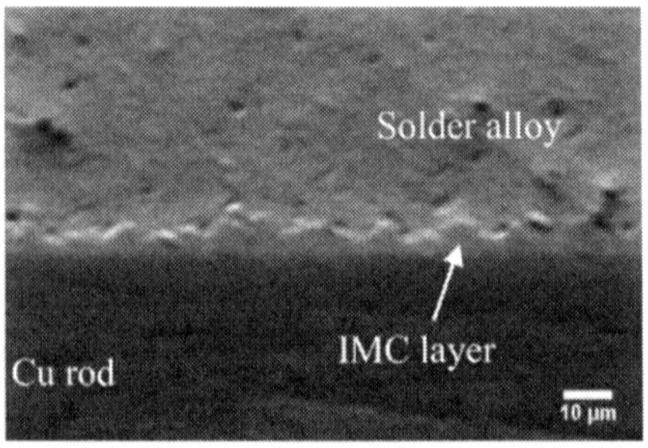

Figure 4. IMC morphology for Sn-3.0Ag-0.5Cu solder and Cu rod without porous Cu interlayer.

Table 1. Elemental analysis of the IMC layer.

Element	Atomic (%)
Cu	52.21
(Ag)	(0.73)
Sn	47.06

Figure 5 shows the IMC layer for solder joint with P15 of porous Cu interlayer soldered at temperature and time of 287°C and 60 s respectively. It is observed that the IMC layer between Cu rod and the solder alloy also appeared rough with scallop-like undulating shape (Figure 5a). EDS analysis confirmed (Table 2) that the elemental composition of the IMC layer is Cu_6Sn_5. The IMC layer which consists of Cu_6Sn_5 is also formed at the interface between porous Cu and solder alloy as shown in Figure 5b. It can be speculated that during the soldering process the porous interlayer have reacted with solder alloy at atomic level which also influence the IMC growth at the porous Cu interface. Similar results are obtained for the solder joint with P15 porous Cu interlayer when soldered at 267°C and soldering time of 60 s.

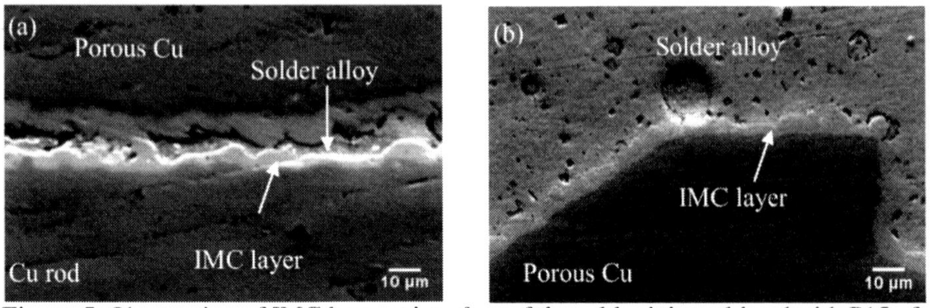

Figure 5. Observation of IMC layer at interface of the solder joint soldered with P15 of porous Cu interlayer (a) solder alloy – rod Cu, (b) solder alloy – porous Cu.

Table 2. Elemental analysis of the IMC layer for sample joined with P15 of porous.

Element	Atomic (%)	
	solder alloy – Cu rod	solder alloy – porous Cu
Cu	57.29	51.42
(Ag)	(0.36)	(0.35)
Sn	42.35	48.23

Figure 6 shows the IMC layer for the solder joint with P25 of porous Cu interlayer soldered at temperature and time of 287°C and 60 s respectively. In this instance, the solder layer is thicker than that of solder joint with P15 of porous Cu interlayer. The scallop-like grains at IMC layer is similarly observed at the interface of the Cu rod with solder alloy (Figure 6a) as well as between the porous Cu interlayer with solder alloy (Figure 6b). The EDS also confirmed the IMC layer consists of Cu_6Sn_5 on both sides of the IMC layer. Similar results are obtained for the solder joint with P25 porous Cu interlayer when soldered at 267°C and soldering time of 60 s except that the solder layer appeared slightly thinner.

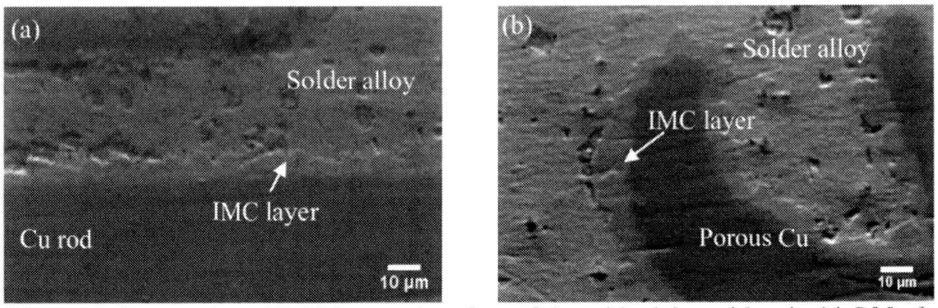

Figure 6. Observation of IMC layer at interface of the solder joint soldered with P25 of porous Cu interlayer (a) solder alloy – Cu rod, (b) solder alloy – porous Cu

Table 3. Elemental analysis of the IMC layer for sample joined with P25 of porous.

Element	Atomic (%)	
	solder alloy – Cu rod	solder alloy – porous Cu
Cu	59.68	52.09
(Ag)	(0.53)	(0.41)
Sn	39.79	47.5

Effect of pore size interlayer on the IMC thickness is shown in Figure 7. It was found that the IMC thickness increased with increasing pore size of porous Cu interlayer. This shows that the molten solder is getting more difficult to penetrate the porous Cu with increasing pore size and finally solidified at the point between porous Cu and solder alloy.

It has also been reported that the cooling rate had a significant effect on the thickness of the IMC layer [6,9]. In the present study, the cooling rate for various solder joints is as shown in Figure 8. Since all the porous Cu has nearly similar weight, it can be deduced that the pore sizes have influence on the cooling rate. It appears that the smaller pore sizes of porous Cu interlayer (P25 and P35) results in longer cooling time for solidification of molten solder than that observed for P15. This slower cooling rates had allowed more time for Cu_6Sn_5 to grow and develop at the interface layer of molten solder. This results in thicker IMC layer for P25 and P35 as compared to P15. In contrast, the cooling time for the solder joint without porous Cu was shorter. As a result a thinner IMC layer was formed compared with solder joint joined with porous Cu interlayer.

The difference in the cooling rate can also be explained by differences in thermal conductivity provided by the porous Cu interlayer inside the solder joint. Thermal conduction through the Cu metal during cooling process will dissipate heat into molten solder that flow through them. This will eventually have a bearing on the solidification of solder alloy which ultimately affect the growth of IMC layer in the solder joint.

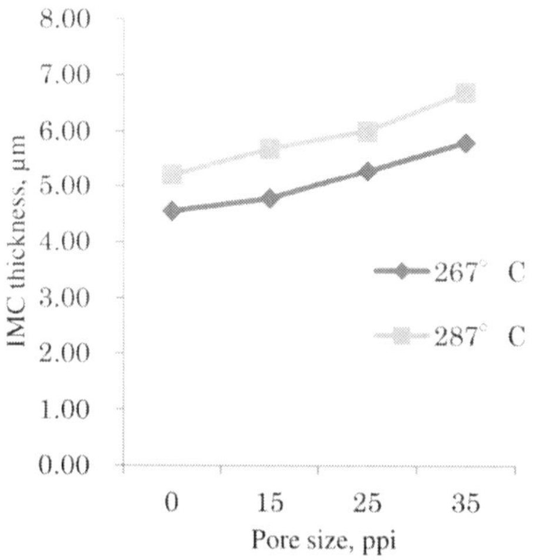

Figure 7. Effect of pore size and soldering temperature on IMC thickness.

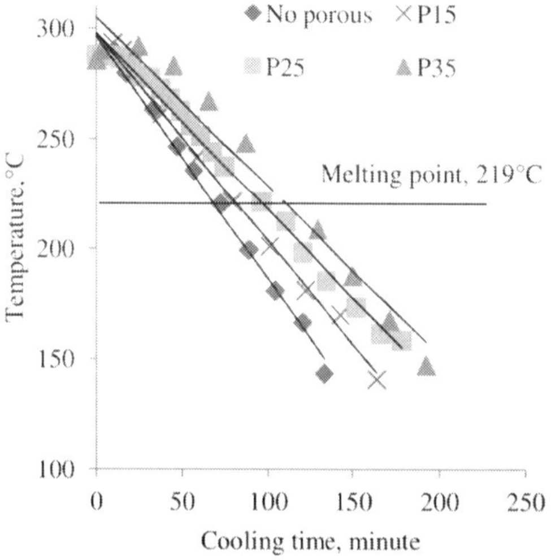

Figure 8. Cooling rate for various solder joint.

4. Conclusions

The effect of adding porous Cu interlayer to Sn-3.0Ag-0.5Cu solder alloy on IMC growth and morphology have been studied. The IMC layer which consists of Cu_6Sn_5 compound was observed at the interface between Cu substrate and Sn-3.0Ag-0.5Cu solder, and also between porous Cu layer and the solder alloy. It was found that increase in soldering temperature and the number of pores inside porous Cu interlayer resulted in increasing thickness of the IMC layer. Results further showed that cooling rate after soldering process have significant effect on the IMC thickness. It was also observed that the cooling time for the sample soldered with the porous Cu interlayer was longer compared to that for the solder joint without the porous interlayer. The cooling rate decreased with increase in number of pores of the porous Cu interlayer.

Acknowledgement

The author would like to acknowledge the financial support of the Fundamental Research Grant Scheme (FRGS; project no. FP024-2012A) of University of Malaya for this work.

References

[1] Harrison M R, Vincent J H and Steen H A H 2001 *Solder. Surf. Mt. Technol.* **13** 21
[2] Abtew M and Selvaduray G 1995 *Mater. Sci. Eng. Rep.* **27** 95
[3] Takemoto T and Miyazaki M 2001 *Mater. Trans.* **42** 7
[4] Huh S H, Kim K S and Suganuma K 2001 *Mater. Trans.***42** 739
[5] Gagliano R A, Ghosh G and Fine M E 2002 *J. Electron. Mater.* **31** 1195
[6] Sigelko J 1999 *J. Electron. Mater.* **28** 1184
[7] Bo L, Yaowu S, Yongping L, Fu G and Zhidong X 2005, *J. Electron. Mater.***34** 217
[8] Yang K S, Chung C H, Lee M T et. Al 2013 Int. J. Heat Mass Transf. **48** 73
[9] Yang W, Felton L E, Messler Jr W 1995 *J. Electron. Mater.* **24** 1465

Flow curves of Sn and Sn-3.5Ag obtained by rotational viscometry using a stainless steel cone

Takahisa Yamazaki, Shinya Oishi, Hirosato Gamou, Toshi-Taka Ikeshoji and Akio Suzumura

Graduate School of Science and Engineering, Tokyo Institute of Technology,

2-12-1 Ookayama, Meguro, Tokyo, Japan, 152-8552

yamazaki.t.ae@m.titech.ac.jp

Abstract. Corrosion of stainless steel in a flow soldering bath by a lead-free solder was investigated using a cone-plate-type rotational viscometer. The rotational torque of the stainless-steel cone in contact with a molten solder was measured at various shear rates. The delicate measured torque was related to the change of the viscosity of the solder owing to dissolution of materials originating from the cone. The estimated viscosity coefficient was ten times greater than the values which have been reported. The result was attributed to the tin content of the solder combined with oxygen from the passive state oxide film on the cone surface. The increase of the viscosity of the silver-containing solder was much greater than in case of pure Sn.

1. Introduction

Alternative Pb-solders, which contain Sn in high concentration ratios, lead to erosion and corrosion damage of stainless steel due to metal flow occurring in the soldering bath. A Sn-Ag solder system is used in flow soldering. The viscosity coefficient of the Sn-3Ag-0.5Cu solder at 523 K was 37 mPa·s, calculated from the data reported by M Jeyakumar et al. [1]. This value was greater than the value of 2.42 mPa·s for Sn-2Cu solder at 533 K by N Zhao et al. [2]. There are three types of rotating bob devices: a parallel-plate rotational device, a cone/plate-type rotational device and a double concentric cylindrical rotational device [3]. The cone/plate-type viscometer creates constant shear stress within the molten metal and on the surface of the cone/plate. Using a rotational cone-plate- type viscometer, the delicate rotational torque change related to erosion and corrosion of the stainless-steel cone was investigated for the Sn-3.5Ag solder.

2. Experimental

Sn-3.5 mass% Ag solder alloys were produced by a mixture of Sn (99.97% purity) drops and Ag (99.99% purity) wire at their eutectic composition. The weight of the mixture was approximately 5.5 mg. The diameter of the cone made from stainless steel 304 with a taper θ of 3° was 21 mm. Additionally, Sn (99.97% purity) melts were also prepared for a comparison of the viscosity changes caused by oxidation of the fluid in contact with the oxide film on top of stainless steel. The viscosity was measured using a graphite cone.

Figure 1 schematically shows the cone/plate viscosity tester used in this experiment. The viscometer was composed of a motor, a shaft, a vortex spring and a pointer that indicated the distortion after

Content from this work may be used under the terms of the Creative Commons Attribution 3.0 licence. Any further distribution of this work must maintain attribution to the author(s) and the title of the work, journal citation and DOI.
Published under licence by IOP Publishing Ltd

rotation. The cone plate was rotated by means of a lever fixed to the spring end. The vortex spring was made of a bimetal serving as a thermometer from 270 to 670 K to calibrate the thermal distortion of the spring at the measurement temperature of 523 K, and its spring constant was 1.117 mN/rad. When the temperature of the graphite crucible reached 623 K, the solder in the crucible was entirely molten. The crucible was then slowly cooled until it reached 523 K. During this time, the heated cone made of stainless steel 304, also kept at 523 K, was placed into the molten solder.

The cone/plate-type viscometer creates a constant shear velocity D on the cone surface, which is related to a constant angular speed ω. D is expressed in equation (1), where V is the speed of the cone surface and y is the separation distance between the cone and the plate, and is inversely proportional to the cone-plate separation in the fluid. The shear rate $\dot{\gamma}$, in the fluid is expressed by $\dot{\gamma} = \partial u / \partial y$, where u is the velocity in the fluid, as shown in Figure 2.

$$D = \frac{V}{y} \tag{1}$$

The shear rate does not yield the radius r from the rotational axis, according to equation (2).

$$D = \frac{r\omega}{r \tan\theta} = \frac{\omega}{\tan\theta} \tag{2}$$

This rotational viscometer can provided various fluid shear rates by changing the rotational speed. The shear stress τ is assumed to be proportional to $\dot{\gamma}$ by $\tau = \mu \partial u / \partial y$, where μ is the dynamic viscosity of a Newtonian fluid. However, the used cone/plate viscometer creates constant τ across the local cone surface according to equation (3).

$$\tau = \mu D \tag{3}$$

The overall torque T_{all}, as measured by the viscometer, is the integral of the shear stresses on the cone surface.

$$T_{all} = \int_0^{\frac{R}{\cos\theta}} r\tau 2\pi l \cos\theta dl = \frac{2}{3}\tau\pi R^3 \frac{1}{\cos\theta} \cong \frac{2}{3}\tau\pi R^3 \tag{4}$$

In equation (4), $\cos\theta \approx 1$, and the viscosity is finally derived from equation (5).

$$\mu = \frac{3T_{all}}{2\pi R^3 D} \tag{5}$$

Based on equation (5), the viscosity of fluid can be estimated from the slope of the curve of shear stress τ versus shear velocity D.

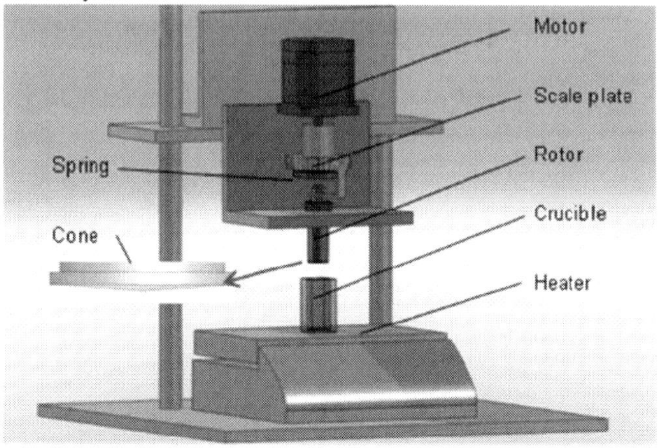

Figure 1. Schematic illustration of the used cone/plate-type rotational viscometer.

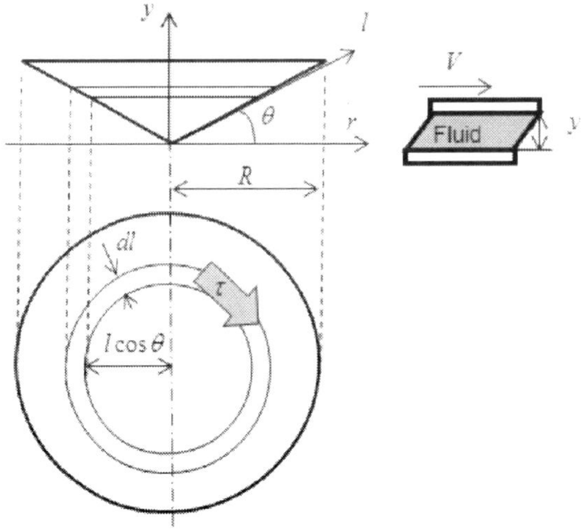

Figure 2. Illustration of the calculation of rotational torque at a constant shear rate.

3. Results

3.1. Flow curves of Sn using graphite and stainless steel cones

Rotational torque data for molten Sn at 573 K using a graphite cone were stably obtained at various rotation speeds of 102, 148, 184, 216, 248, 289, 330, 369, 402, 435 and 468 rpm (204, 296, 368, 432, 498, 577, 659,737, 803, 869 and 935 s^{-1} in shear velocity), as shown in Figure 3. The torque values were obtained avoiding oxidation of molten Sn by the use of excess flux. The viscosity coefficient was estimated to be 3 mPa·s from the linear relationship between shear stress τ and shear velocity D as in a Newtonian fluid (Figure 2). At the rotation speeds listed above, the flow of the molten solder was lamellar with a shear rate of below 800 s^{-1}. The derived viscosity value is almost identical to the reported value of 1.56 mPa·s[4] employing this simple measurement method.

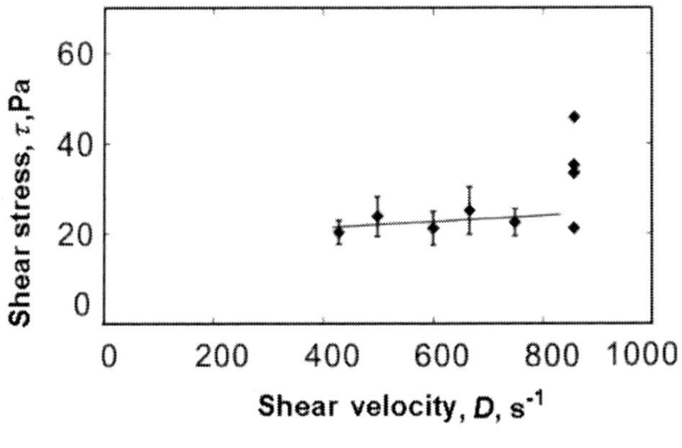

Figure 3. Flow curve of pure Sn, obtained using a graphite cone at 573 K.

Rotational torque data for molten Sn at 573 K using a stainless steel cone were reliably obtained at various rotation speeds of 184, 216, 248, 289, 330, 369 and 402 rpm. The viscosity coefficient was estimated to be 100 mPa·s from the linear relationship between τ and D as in a Newtonian fluid (Figure 4). At these rotation speeds, the flow of the molten solder was lamellar and was maintained below a shear rate of 800 s^{-1}. This viscosity coefficient was greater than the previously reported value [4] due to the presence of oxygen.

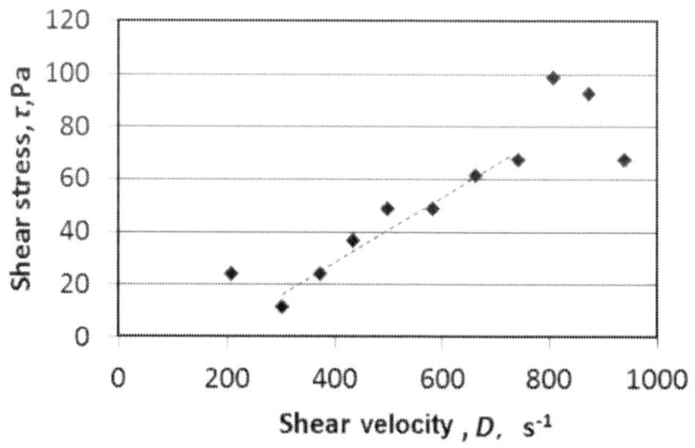

Figure 4. Flow curve of Sn, obtained using a stainless-steel cone at 523 K.

3.2. Flow curve for Sn-3.5Ag

Rotational torque data for Sn-3.5Ag solder at 523 K were stably obtained at various rotation speeds of 184, 216, 248, 289, 330, 369 and 402 rpm. High shear stresses were observed at low rotational speeds of 184 and 216 rpm. The viscosity coefficient was estimated to be 256 mPa·s from the slope of the flow curve shown by the broken line between 500 and 800 s^{-1} (Figure 5). At these rotation speeds, the flow of the molten solder was lamellar and was maintained below a shear velocity of less than 800 s^{-1}. At the slow rotational speeds of 184 and 216 rpm, large values of shear stress were obtained, similar to those previously reported [1,5].

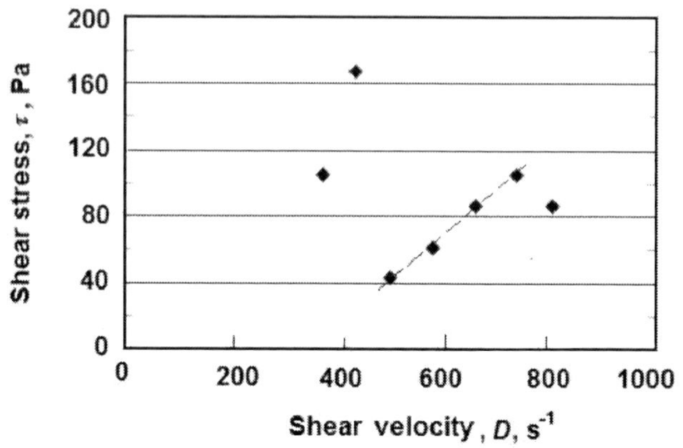

Figure 5. Flow curve of Sn-3.5Ag obtained using a stainless-steel cone at 523 K.

We attempted to measure speeds for accelerated life testing for erosion and corrosion tests at higher shear stress at the entire local cone surface in lamellar flow. The selected speed was 500 s^{-1} (250 rpm) because stable rotation could be achieved.

3.3. Change in rotating torque with time

The measurements of the rotational torque could be started one minute after immersion of the stainless steel cone into the solder. Measurements were taken every 30 min with a resolution of 2×10^{-5} N · m. Once an increase in the torque was observed, the cone was removed from the graphite crucible. The cones and solder alloys were then cooled under ambient conditions. Figure 6 shows the measured rotational torque deduced from the shear resistance of the molten solder with respect to time at 523 K at the velocity of 250 rpm. The initial torque was 0.11 mN · m when the cone was immersed in the solder, and an increase in the rotational torque was observed for approximately 500 min after immersion. Then, the torque quickly reached 0.15mN · m.

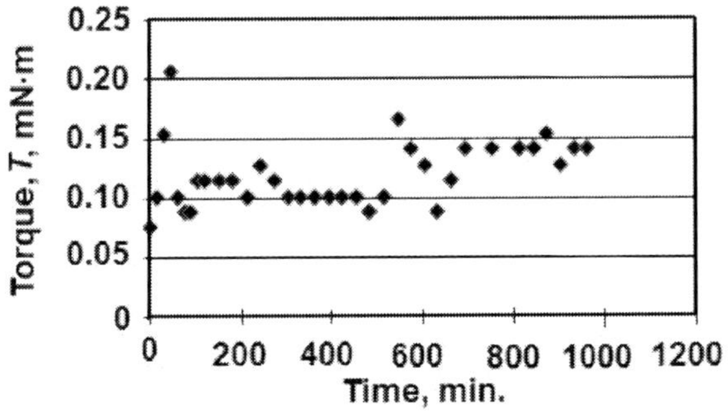

Figure 6. Measured rotational torque deduced from the shear resistance of the molten Sn-3.5Ag solder with respect to time at 523 K.

3.4. Observation of cone plate after rotating torque change test

Cone plates were cross-sectioned after erosion and corrosion tests to observe if any solder alloy had adhered to the stainless steel. A scanning electron microscope (SEM) equipped with energy-dispersive X-ray spectroscopy (EDX) analysis examined the surface of the stainless-steel cone that had been in contact with the solder. The optical image (a) and SEM image (b) of Figure 7 show the surface damage to the apex of the cone. The line analysis for (c) Ag, (d) Cr, (e) Ni, (f) Sn and (g) Fe in Figure 7(b) were performed from the apex at point A to point B, corresponding to approximately 40 μm. Figure 7(f) shows the adhesion of Sn at about 5 μm from A and around B. The dark area in Figure 7 (b) is considered to be related to the distribution of (c) Ag in the length from 10 to 20 μm.

Figure 8 shows the EPMA maps corresponding to the cross-section of Figure 7(a) at a ring radius of $l = 9$ mm where the surface lost luster. In other areas of the cone, adherent materials were not observed. Adherent Sn could not form an Fe-Sn alloy because iron was not detected. Elemental Sn was found inside the stainless-steel cone. The composition of the damaged surface was 73.41 mass% of Fe, 19.76 mass% of Cr, 6.10 mass% of Ni and 0.75 mass% of Sn. Cr atoms remained at the surface and Sn atoms diffused in the cone, according to the detection of 0.03 mass% of Sn inside the cone. The compositional change of the surface was attributed to the loss of oxygen dissolved into the solder. This phenomenon was caused by the adhesion of Sn atoms, but also by the dissolution of Fe and Ni atoms from stainless steel into the solder, based on a decrease of the Ni concentration from 8 mass% to 5-6 mass%. The elution of Ni was considered to be promoted by a high shear velocity and friction.

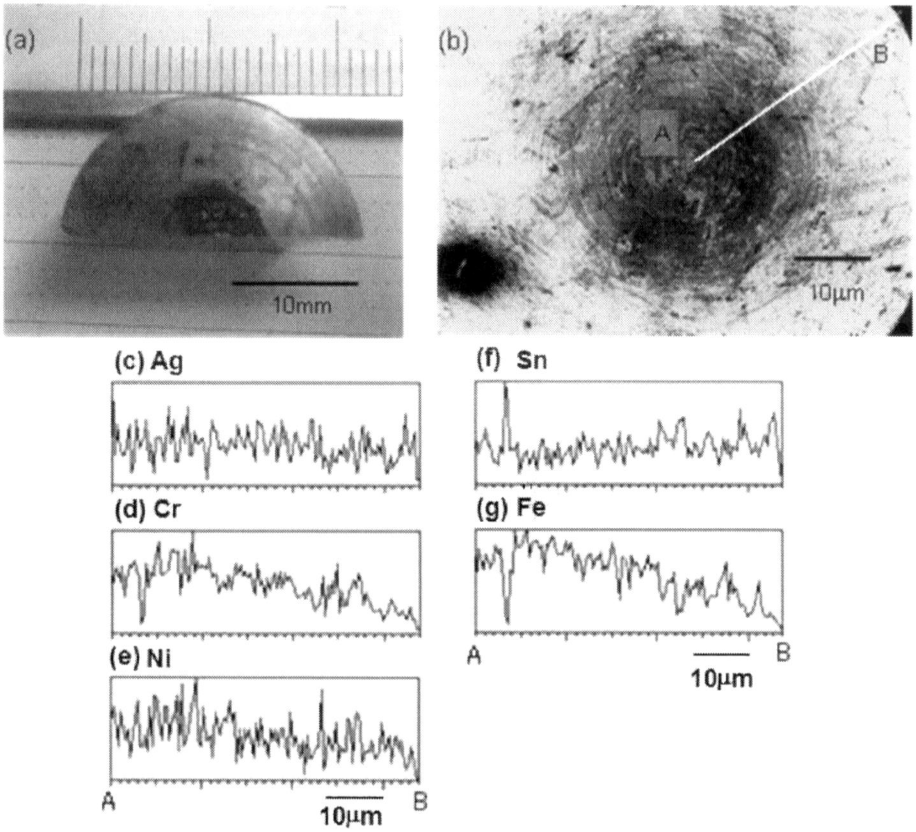

Figure 7. Energy-dispersive X-ray spectroanalysis of the conical surface after the erosion and corrosion test. (a) Optical and (b) SEM images of sectioned cone after the test. Line analyses of (c) Ag, (d) Cr, (e) Ni, (f) Sn and (g) Fe were performed from the apex at point A of the cone to point B, covering a distance of approximately 40 μm.

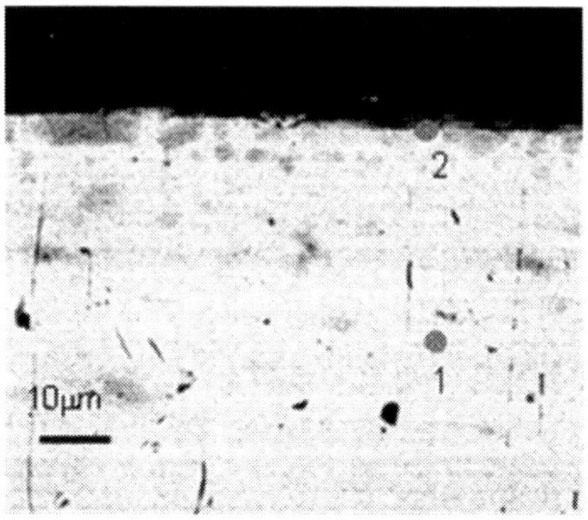

Figure 8. EDX analysis of Fe, Cr, Ni, Sn, and Ag and back-scattering electron images of the sectioned cone after testing.

Table 1. EDX analysis of Fe, Cr, Ni, Sn and Ag at the points 1and 2 indicated in Figure 8.

	Fe mass%	Cr mass%	Ni mass%	Sn mass%	Ag mass%
1	70.97	18.12	10.18	0.03	Not detected
2	73.41	19.76	6.10	0.75	Not detected

3.5. Differential thermal analysis (DTA)

A difference between the DTA curves of the solders after 360 and 840 min of the erosion and corrosion tests was verified by a rise in the liquidus line of the solder by approximately 3 K (Figure 9). The DTA-plot was obtained using a heating rate of 10 K per minute under an argon atmosphere. The eutectic point was detected at 220.48 °C (493.63 K) according to the DTA curves recorded after 360 and 840 min testing time. The DTA-plot revealed that the liquidus line rose with time and reached 228 °C (501 K) after 840 min. It was considered that this rise is due to the elution of nickel to the solder, according to the Ni-Sn binary phase diagram.

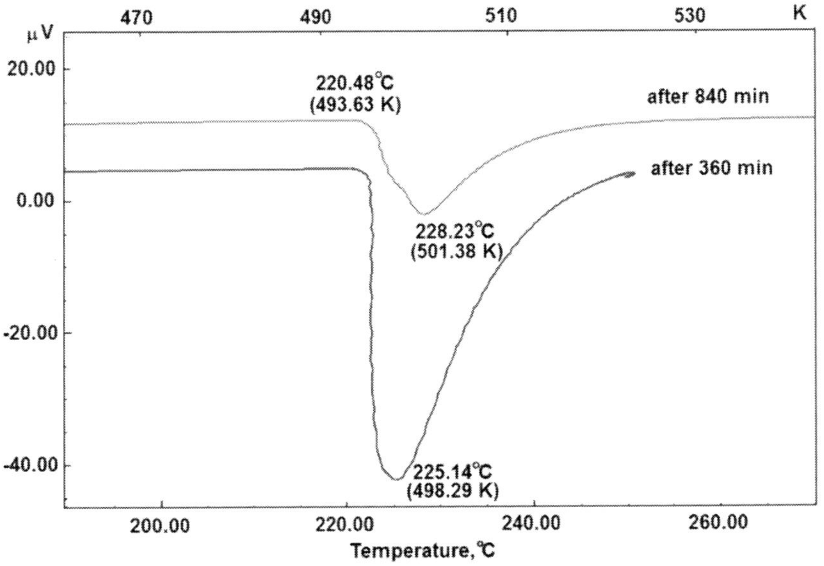

Figure 9. DTA analyses of the solders after erosion and corrosion testing.

4. Discussion

In the case of the inert graphite cone, molten Sn dissolved oxygen adsorbed at the cone surface and the viscosity increased beyond than that of pure Sn. In the case of the stainless steel cone, adhesion of Sn atoms first occurred at the passive oxide film on top of the cone surface but, subsequently, Sn-Ag solder then dissolved oxygen atoms gradually for a while. Therefore, the viscosity of the Sn-Ag solder containing oxygen increased. The large value of 256 mPa·s using stainless steel cone was attributed to the silver content combined with the presence of oxygen around the conical apex area at low shear velocities. The flow curve of molten Sn-3.5Ag solder was not linear at low shear rates, while the viscosity of molten pure Sn was linear [1,5]. Adhesion of Sn after erosion and corrosion tests was observed at the apex of the cone where the shear velocity was low. The sudden increase in torque during the tests for the Sn-3.5Ag molten solder was attributed to the fracture of the passive oxide film by shear stress. When the cone was immersed in molten Sn-3.5Ag alloy, oxygen atoms were unable to penetrate into the cone and to recover from the passive layer because the concentration of dissolved oxygen atom in the molten solder was less than that in the stainless steel [6-10]. As for the elution of stainless steel into the molten solder, the concentration of dissolved oxygen in the oxide layer at 523 K has an effect on the elution speed. As a result, Ni atoms eluted into the Sn-Ag solder, and Sn atoms diffused into the cone after testing for 840 min. Local corrosion caused damage to stainless steel after the protective oxide film disappeared from the cone at a 9 mm radius where the shear rate was higher.

5. Conclusions

The viscosity of solders was investigated using the change in torque during rotating rheometry tests with the cones of stainless steel 304. An increase in the rotational torque was observed approximately 500 min after dipping the cone into the molten solder. In terms of the elution of stainless steel, the effect of the consistency of dissolved oxygen due to the high viscosity coefficient of 256 mPa·s for Sn-3.5Ag solder containing oxygen was examined.

Adhesion of Sn atoms occurred at the passive oxide film on top of the cone surface. Sn-Ag solder then dissolved oxygen, and the viscosity of the Sn-Ag solder containing oxygen increased. The increase in torque during the erosion and corrosion test for Sn-3.5Ag molten solder was attributed to the damage of the cone surface. Stainless steel is prone to erosion because its passivity is not recovered under the conditions present in a molten metal.

Ni atoms eluted into the Sn-Ag solder, and Sn atoms diffused into the cone. Local corrosion causes damage to stainless steel after the oxide film disappeared from the cone at about 9 mm in radius.

Acknowledgments

The authors wish to acknowledge assistance and technical support from the Center for Advanced Materials Analysis, Tokyo Institute of Technology.

References

[1] Jeyakumar M, Hamed M, and Shankar S 2011 *J. Non-Newtonian Fluid Mechanics* **166** 831
[2] Zhao N, Pan X M, Yu D Q, Ma H T, and Wang L 2009 *J. Electronic Materials* **38**-6 828
[3] JIS Z 8803 Method for Viscosity Measurement of Liquid (2011) 1-40.
[4] Assael M J, Kalyva A E, Antoniadis K D, Banish R M, Egry I, Wu J, Kaschnitz E and Wakeham W A Aug 2010, *J. Phys. Chem. Ref. Data* **39**-3 033105-1.
[5] Malika M M, Jeyakumara M, Hameda M S, Walkerb M J, Shankara S 2010 *J. of Non-Newtonian Fluid Mech.* **165** 733
[6] Huang Y, Chen S and Gierlotka W 2007 *J. Mater. Res.* **22**-10 2924
[7] Ricci E, Giurano D, Grosso I, Lanata T, Amore S, Novakoovic R and Arato E 2009 *J. Chem. Eng. Data* **54** 1660
[8] Zang L, Yuan Z, Xu H and Xu B 2011 *Applied Surface Science*, **257** 4877
[9] Japan Society of Corrosion Engineering, Fushoku Boshoku Databook, Maruzen(1995-6) p. 316.
[10] Yamazaki T, Nishiyama M, Ikeshoji T-T, Suzumura A: IBSC 2012 (5th International Brazing and Solderng Conference 2012 USA American Welding Society.)

IOP Conf. Series: Materials Science and Engineering **61** (2014) 012022 doi:10.1088/1757-899X/61/1/012022

Effect of intersplat interface bonding on the microstructure of plasma-sprayed Al_2O_3 coating

Er-Juan Yang[1], Chang-Jiu Li[1,*], Guan-Jun Yang[1], Cheng-Xin Li[1] and Makoto Takahashi[2]

[1]State Key Laboratory for Mechanical Behaviour of Materials, School of Materials Science and Engineering, Xi'an Jiaotong University, Xi'an, Shannxi 710049, P. R. China

[2]Japan Welding and Joining Research Institute, Osaka University, Ibaraki 565-0871, Osaka , Japan

*: E-mail: licj@mail.xjtu.edu.cn

Abstract. The interface bonding between lamellae dominates the properties and performance of plasma-sprayed ceramic coatings. In this study, the interlamellar interface bonding and its effect on the splat microstructure were examined using TEM analysis of the microstructure of a plasma-sprayed Al_2O_3 coating. The obtained results revealed that the intersplat interface microstructure depends significantly on the intersplat bonding. An amorphous phase was observed in the interface region in which the bonding was formed, while the γ-Al_2O_3 phase was observed at the interface where no bonding was formed. In addition, it was found that the interface bonding significantly influenced the interface microstructure of the coating. After heat treatment, the phase of the bonding between adjacent splats was transformed from amorphous to γ-Al_2O_3. In the interface region in which the amorphous phase recrystallization occurred, nanosized pores evolved owing to the volume shrinkage accompanying the transformation of alumina from the amorphous to the γ-Al_2O_3 phase.

1. Introduction

Plasma spraying is a flexible process for producing thick coatings (thickness ranging from micrometers to millimeters) at a high deposition rate and has been employed extensively in different industrial fields. During plasma spraying, molten or semimolten particles successively impact on the substrate, followed by rapid spreading and subsequent solidification of the melt, thereby forming a deposit stacked with splats [1-4]. A splat is the basic unit constituting a plasma spray coating. The successive stacking of splats is required to deposit a thermal spray coating with a certain thickness. Thus, the resulting coating possesses a lamellar structure. It has been demonstrated that the intersplat bonding dominates the mechanical and physical properties such as strength and thermal or electrical conductivity of certain materials [5]. Therefore, bonding formation between splats is an important factor that affects the stacking of splats to form a coating. Accordingly, a systematic quantitative investigation of splat bonding was carried out by utilizing the bonding ratio [6]. It was found that the bonding ratio changes from ~10% in the case of D-gun alumina to a maximum value of ~1/3 for optimized plasma-sprayed ceramic coatings deposited by following a conventional routine [7]. This accounts qualitatively for the fact that most of the coating properties such as Young's modulus [7, 8],

Content from this work may be used under the terms of the Creative Commons Attribution 3.0 licence. Any further distribution of this work must maintain attribution to the author(s) and the title of the work, journal citation and DOI.
Published under licence by IOP Publishing Ltd

fracture toughness [9], thermal conductivity [5, 10], and electrical conductivity [5, 11] are found to range from one-tenth to one-third of those of bulk materials. Moreover, it is evident that the coating properties also depend on the crystalline structure of splats.

Aluminum oxide (Al_2O_3) coatings have been widely used as dielectric coatings, protective coatings, and corrosion-resistance barriers. However, aluminum oxide is a complex material even in the massive bulk form as it exhibits various crystalline structures. Several transient metastable phases (χ, κ, γ, δ, and θ) may evolve and only one thermodynamically stable phase (α-Al_2O_3 in a hexagonal close-packed (hcp) lattice) has been observed [12]. As mentioned previously, in a thermally sprayed ceramic coating, less than one-third splat interface area is bonded together [7] and the alumina coating mainly consists of metastable phases [10, 13]. This is because after the flattening of droplets, the cooling rate is dominated by the contact conditions between the droplets and underlying substrate. Assuming that the contact of the melt with solid splats before solidification is positively related to interface bonding, the cooling rate would differ in the bonded and unbonded zones, possibly resulting in different microstructures. Thus, it can be naturally considered that a local splat bonding state corresponds to the droplet–splat contact condition prior to solidification of the molten splat and thus influences the phase formation upon crystallization. However, little attention has been paid to the effect of splat bonding on the local splat microstructure. In this study, an alumina coating was deposited by atmospheric plasma spraying (APS), and high-resolution transmission electron microscopy (HR-TEM) analysis was performed to examine the local microstructures near the interfaces in both the bonded and unbonded zones. Further, microstructure changes resulting from heat treatment were also investigated.

2. Materials and methods

Commercial fuse-crushed Al_2O_3 powder with a particle size range of -45 to 15 μm was used as the feedstock. An Al_2O_3 coating was deposited on grit-blasted steel by atmospheric plasma spraying (APS, 80 kW class, Jiujiang, China) at a plasma arc power of 33 kW. During spraying, an air jet attached to the torch and oriented toward the substrate was employed by following a conventional routine to cool the sample by suppressing any rise in the sample temperature. Subsequently, post-spray heat treatment of the sample was carried out at 1073 K for 12 h.

The deposits were characterized using a variety of techniques, including X-ray diffraction (XRD), scanning electron microscopy (SEM), and HR-TEM (JEM-2100F JEOL) carried out at 200 kV. In order to distinguish the deposition direction in HR-TEM specimens, their substrate side was marked during sample preparation.

Vickers microhardness of the coating was measured from a polished cross-section at a load of 2.92 N and a loading time of 30 s. Young's modulus was measured through the Knoop indentation approach as proposed in literature [14, 15] at a load of 9.8 N and a loading time of 50 s.

3. Results and discussion

3.1. Effect of interface bonding on interface microstructure

3.1.1. Microstructure characterization at unbonded intersplat interface in APS-sprayed alumina coating. A bright-field TEM image of a region near the unbonded lamellar interface zone in APS-sprayed alumina coatings is shown in Figure 1(a). A ~20–40-nm-wide crack at the splat interface can be easily observed. Such crack was presented as the unbonded splat interface. Moreover, the splat showed a columnar substructure. The columnar grains were perpendicular to the splat interface. Global pores possibly resulting from gas entrapment were also observed. Selected area electron diffraction (SAED) patterns taken from regions in two adjacent splats across the unbonded interface marked "A" and "B" in Figure 1(a) are shown in Figure 1(b) and 1(c), respectively. These patterns clearly indicate that both splats have a γ-Al_2O_3 lattice structure. This is in good agreement with the results widely reported in the literature [10, 13]. As expected on the basis of common knowledge on thermally sprayed coatings, substantial lamellar interfaces were present in the as-

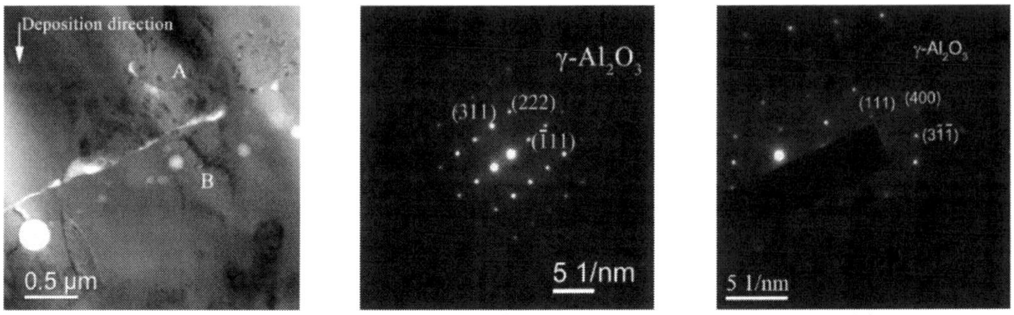

(a) (b) (c)

Figure 1. (a) Bright-field TEM image of an unbonded zone in an APS-sprayed alumina coating; (b) selected area electron diffraction (SAED) pattern corresponding to the region marked "A" in (a), which is in the unbonded zone; and (c) SAED pattern corresponding to the region marked "B" in (a), which is in the unbonded zone.

sprayed alumina coatings as unbonded interfaces, as shown in Figure 1 (a). An investigation of the crystalline structure of two splats immediately across an unbonded interface confirmed the above-mentioned results.

3.1.2 Microstructure characterization at bonded intersplat interface in APS-sprayed alumina coating. A bright-field TEM image of and SAED patterns corresponding to a region near the bonded interface in an alumina coating are shown in Figure 2. At the bottom of the image shown in Figure 2(a), the region marked "C" clearly shows columnar grains, which grew to the splat surface and became the interface between two splats. The columnar grains were confirmed to have a γ phase. The SAED pattern corresponding to the region marked "B" in Figure 2(a) indicates the formation of an amorphous phase, as shown in Figure 2(c). The amorphous phase layer within a splat reached up to several hundreds of nanometers in thickness with a nanocrystal γ-Al_2O_3 phase extended throughout the remaining part of the splat, as indicated by the SAED pattern in Figure 2(b), which corresponds to the region marked "A" in Figure 2(a). The origin of the amorphous phase was attributed to the cooling rate in the bonded zone being higher than that in the unbonded zone owing to good contact of the molten alumina. The formation of nanocrystal γ-Al_2O_3 in the remaining part of the splat may be due to the decreased cooling rate resulting from droplet solidification by the release of fusion latent heat.

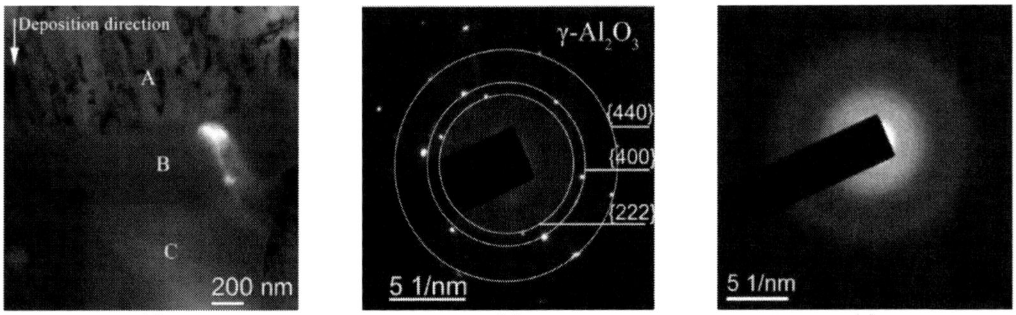

(a) (b) (c)

Figure 2. (a) Bright-field TEM image of a bonded zone in an APS-sprayed alumina coating; (b) SAED pattern corresponding to the region marked "A" in (a); and (c) SAED pattern corresponding to the region marked "B" in (a).

Amorphous structures are frequently observed in plasma-sprayed ceramic coatings owing to the rapid cooling rate of molten microsized powders [16, 17]. However, only a few studies have focused on the relationship between amorphous phase formation and intersplat bonding features. The results presented herein reveal that splat interface bonding significantly influences phase formation in the splat across the interface between bonded and unbonded regions. As mentioned above, in APS-sprayed alumina coatings, the γ phase appears in Al_2O_3 splats over the unbonded interface while the amorphous phase appears in those over the bonded interface area.

3.2. Changes in microstructure at bonded intersplat interface through heat treatment

It has been confirmed that in APS-sprayed alumina coatings, the γ phase appears in Al_2O_3 splats over the unbonded interface while the amorphous phase appears in those over the bonded interface area. This intersplat bonding features may change when the alumina coating is exposed to high-temperature conditions. Thus, heat treatment was carried out to investigate the changes in the microstructure of intersplat bonded zones resulting from high-temperature thermal exposure. The heat treatment was carried out at 1073 K for 12 h. During the treatment, the γ phase of the alumina coating was stable [13] and the amorphous alumina phase could be crystallized [18].

A bright-field TEM image of a region near the bonded interface of an APS-sprayed alumina coating subjected to heat treatment at 1073 K for 12 h is shown in Figure 3(a). In the regions at the middle and bottom of the image marked "A" and "C," respectively, the splat internal columnar sub-structure is still evident. This columnar substructure grew to the splat surface and became the interface between two splats. The columnar grains were confirmed to have the γ phase. No columnar substructure was observed between the two internal columnar substructure splats. However, some nanosized pores were observed.

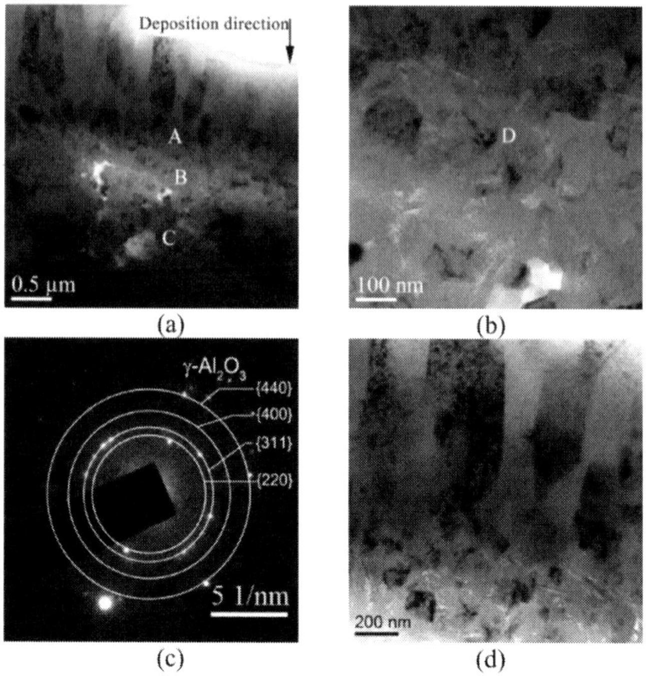

Figure 3. (a) Bright-field TEM image of the bonded zone in an APS-sprayed alumina coating subjected to heat treatment at 1073 K for 12 h; (b) magnified bright-field TEM image corresponding to the region marked "B" in (a); (c) SAED pattern corresponding to the region marked "B" in (a); and (d) magnified bright-field TEM image corresponding to the region marked "A" in (a).

In order to further examine the microstructure in the region marked "B," a magnified bright-field TEM image was obtained and is shown in Figure 3(b). A number of nanocrystalline grains can be observed in this figure. Furthermore, many nanosized pores around these nanocrystals can also be observed. The SAED pattern corresponding to the region marked "D" in Figure 3(b) is shown in Figure 3(c). The pattern clearly indicates that the nanocrystals had a γ-Al_2O_3 lattice structure. The present observations reveal that the amorphous phase of the bonded zones in the APS-sprayed alumina coating was crystallized to nanocrystalline γ-Al_2O_3 after heat treatment at 1073 K for 12 h, and this phase transformation resulted in the formation of nanosized pores. As shown by a molecular dynamics simulation [19], the amorphous phase follows the same trend in the distribution of Al coordination numbers and ring distributions as that in the liquid phase. The density of both phases is almost equal at close to 3.175 g/cm^3 [19, 20]. Further, the density of γ-Al_2O_3 is 3.66 g/cm^3 [19, 21]. During the heat treatment at 1073 K for 12 h, the amorphous phase transformed to γ-Al_2O_3, resulting in a density difference of 3.175 and 3.66 g/cm^3 for amorphous alumina and γ-Al_2O_3, respectively. Thus, the dynamics of the heating process induced a volume shrinkage of about 13% and led to the formation of nanosized pores. The microstructure at the bonded intersplat interface after the heat treatment is shown in Figure 3(d). It can be concluded that the intersplat bonding phase changed from amorphous to nanosized γ-Al_2O_3, with a less dense microstructure.

3.3. Microhardness and Young's modulus for as-sprayed and heat-treated alumina coatings

Figure 4 shows the microhardness of the as-sprayed and heat-treated alumina coatings. The as-sprayed alumina coating exhibited a mean hardness of 840 $Hv_{0.3}$. However, the microhardness of Al_2O_3 coatings increased to 982 $Hv_{0.3}$ after the heat treatment.

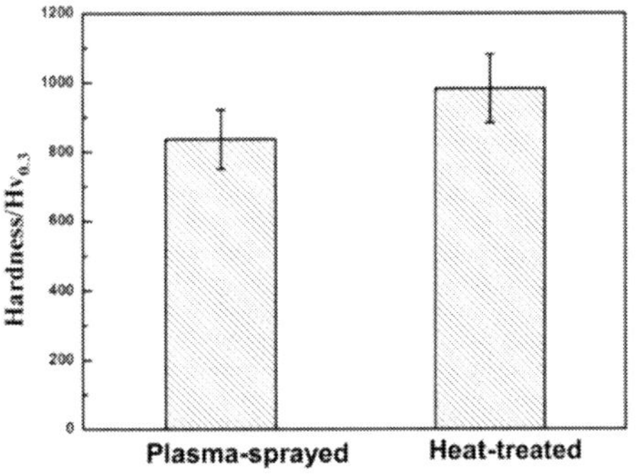

Figure 4. Comparison between microhardness of as-sprayed Al_2O_3 and that heat-treated at 1073 K for 12 h.

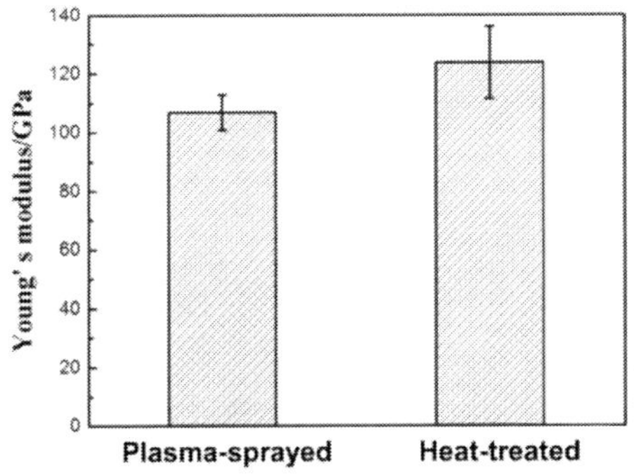

Figure 5. Comparison between Young's modulus of as-sprayed Al_2O_3 and that heat-treated at 1073 K for 12 h.

The Knoop indentation method [14, 15] was employed to measure Young's modulus of coatings deposited at different temperatures. The cross-sections of the samples were prepolished and subjected to Knoop indentation. The test was performed at a load of 1000 gf and loading time of 50 s to ensure sufficient deformation of the coating so that it yields in the direction perpendicular to the lamellae. Accordingly, obtained Young's modulus was that for the direction perpendicular to the lamellar interface. Young's modulus of the as-sprayed and heat-treated alumina coatings is shown in Figure 5. Clearly, Young's modulus of the as-sprayed alumina coating was about 107 GPa, and increased to 124 GPa after heat treatment. The above-mentioned results show that although after heat treatment, the crystalline structure near the interface changed from amorphous to γ-Al_2O_3, only a little increase occurred in coating hardness and Young's modulus. This is because the interface bonding was not altered under the annealing conditions employed and because the region in which microstructure change occurred was limited to a small area. However, when interface areas are required to transfer mechanical or thermal loading, such changes in the interface microstructure may significantly influence the coating load-bearing ability.

4. Conclusions

The influence of interface bonding on the local microstructure near the interface of the as-sprayed and heat-treated alumina coatings was clarified through HR-TEM observations. The obtained results showed that over unbonded zones, splats exhibit a γ-Al_2O_3 phase. An amorphous phase was observed in the splat over the bonded interface area. The amorphous phase layer within a splat was several hundred nanometers thick with nanocrystal γ-Al_2O_3 extended throughout the splat. After heat treatment, the phase of the bonding between adjacent splats transformed from amorphous to nanosized γ-Al_2O_3, thereby resulting in amorphous phase recrystallization during heat treatment. Moreover, nanosized pores were observed in crystallized γ-Al_2O_3; these pores were formed because of the difference in the densities of the two phases involved in transformation. The microhardness and Young's modulus of an as-sprayed alumina coating were 840 $Hv_{0.3}$ and 107 GPa, respectively, while these values increased to 982 $Hv_{0.3}$ and 124 GPa, respectively, after heat treatment. The small change in the values was attributed to the limited improvement in interface bonding after heat treatment.

Acknowledgements

The work was supported by the National Basic Research Program (Grant No. 2012CB625100) and the National Natural Science Foundation of China (Grant No. 51171144).

References

[1]. Dhiman R, McDonald A G and Chandra S 2007 *Surf. Coat. Technol.* **20** 7789
[2]. Li H, Costil S, Liao H L, Li C J, Planche M and Coddet C 2006 *Surf. Coat. Technol.* **200** 5435
[3]. Li L, Vaidya A, Sampath S, Xiong H and Zheng Li 2006 *J. Therm. Spray. Technol.* **15** 97
[4]. Sampath S and Jiang X 2001 *Mater. Sci. Eng.* A **304-306** 144
[5]. Li C J and Ohmori A 2002 *J. Therm. Spray Technol.* **11** 365
[6]. Li C J, Yang G J and Li C X 2013 *J. Thermal Spray Technol.* **22** 131
[7]. Li C J and Ohmori A 1996 *Surf. Coat. Technol.* **82** 254
[8]. Kuroda S and Clyne T W 1991, *Thin Solid Films* **200** 49
[9]. Li C J, Wang W Z and He Y 2004 *J. Therm. Spray. Technol.* **13** 425
[10].McPherson R 1973 *J. Mater. Sci.* **8** 851
[11].Goberman D, SohnY H, Shaw L, Jordan E and Gell M 2002 *Acta Mater.* **50** 1141
[12].Holleck H 1986. *J Vac. Sci. Technol.* A **4** 2661
[13].Damani R J and Mkroczy P 2000 *J. Eur. Ceram. Soc.* **20** 867
[14].Marshall D B, Noma T and Evans A G 1982 *J. Am. Ceram. Soc.* **65** 175
[15].Kroupa F and Dubsky J 1999 *Scr. Mater.* **40** 1249
[16].Tarasi F, Medraj M and Dolatabadi A, Oberste-Berghaus J and Moreau C 2013 *Surf. Coat. Technol.* **220** 191
[17].Kratschmer T and Aneziris C G 2011 *Ceram. Int.* **37** 181
[18].Tane M, Nakano S, Nakamura R, Ogi H, Ishimaru M, Kimizuka H and Nakajima H 2011 *Acta Mater* 59 4631
[19].Gutierrez G and Johansson B 2002, *Phys. Rev. B* **65** 104202
[20].Gutierrez G, Belonoshko A B, Ahuja R and Johansson B 2000 *Phys. Rev. E* **61** 2723
[21].Levin I and Brandon D 1998 *J. Am. Ceram. Soc.* **81** 1995

Interfacial state and characteristics of cold-sprayed copper coatings on aluminum substrate

Yuichiro Yamauchi and Shinya Miyaji
NHK SPRING CO., LTD. 3-10 Fukuura, Kanazawa-ku, Yokohama 236-0004, Japan

E-mail: y.yamauchi@nhkspg.co.jp

Abstract. Cold spray is a new process for bonding copper and aluminum. In this paper, the interfacial states of cold-sprayed copper on aluminum substrate were evaluated and compared with those resulting from hot extruding. No critical intermetallic layers were observed at the cold-sprayed interfaces and good electrical and mechanical properties were achieved. The effect of annealing on the characteristics of cold-sprayed bonding material was also examined. By annealing at appropriate temperature, a continuous layer of an intermetallic compound was formed and the electrical conductivity and bending strength were improved. This indicates that a good bonding state is obtained by using cold-spray technology in combination with subsequent thermal processing.

1. Introduction

Copper-aluminum bonding material is expected to be used for conductive parts, such as electrodes and connectors for electric vehicles and heavy electric machineries, and several other applications. The hot extruding method has been predominantly used as the process for bonding copper and aluminum. However, due to the formation of intermetallic compounds through the high-temperature process, it is difficult to obtain high mechanical strength and good electrical conductivity.

Recently, cold-spray technology has emerged as an intermetallic bonding method. In this process, compressed gas is injected into a Laval-barrel nozzle to produce a supersonic jet. At the same time, metal particles are fed by a turntable with nitrogen gas. Then, the metal particles are injected into the supersonic jet of compressed gas and accelerated to a high velocity of nearly Mach-2 (300–1,200 m/s) [1,2]. The accelerated particles are sprayed toward the substrate and plastic deformation occurs when colliding with the substrate. A bonding mechanism is inferred when the metal particles collide with a metal substrate. The metal bonding occurs in addition to the anchor effect caused by plastic deformation [3]. Since bonding occurs below the melting point, the bonding materials have the following characteristics, among others: no phase transformation, little oxidization, small thermal stress.

Previous studies reported that the particle deposition depends on the impact velocity. It is also suggested that the impact velocity must be higher than a certain critical velocity to obtain good bonding conditions. Below the critical velocity, colliding particles would cause only erosion of the substrate [4,5]. It is also proposed that the critical velocity depends on the combination of the particle and substrate materials [6,7]. The critical velocity is approximately 600 m/s in case of depositing copper particles on aluminum substrate. Spraying of copper particles on aluminum substrate is a relatively easy deposition process.

Content from this work may be used under the terms of the Creative Commons Attribution 3.0 licence. Any further distribution of this work must maintain attribution to the author(s) and the title of the work, journal citation and DOI.
Published under licence by IOP Publishing Ltd

In this paper, the focus is on the interface states generated by cold spraying and hot extruding, and a comparison of the resulting mechanical and electrical properties is provided.

2. Experimental procedure

A commercially available cold-spray system (PCS-305, Plasma Giken Co. Ltd., Japan), as shown in Figure 1, was employed to make copper coatings by using nitrogen gas. The cold spray parameters are listed in Table 1.

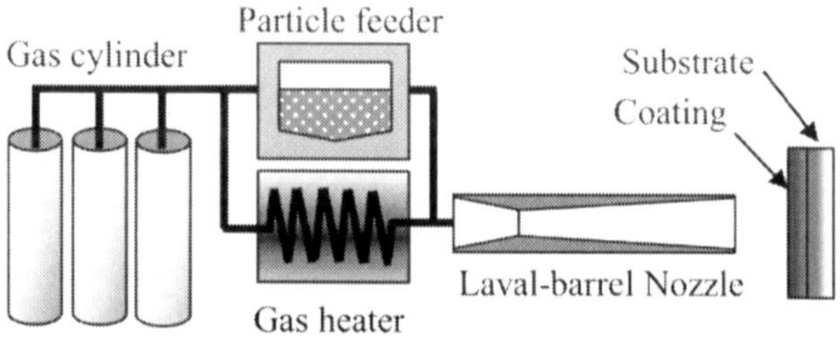

Figure 1. Schematics of the cold-spray system.

Table 1. Cold spray conditions.

Gas type	N2
Main gas temperature [K]	873
Main gas pressure [MPa]	3
Spray distance [mm]	25
Particle feed rate [g/min]	300

In this study, a pure aluminum block (100 × 100 × 50 mm^3) was used as substrate material. Non-spherical copper particles with an average particle size of 35 μm and 99.9% purity, as seen in Figure 2, were used to deposit the coatings up to 50 mm thick onto the substrate.

The specimens for cross-sectional observations of the surface were prepared by electrical discharge wire processing. Shape and size of the cross-sectional samples are shown in Figure 3 (A). Next, the specimens were examined by scanning electron microscopy (SEM, Carl Zeiss, Sigma, Germany) after being filled with resin and subsequent optical polishing. Furthermore, the samples were observed by transmission electron microscopy (TEM, JEOL, AMR-200F, Japan). In order to examine the possible generation of intermetallic compounds at the interface, sample (A) was also heat treated in vacuum. Annealing was conducted under various conditions, at temperatures ranging from 523 K to 773 K and holding times from 4 h to 300 h.

The specimens for the investigation of the mechanical and electrical properties were made with corresponding sizes and shapes as shown in Figure 3 (B) and (C). Samples of type (B) were examined by three-point bending tests after conductivity measurements using specimens of size 2 × 2 × 80 mm^3. Measurements of the electrical conductivity were carried out using the four-terminal method shown in Figure 4. The measurement distance was set at 35 mm when measuring copper conductivity. In case of measuring the conductivity of copper-aluminum including the bonding surface, the distance was set at 66 mm. As shown in Figure 5, the three-point bending tests were conducted with a displacement rate of 1 mm/min and fulcrum-gap distance of 30 mm. Samples of shape (C) were used for tensile tests, as shown in Figure 6.

The hot-extruded samples serving as comparison material were prepared from commercially available materials.

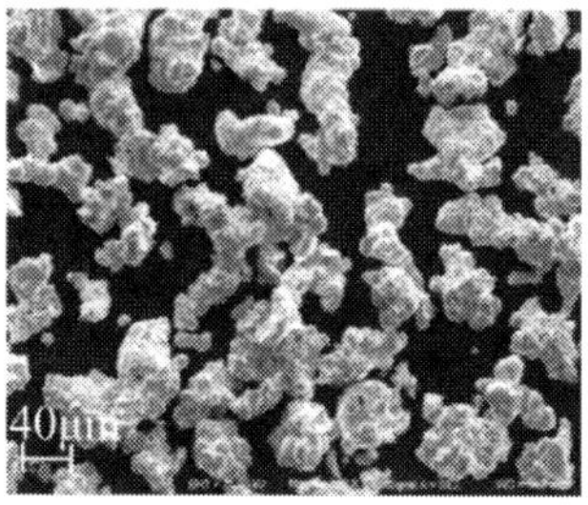

Figure 2. Morphology of copper particles.

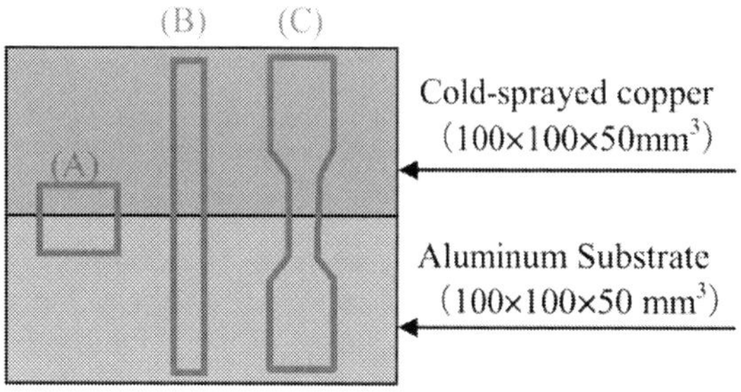

Figure 3. Shape and size of the specimens.

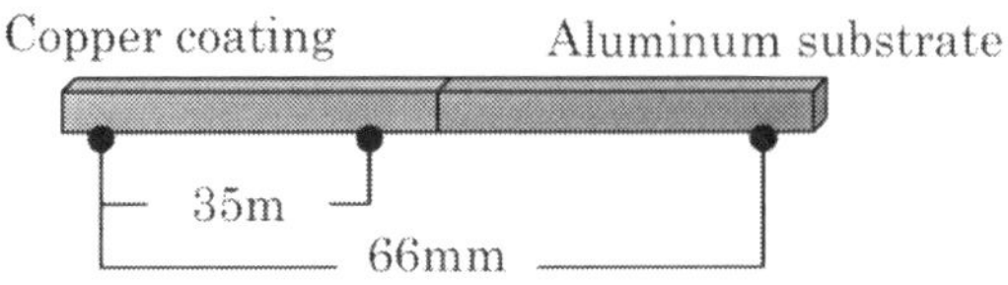

Figure 4. Layout of the conductivity measurements.

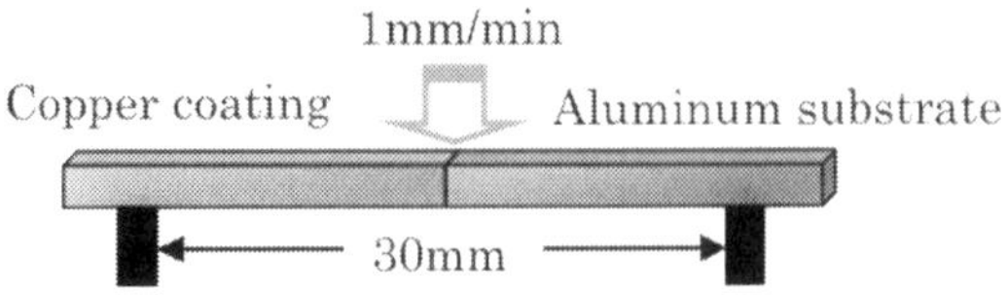

Figure 5. Schematic diagram of the bending test.

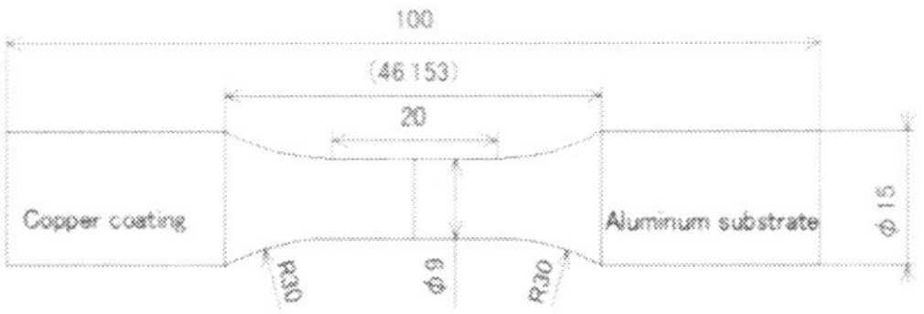

Figure 6. Shape of the tensile-test specimen.

3. Results and discussion

3.1. Cross-sectional observations

Figure 7 shows SEM images of the hot-extruded interface (A) and cold-sprayed interface (B). After the hot extruding process, discontinuous intermetallic compounds were observed at the bonding interface. On the other hand, employing the cold spray process, an intermetallic compound was not clearly observed in the SEM images. Therefore, the cold-sprayed interface was additionally examined by TEM and a continuous intermetallic layer of 200 nm in thickness was observed, as shown in Figure 8. Thus, these results clearly show that the formation of an intermetallic compound in the cold spray bonding process is suppressed compared with in the hot extruding process. Cold-sprayed interfaces were also observed after annealing, shown in Figure 9. As a result, the intermetallic layer gradually grew by high-temperature annealing in a uniform and continuous manner. Therefore, a stable bonding between aluminum and copper was not created in the hot extruding process, whereas it was confirmed that the cold sprayed interface was in a stable bonding state. Figure 10 shows the growth curve of the intermetallic compound at 523 K. According to this graph, the intermetallic compound is expected to grow up to a thickness of 15 μm after 2,000 h.

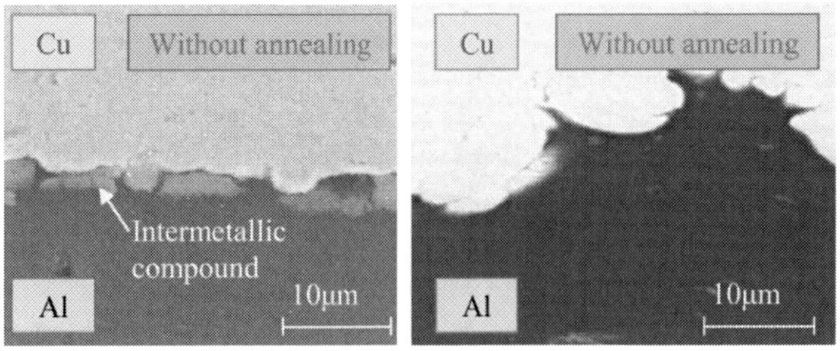

(A) Hot extruding interface (B) Cold sprayed interface

Figure 7. SEM images of bonding interface.

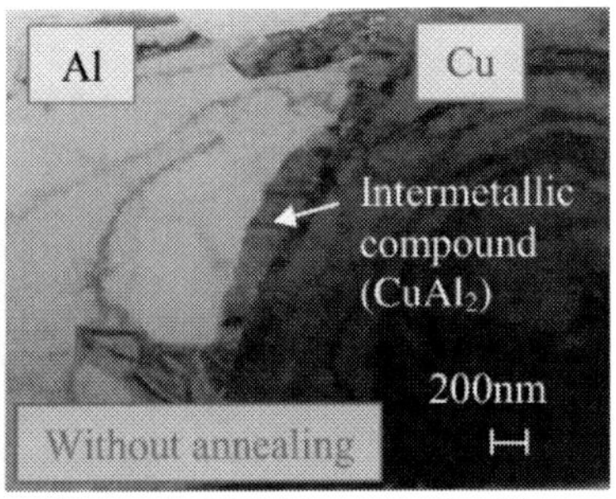

Figure 8. TEM image of cold-sprayed interface.

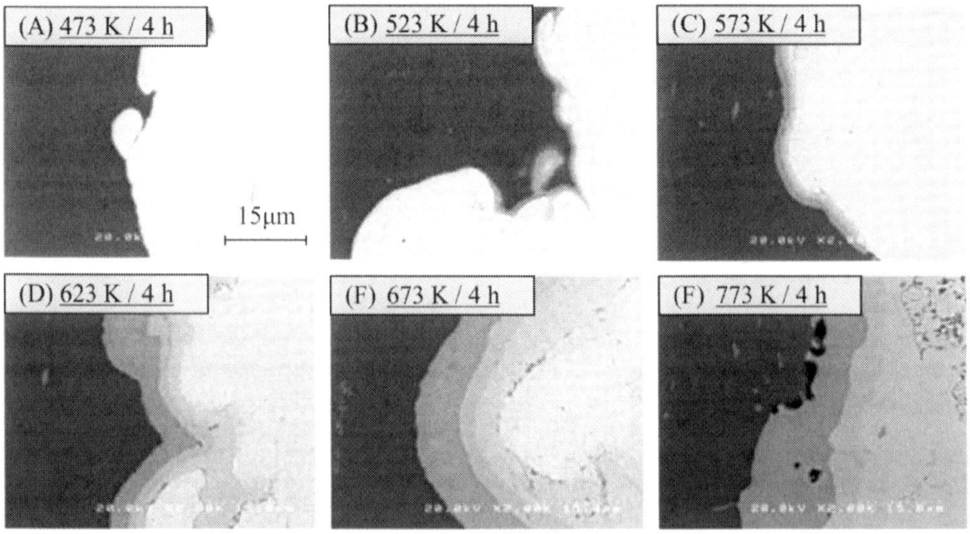

Figure 9. Growth process of the intermetallic compound at the cold-sprayed interface.

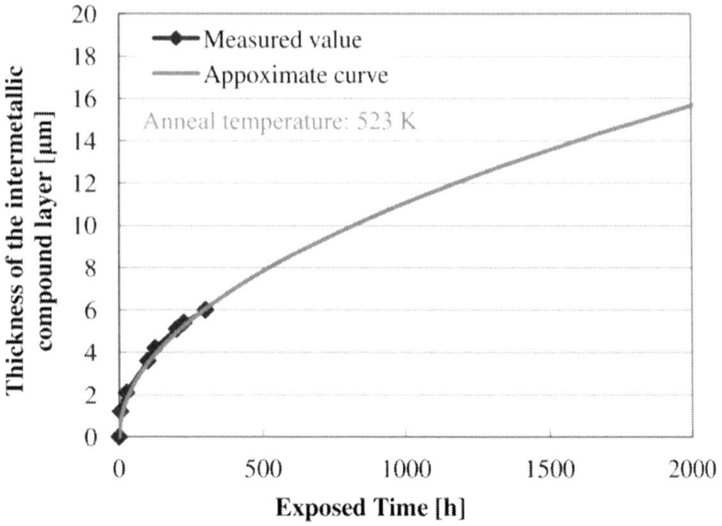

Figure 10. Growth curve of the intermetallic compounds at the cold-sprayed interface.

3.2. Mechanical properties

Figure 11 shows results of the three-point bending test of cold-sprayed copper on aluminum material. It was found that the deposited cold spray material was ruptured in the copper coating. However, if annealed at 623 K for 4 h, this material deformed plastically without fracturing. In addition, if annealed at 673 K for 4 h, the material ruptured at the copper-to-aluminum interface. It is conjectured that brittle fracture occurred by dislocations in the copper coatings without annealing. In the annealed materials, ductility was generated in the copper coatings by dislocation recovery at appropriate lower annealing temperatures, whereas interfacial fracture occurred due to the growth of intermetallic compounds by high temperature annealing.

In order to obtain further insight into the mechanical behavior, the cold-sprayed bonding material that was annealed at 623 K for 4 h was evaluated by tensile tests. The bonding interface was not ruptured and the breaking point was within the aluminum substrate, as shown in Figure 12. A mechanical strength of about 80 MPa was achieved. This breaking strength is different from the result shown in Figure 11 because of the different geometry between the bending and the tensile test.

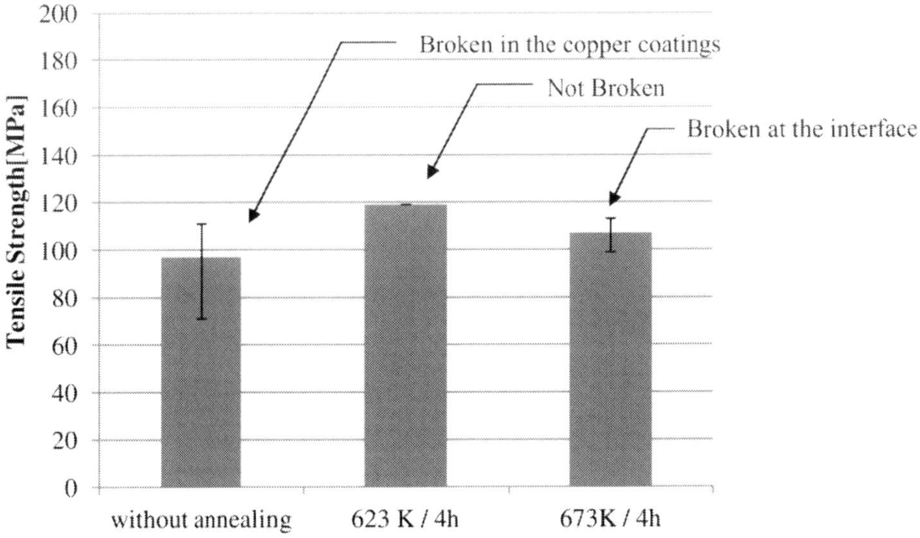

Figure 11. Test results of three-point bending tests.

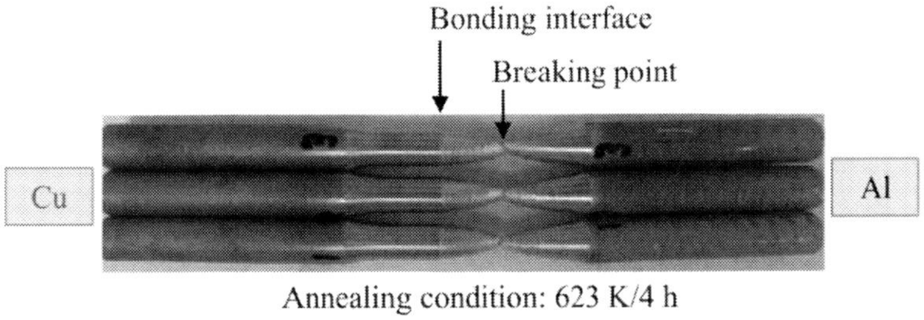

Figure 12. Photo of a specimen after tensile test.

3.3. Electrical properties

The evaluation of the electric conductivity was carried out by using the four-terminal method and the result is shown in Figure 13. The cold-sprayed copper-aluminum materials have excellent electrical conductivity, close to theoretical bonding limit. This result evidences that the thin intermetallic layer at the bonding interface has no critical effect on electrical conductivity.

In addition, the effect of the various annealing conditions on the electrical characteristics of cold-sprayed copper-aluminum bonding materials was evaluated and the result is shown in Figure 14. The figure illustrates the conductivity changes due to annealing of the copper coatings and copper-aluminum bonding material. The conductivity without annealing was fixed at 1. Consequently, the conductivity increased on higher annealing temperature in the copper coatings until 673 K, but not further at 723K. This phenomenon can be explained by the recrystallization of copper. In addition, the conductivity of the copper and aluminum bonding increased with higher annealing temperature. However, the increase rate is slightly smaller in the copper-aluminum junction than in the pure copper coatings because of the intermetallic layer grown at the interface between copper and aluminum. Note that the conductivity increased in the copper coatings by annealing.

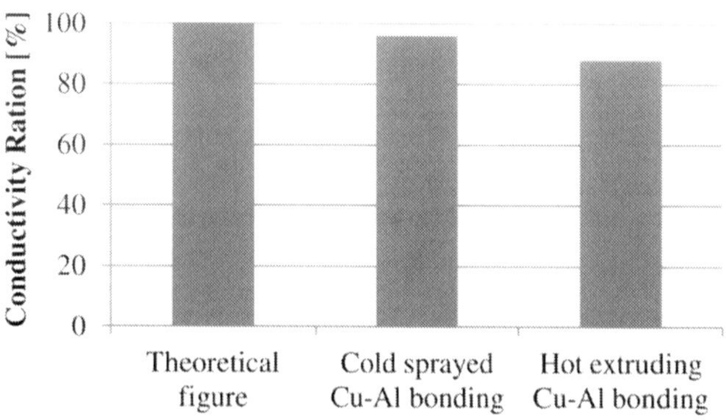

Figure 13. Comparison of the conductivities of cold-sprayed and hot-extruded Cu-Al to the theoretically expected conductivity.

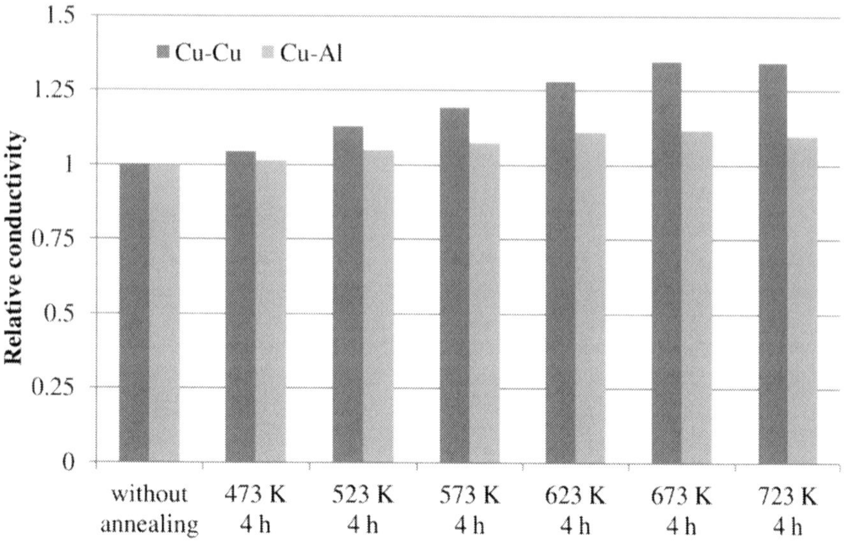

Figure 14. The conductivity changes induced by annealing.

4. Conclusions

In this study, the mechanical and electrical properties of cold-sprayed copper and aluminum bonding were evaluated with a focus on the interface and compared to those of the materials obtained by hot extruding. According to the experimental results obtained in this study, the following conclusions can be drawn:

- No critical intermetallic compound was observed in the cold-sprayed material, whereas a discontinuous intermetallic compound was observed at the hot-extruded interface.
- At the cold-sprayed interfaces, a continuous and uniform intermetallic layer was observed by TEM.
- It is estimated that the growth curve of the intermetallic layer is parabolic and that a good bonding interface is obtained.
- Using the cold spray process, superior mechanical and electrical properties can be achieved by appropriate annealing compared with the hot extrusion process.

References

[1] Alkimov A.P, Papyrin A.N et al., 1994, U.S.Patent **5**, 302, 414
[2] Huang R and Fukanuma H, 2012, *Journal of Thermal Spray Technology*, **21**(3-4), 541
[3] Ogawa K et al., 2009, *Thermal spray technology* **18**(3), 331-342
[4] Fukanuma H et al., 2005, *thin solid films,* **489**, 79-85
[5] Papyrin A.N et al., 2003, *Thermal spray 2003*, **1**, 27-35
[6] McCartney D.G et al., 2005, *J. Therm. spray technol.* **14**(1), 109-116
[7] Smith M.F et al., 1999, *Thermal spray technology*, **8**(4), 576-582

Effects of coating thickness and interfacial roughness on cracking and delamination strength of WC–Co coating measured by ring compression test

Masahiko Kato[1*], Mahmoud Nazul[2], Takeshi Itti[3], Hiroyuki Akebono[1], Atsushi Sugeta[1] and Eiji Mitani[4]

[1] Major of Mechanical System Engineering, Hiroshima University, 1-4-1 Kagamiyama, Higashi-Hiroshima, Hiroshima 739-2115, Japan
[2] Bachelor Student, Hiroshima University, 1-4-1 Kagamiyama, Higashi-Hiroshima, Hiroshima 739-2115, Japan
[3] Graduate Student, Hiroshima University, 1-4-1 Kagamiyama, Higashi-Hiroshima, Hiroshima 739-2115, Japan
[4] SSC Co., Ltd., 26-216 Ushiroji, Tomo-Chou, Fukuyama, Hiroshima 720-0202, Japan

*E-mail:mkato@hiroshima-u.ac.jp

Abstract. The effects of coating thickness and interfacial roughness on the interfacial fracture toughness of tungsten carbide–cobalt (WC–Co) coatings were evaluated using a ring compression test. WC–Co powder was sprayed on steel (JIS:SS400) rings by a high-velocity air-fuel method in coatings with various thicknesses and values of interfacial roughness. The ring compression test was carried out, and the cracking and delamination behavior of the coatings was observed using charge-coupled-device cameras. The results showed that cracking perpendicular to the loading direction occurred in the coatings during the ring compression test, and the cracking strength obtained from the ring compression test decreased slightly with increasing coating thickness, but was independent of the interfacial roughness. Upon further increase of the compression load, the coatings delaminated from the substrate. The interfacial fracture toughness calculated from the delamination of the coatings during the ring compression test decreased with increasing coating thickness and increased with increasing interfacial roughness.

1. Introduction

In order to improve wear properties of materials for mechanical parts, the tungsten carbide–cobalt (WC–Co) is frequently coated on the surface of the materials by spraying because of its superior wear resistance. However, the improvement effect disappears if the coating delaminates. Therefore, it is very important not only to develop spraying methods that yield coatings with high adhesion strength, but also to quantitatively evaluate the delamination strength of the coatings.

Many methods have been proposed to study the cracking and delamination strength of coatings [1-5]. Kendall developed a tensile testing method to evaluate the delamination strength of brittle coatings on elastic substrates with rectangular (flat) shapes by applying tensile loads to the substrates, and equations were derived to calculate the interfacial fracture toughness [6]. Nakasa et al. extended this method to elastic–plastic substrates, and Zhang et al. further developed this method for specimens subjected to four-point bending [7,8]. In practical applications, WC–Co is frequently sprayed on

Content from this work may be used under the terms of the Creative Commons Attribution 3.0 licence. Any further distribution of this work must maintain attribution to the author(s) and the title of the work, journal citation and DOI.
Published under licence by IOP Publishing Ltd

columnar or cylindrical substrates to serve as sliding bearings. However, the cracking and delamination strength of the coating sprayed on columnar or cylindrical substrates cannot be evaluated by the tensile testing method, and evaluation method of cracking and delamination strength of coatings sprayed on columnar or cylindrical substrates has not been fully established.

The coating thickness and interfacial roughness are important factors in coating design. In this study, we applied a compression load to WC–Co coating sprayed on ring-shaped substrates to evaluate the effects of the two factors on the cracking and delamination strength of the coating. The validity of this method was also assessed.

2. Experimental procedure

The substrate used in this study was a mild steel (JIS SS400) ring with an outer diameter of 100 mm and a thickness, B_2, of 4 mm. To investigate the influence of interfacial roughness, the outer surfaces of the substrates were grit-blasted with alumina powder or polished using sandpaper until the surface roughness, R_a, reached 0.062, 0.485, 3.34, and 4.50 μm. After blasting the surface of the substrate, coatings of WC–8.8 mass% Co (grain size of powder: 10–38 μm) with different thicknesses, B_1, of 0.1, 0.2, 0.5, and 1.0 mm were sprayed on the substrate to investigate the effect of coating thickness. The WC–Co coatings were sprayed using a high-velocity air-fuel (HVAF) method (pressure: 0.41 MPa; distance: 200 mm; fuel supply: 150 L/h; air flow: 12 m^3/min)

A nano-indenter (MZT-4, Akashi Co Ltd.) with a Berkovich indenter was used to measure the indentation hardness, H_1, and the indentation modulus, E_1, of the WC–Co coatings. The indentation test was carried out on the surface of a coating under a maximum load, P_{max}, of 1000 mN, with an indentation speed, v, of 33.3 mN/s. The indentation hardness and the indentation modulus of the WC–Co coatings were obtained according to the procedure in ISO14577-1 [9].

An X-ray diffractometer (MXP-3, MAC Science Co. Ltd.) was used to measure the residual stress in the coating, σ_{r1}. Characteristic X-rays from a Cr target (Cr Kα; tube voltage: 40 kV; tube current:40 mA) were selected and the residual stress was measured based on the $\sin^2(\psi - 2\theta)$ method (θ: diffraction angle; ψ: angle between the normal to the diffracting plane and the sample surface; diffraction plane: WC(112); elastic modulus: 700 GPa [10]; Poisson's ratio: 0.20 [10]).

A ring compression test was carried out using a universal tension-and-compression testing machine (AG-50kNI, Shimadzu). Figure 1 shows the schematic diagram of the experimental setup for the ring compression test. The specimen was placed on the testing machine and the compression load, P, was gradually increased. The displacement, δ, was determined from the cross-head displacement. Cameras

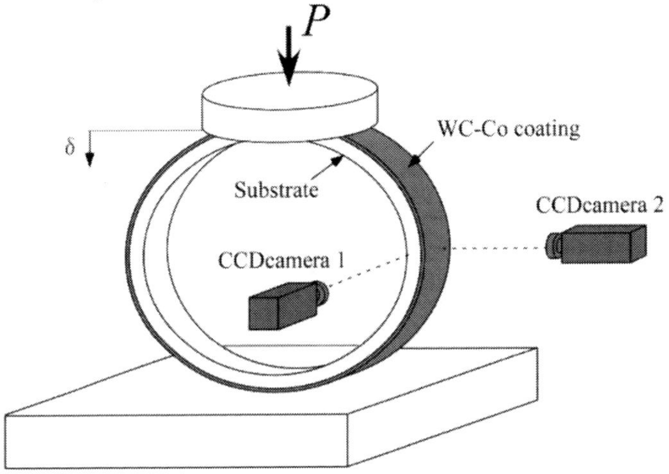

Figure 1. Schematic illustration of ring compression test.

1 and 2 captured the side and surface images of the specimen, respectively. From the images taken by camera 1, the radius of curvature of the specimen, ρ, was calculated.

The interfacial fracture toughness, G_C, was evaluated by expanding the equations derived by Kendall, Nakasa et al., and Zhang et al. [6-8]. G_C can be described by the following equation [11]:

$$G_C = \frac{E_{c1}}{6}\left(\frac{1}{\rho_{AB,d}} - \frac{1}{\rho_0}\right)^2 \left\{(\eta_{0,d} + B_1)^3 - \eta_{0,d}^{\ 3}\right\} - \frac{E_{c2}}{3}\left(\frac{1}{\rho_{C,d}} - \frac{1}{\rho_0}\right)^2 \left\{(1-\beta)\varepsilon_{ys2}^{\ 3}\rho_{C,d}^{\ 3} + \beta\left(\frac{B_2}{2}\right)^3\right\}$$
$$+ \frac{E_{c2}}{6}\left(\frac{1}{\rho_{AB,d}} - \frac{1}{\rho_0}\right)^2 \left\{2(1-\beta)\varepsilon_{ys2}^{\ 3}\rho_{AB,d}^{\ 3} + \beta(B_2 - \eta_{0,d})^3 + \beta\eta_{0,d}^{\ 3}\right\} + \frac{P_d r_x}{2b}\left(\frac{1}{\rho_{C,d}} - \frac{1}{\rho_{AB,d}}\right), \quad (1)$$

where E_{e1} and E_{e2} are the elastic constants of the coating and substrate, respectively; $\rho_{AB,d}$, $\rho_{C,d}$, and ρ_0 are the radius of curvature just before delamination, at delamination, and before loading (initial radius of curvature), respectively; $\eta_{0,\delta}$ is the distance between the interface and neutral axis at delamination; β is the ratio of the plastic constant to the elastic constant of the substrate (elastic–plastic property of the substrate is approximated to linear hardening material); ε_{ys2} is the yield strength of the substrate; P_d is load at delamination; and r_x is the radius of the neutral axis of the substrate.

3. Results and discussion

3.1. *Indentation modulus and indentation hardness*

To investigate the effect of the coating thickness, B_1, on the indentation hardness and the indentation modulus of the WC–Co coatings, indentation tests were carried out on the coatings with thicknesses ranging from 0.1 to 1.0 μm. The results are shown in Figure 2 (open circles: average value of 10 measurements; error bars: maximum and minimum values). The indentation hardness and the indentation modulus of the WC–Co coatings were both independent of the coating thickness.

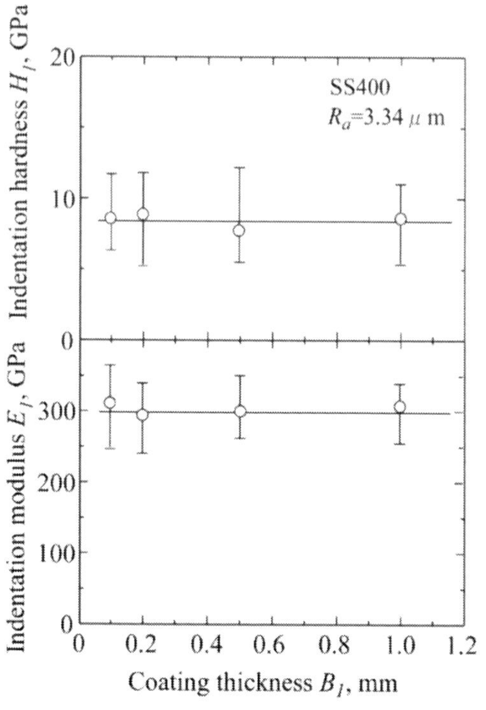

Figure 2. Effect of coating thickness on indentation modulus and indentation hardness.

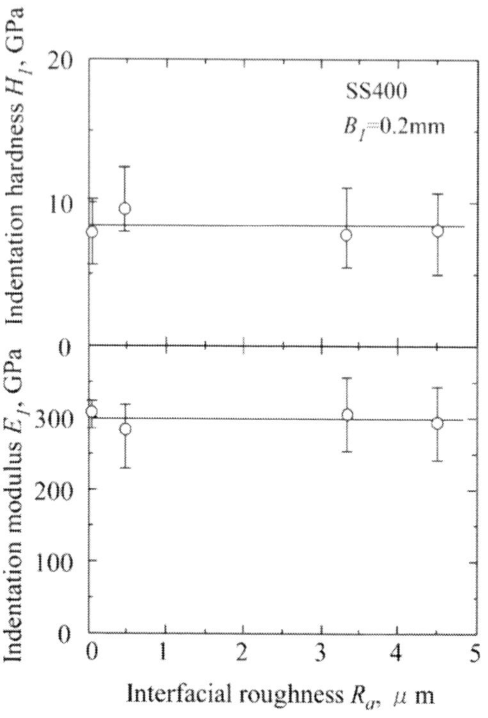

Figure 3. Effect of interfacial roughness on indentation modulus and indentation hardness.

Figure 3 shows the effect of the interfacial roughness, R_a, on the indentation hardness and indentation modulus of the WC–Co coatings. Both parameters had almost constant values irrespective of increases in the interfacial roughness. This means that
the indentation hardness and the indentation modulus of the surface of the WC–Co coatings were both independent of not only the coating thickness but also interfacial roughness.

3.2. *Residual stress*

The residual stress in the coating, σ_{r1}, shown in Figure 4(a). Compressive residual stress is introduced in the coating, and the stress becomes more compressive with increasing coating thickness. This tendency qualitatively agrees with the results reported by Stokes et al. [12]. In the coating, the cooling and quenching stresses (residual stresses) arose because of the thermal mismatch during cooling after spraying was finished, as well as the impact and quenching of individual lamella [12,13]. Stokes *et al.* analytically explained the effect of coating thickness on the residual stresses [12]. On the other hand, the residual stress of the specimens with different interfacial roughness was around -600 MPa, independent of the interfacial roughness, as shown in Figure 4(b).

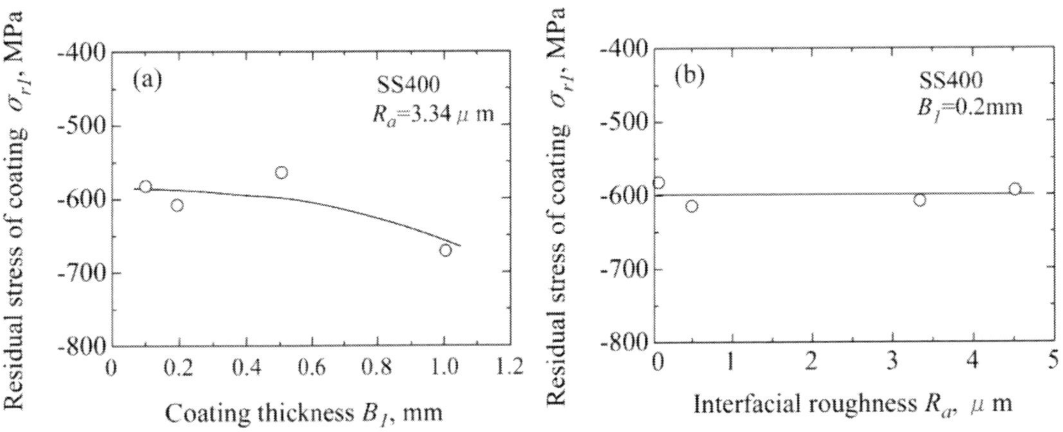

Figure 4. Effects of (a) coating thickness and (b) interfacial roughness on residual stress.

3.3. *Ring compression test*

3.3.1. Cracking and delamination of coating. Figure 5 shows a typical example of the coating surface during the ring compression test. When the load reaches a critical value after the beginning of the plastic deformation of the substrate, cracking of the coating occurs perpendicular to the loading direction (Figure 5b). The number of cracks increases with increasing load (Figure 5c–e), and the coating delaminates from the substrate when the load reaches the critical value, P_d (Figure 5f).

The surfaces of the coatings with different thicknesses at the beginning of delamination are shown in Figure 6. When the coating is thin (Figure 6a), many cracks are introduced in the coating and small area of the coating delaminates. The number of cracks decreases with increasing coating thickness. When the coating becomes thick (Figure 6c and d), only one or two cracks are observed. The coating does not peel off but is lift up from the substrate. The surfaces of the specimens with different interfacial roughness just before delamination are shown in Figure 7. Almost the same cracking behavior is observed for every specimen.

3.3.2. Cracking strength of the coating. The cracking strength, σ_{1C}, was calculated from the following equation:

$$\sigma_{1C} = E_1\left(\frac{1}{\rho_i} - \frac{1}{\rho_0}\right)\left(B_1 - \frac{B_2}{2}\right), \tag{2}$$

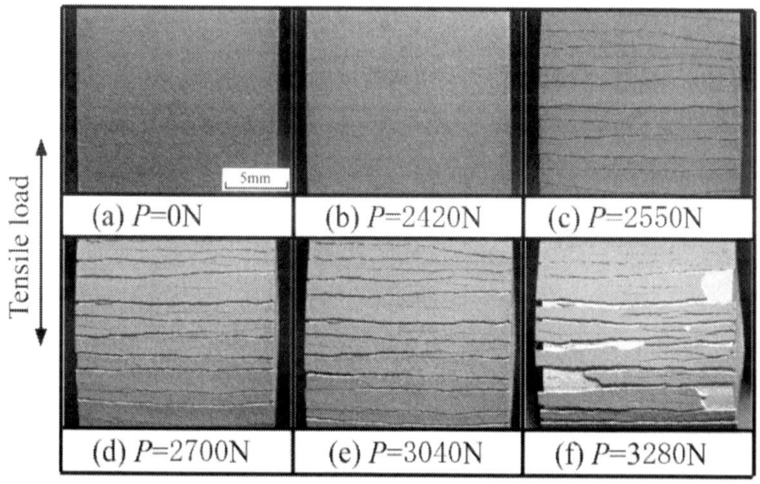

Figure 5. Surface of WC–Co coating during ring compression test (substrate: SS400; B_l = 0.2 mm, R_a = 3.4 μm).

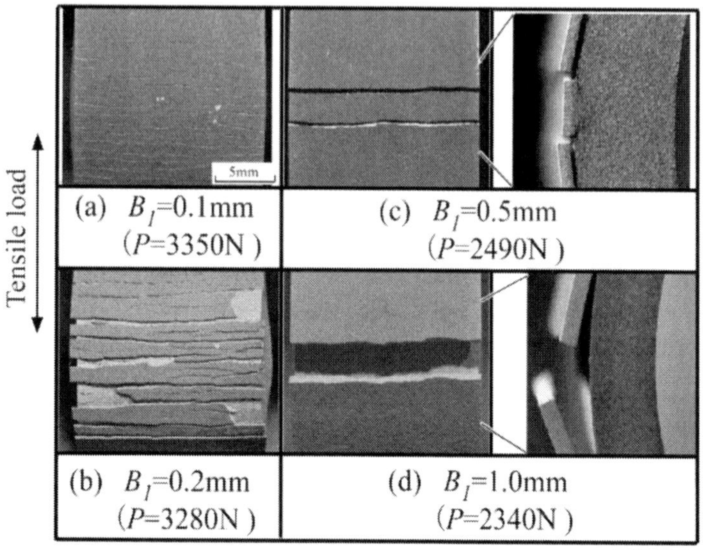

Figure 6. Surface of WC–Co coating at delamination (substrate: SS400; R_a = 3.4 μm).

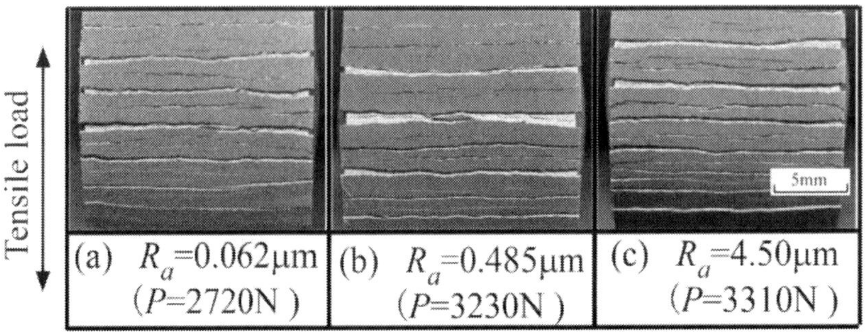

Figure 7. Surface of WC–Co coating at delamination (substrate: SS400; B_1 = 0.2 mm).

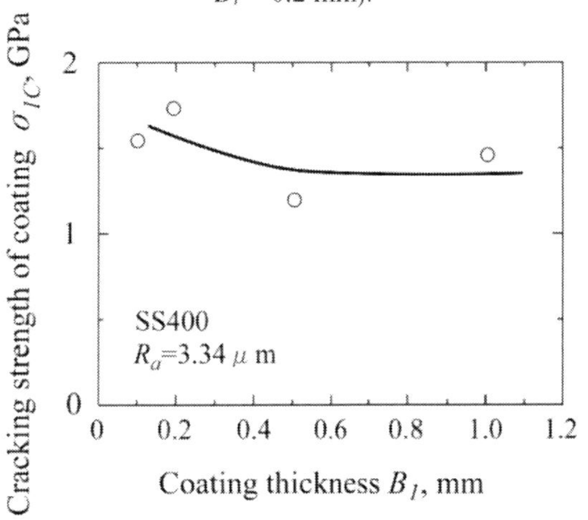

Figure 8. Relationship between coating thickness and cracking strength of coating.

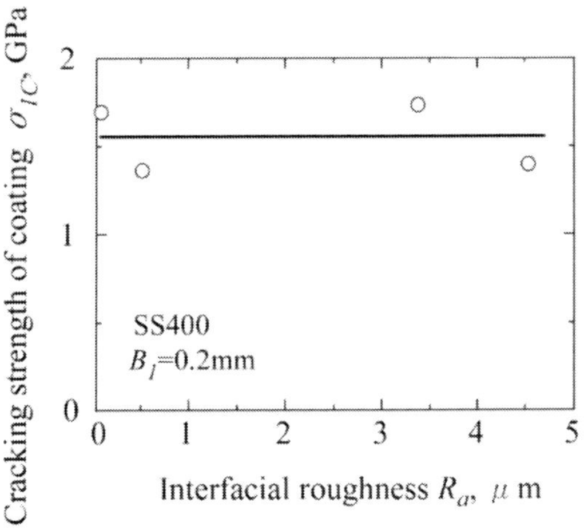

Figure 9. Relationship between interfacial roughness and cracking strength of coating.

where ρ_i is the radius of curvature at crack initiation. The crack initiation was detected using the images taken by camera 2 shown in Figure 1.

The relationship between the coating thickness and the cracking strength of the coating is shown in Figure 8. A small decrease in cracking strength is observed with increasing coating thickness.

Figure 9 shows the relationship between the interfacial roughness and cracking strength of the coating. The cracking strength ranges between 1.3 and 1.8 GPa, and the obvious effect of the interfacial roughness is not observed. According to the results reported by Stokes et al., the amount of the residual stress changes with increasing coating thickness [12]. On the other hand, the influence of interfacial roughness on the residual stress is considered to be negligibly small as the residual stress is almost constant under the same coating thickness. Therefore, it is implied that main factor affecting the cracking strength of our specimens is the residual stress, which causes damage to the coatings and changes the stress distribution in the specimens.

3.3.3. Interfacial fracture toughness. The interfacial fracture toughness of the specimen with various coating thicknesses calculated using Equation (1) is shown in Figure 10. E_{e1} and E_{e2} were measured using the nano-indenter, and β was obtained from the load-displacement curve. The interfacial fracture toughness decreases with increasing coating thickness. Similar to the cracking strength, it can be said that the interfacial fracture toughness of the specimens is affected by the residual stress.

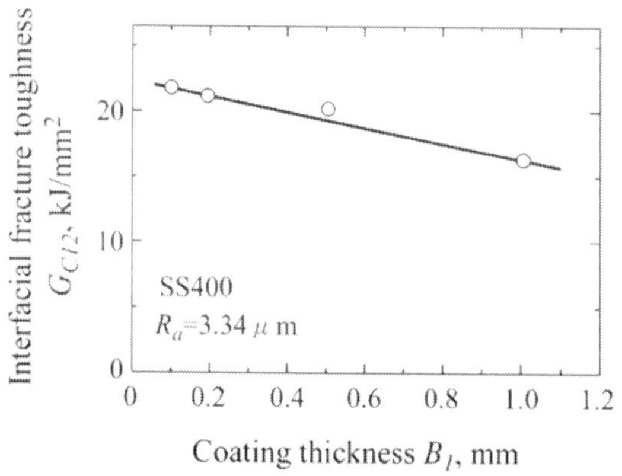

Figure 10. Relationship between coating thickness and interfacial fracture toughness.

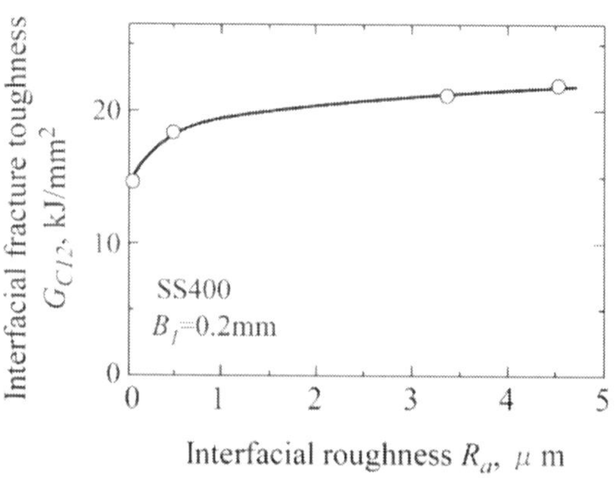

Figure 11. Relationship between interfacial roughness and interfacial fracture toughness.

According to the stress analysis at the interface, both tensile and shear stresses are introduced (mixed mode) [14]. In Figure 6, the delaminated coating with a thickness (d) of 1.0 mm recovered almost elastically. The recovery force (mode 1 component) increase with increasing coating thickness based on the theory of elasticity, and hence the mode ratio (mode 1/mode 2) increases with increasing coating thickness. Therefore, it is considered that the change of the mode ratio is another factor affecting the interfacial fracture toughness of the specimens.

The interfacial fracture toughness increases with increasing interfacial roughness, and the effect is shown in Figure 11. These results qualitatively agree with those reported by Fukumoto et al. [15], who noted that the increase resulted from both the anchor effect and the physical adhesion effect. Moreover, the value of the interfacial fracture toughness obtained in this study (15–22 kJ/m^2) is close to that measured by a four-point bending test [8].

From this study, we conclude that the ring compression test is a valid means to evaluate the effects of coating thickness and interfacial roughness on the cracking and interfacial fracture toughness of WC–Co spray-coated using the HVAF method.

4. Conclusions

To evaluate the effect of coating thickness and interfacial roughness on the cracking strength and interfacial fracture toughness of WC–Co coatings, a ring compression test was carried out on the WC–8.8 mass% Co coating formed on mild steel substrate using the high-velocity air-fuel (HVAF) spraying method. The results are summarized as follows:

(1) The indentation hardness and indentation modulus measured by the nano-indenter were independent of both the coating thickness and interfacial roughness

(2) The residual stress in the coating, measured by X-ray diffraction, was compressive, and the amount of stress increased with increasing coating thickness.

(3) During the ring compression test, cracking perpendicular to loading direction and delamination of the coatings occurred. The number of cracks decreased with increasing coating thickness.

(4) The cracking strength decreased slightly with increasing coating thickness, but was constant irrespective of interfacial roughness.

(5) The interfacial fracture toughness decreased with increasing coating thickness and increased with increasing interfacial roughness. The measured interfacial fracture toughness ranged between 15 and 22 kJ/m^2.

(6) The ring compression test is applicable to the evaluation of the effects of coating thickness and interfacial roughness on the cracking and interfacial fracture toughness of WC–Co coatings sprayed using the HVAF method.

References

[1] ISO 14916 1999 "Thermal spraying - Determination of tensile adhesive strength"
[2] ISO 4624 2002 "Paints and varnishes -- Pull-off test for adhesion"
[3] ISO 1520 2006 "Paints and varnishes -- Cupping test"
[4] ISO 1519 2011 "Paints and varnishes -- Bend test (cylindrical mandrel)"
[5] ISO 2409 2013 "Paints and varnishes -- Cross-cut test"
[6] Kendall K 1975 *Proceedings of the Royal Soc. Lond.* A.344 287
[7] Nakasa K, Takata S, Ichigo H 1995 *J. Soc. Mater. Sci. Japan* **44** 321(in Japanese)
[8] Zhang D, Kato M, Nakasa K 1999 *J. Soc. Mater. Sci. Japan* **48** 636(in Japanese)
[9] ISO 14577-1 2002 "Metallic materials -Instrumented indentation test for Hardness and materials parameters- Part 1: Test method"
[10] Pina J, Dias A, Lebrun J L 2003 *Mater. Sci. Eng.* **A347** 21
[11] Kato M, Itti T, Akebono H, Sugeta A, 2011 *Proc. 15th Symp. on Frac. Mech.* 274(in Japanese)
[12] Stokes J, Looney L 2004 *Surf. and Coat. Tech.* **177-178** 18
[13] Clyne T W, Gill S C 1996 *J. Therm. Spray Tech.* **5** 401
[14] Toya M, Ono T, Miyawaki T, Kirioka K 1991 *J. Japan Soc. Mech. Eng.* **A57** 1822(in Japanese)
[15] Fukumoto m, Murakami H, Okane I, 1994 *J. Japan Therm. Spray Soc.* **31** 130(in Japanese)

Surface modification of graphite and ceramics with metals using induction heating

Toshi-Taka Ikeshoji[1], Akiko Imoto[1], Akio Suzumura[1], Mana Katori[1,2], Takahisa Yamazaki[1], Masahiro Sakamoto[3] and Satoshi Sakimichi[3]

[1] Mechanical and Aerospace Engineering, Tokyo Institute of Technology, 2-12-1 O-okayama, Meguro-Ku, Tokyo, 152-8552, Japan
[2] LMS, A Siemens Business
[3] DENSO CORPORATION

ikeshoji.t.aa@m.titech.ac.jp

Abstract. In order to join metals to graphite or ceramics by soldering or brazing, a new surface modification method using induction heating was developed for graphite and ceramics. Such source metals as Cu, Ni, Cr, etc. were induction-heated in vacuum atmosphere and making deposited films on the deposition substrate, or the target substrate: graphite, AlN, Si_3N_4. The applicability of this method was investigated and the deposited layer was analysed by SEM observation, Auger electron spectrum analysis, X-ray diffractometry, and EPMA. By comparison of ambient vacuum pressure during deposition and the saturated vaopr pressure of source metals, this method was considered to utilize the sublimation phenomenon.

1. Introduction

Recently, the power electronics module with higher voltage and larger current is demanded, and it is exposed to be used the ceramics semiconductors, *e.g.* SiC, AlN, or GaN, which should be cooled to proper temperature [1]. One of the cooling approaches is soldering ceramics semiconductors chip to graphite with high heat conductivity. The ceramics and graphite, however, have low wettability to molten liquid metals [2] and are difficult to join by soldering. Therefore the surface modification of their surface to wet with the molten metals is required. In this research, a new method for the metalizing ceramics and graphite is proposed. The ceramics and graphite are heated with the source metals simultaneously in the alternating magnetic field in vacuum. They are heated up at the temperature lower than the melting point of the source metals. In this work, the possibility of metallizing by the induction heating (IH) is newly clarified for the various combinations of the target materials and the source metals.

2. Experimental details

Deposition substrates were graphite, AlN and Si_3N_4 with the dimensions of 10×10×3 mm, and source metals were Al, Cu, Cr, Ni, Ti and 18-8 austenitic stainless steel (in the following, SUS304). Both deposition substrates and source metals were cleaned in the ultrasonic cleaner with ethanol bath. As shown in figure 1, a target plate was put on a metal source plate, where a W foil spacer was inserted between the mating surfaces. For SUS304 source heated lower than 1623 K, no spacer was inserted. The specimens were heated by induction heating in the vacuum atmosphere of 1.33×10^{-2} Pa by applying approximately 10 kHz AC current. The heating temperature and the holding time are shown in table 1.

Content from this work may be used under the terms of the Creative Commons Attribution 3.0 licence. Any further distribution of this work must maintain attribution to the author(s) and the title of the work, journal citation and DOI.
Published under licence by IOP Publishing Ltd

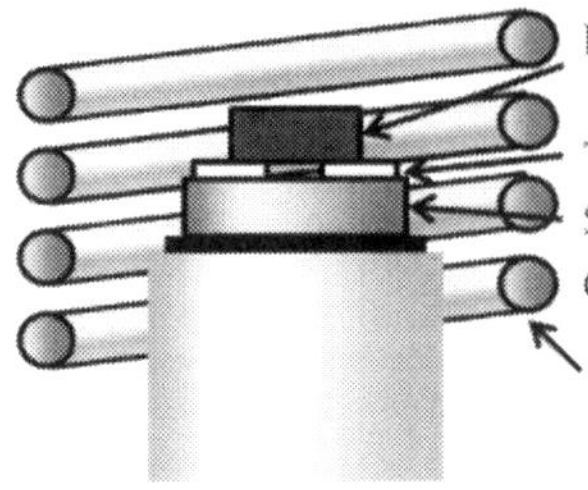

Figure 1. Schematics of experimental apparatus.

Table 1. Conditions of inductions heating.

Source Metals	Melting Point, K	Holding Temp., K	Holding Time, s
Al	933	893	2700, 3600
Cu	1358	1203	900, 3600
Ni	1728	1473, 1623	1800
Cr	2180	1473	900
Ti	1941	1623	1800
SUS304*	1673 --1723	1473	60, 300

* : 18-8 austenitic Stainless Steel,

For comparison, the specimens in the same configuration were heated in a vacuum furnace by radiation heating.

The deposited layer on the target surface was observed by SEM, and analyzed using the Auger electron spectroscopy (AES) and EPMA. Some deposited layer was analyzed using XRD.

3. Results

The appearance of deposited layer of SUS304 source and Cu is compared by heating methods in figure 2. On the case of SUS304 source, both the deposited surfaces by induction heating and radiation heating were covered by the light gray material with slight difference. On the other hand, by Cu source, apparently, the amount of the deposition was different by heating methods. The deposited layer by induction heating seemed thicker. These results suggest that the deposition ability will different among the combinations of deposition substrates and source metals.

It is shown in table 2 that the deposition ability by this induction heating method with variations in deposition substrates and source metals. By Al source, no deposition was observed on any deposition substrates in appearance. By Cu source and Ni source, pure Cu and pure Ni were deposited on all the deposition substrates respectively, which was identified by AES. By Cr source, pure Cr was deposited on graphite plate, which was identified by XRD. By Ti source, XRD revealed that pure Ti and TiC were deposited on graphite substrate (figure 3(a)). By SUS304 source, Cr_7C_3 was deposited on graphite surface. It was identified by XRD (figure 3(b)). EPMA analysis indicated the compositions of the deposited layer by SUS304 source was 42.2Fe-37.7Cr-9.7Mn (in mass%) without Ni.

Secondary electron images of the deposited surfaces are shown in figure 4. On the case of Cu source, the round mounds were observed. It was considered that the deposited particles grew and coalesce each other. On the case of Cr source and Ti, the growth of the crystals was apparently observed. By Ni source and SUS304, the fine and random morphology was formed.

Figure 5 shows the difference in morphology among the combinations of sources and substrates. Both Cu source and Ni was not deposited in the form of crystal growth on all substrate. But on the case of Cu source on Si_3N_4, short whiskers grew.

4. Discussions

The deposition ability among the combinations of source metals and deposition substrates attempted in this research could be explained by the comparison of ambient pressure and saturated vapor pressure. In figure 6, the curves of saturated vapor pressure of source metals with temperature are plotted. The

Table 2. Applicability of deposition and products in deposited layers.

		Deposition substrates		
		Graphite	AlN	Si_3N_4
Source metals	Al	Not deposited	Not deposited	Not deposited
	Cu	Pure Cu [1]	Pure Cu [1]	Pure Cu [1]
	Ni	Pure Ni [1]	Pure Ni [1]	Pure Ni [1]
	Cr	Pure Cr [2]	--	--
	Ti	Pure Ti, TiC [2]	--	--
	SUS304*	Cr_7C_3 [1,2,3]	--	--

* : 18-8 austenitic stainless steel, 1 : AES, 2 : XRD, 3 : EPMA

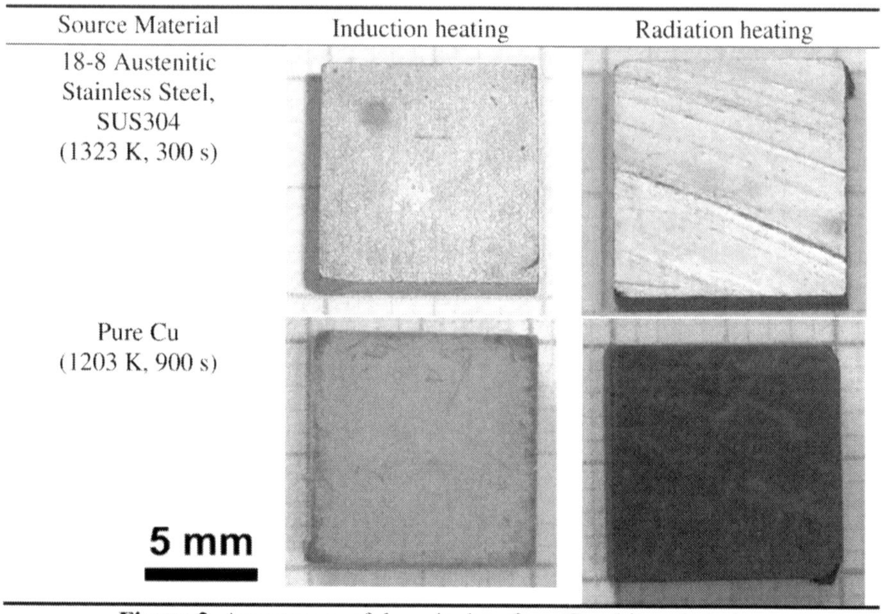

Figure 2. Appearance of deposited surface. Graphite substrates.

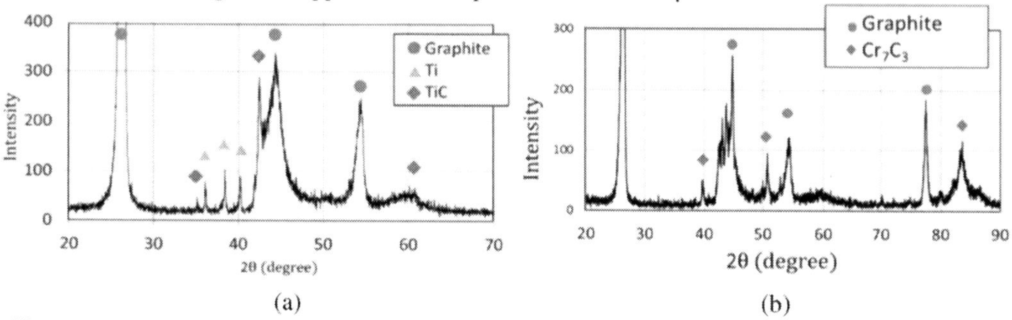

Figure 3. X-ray diffractometry spectrum of the deposited layer on graphite substrate by (a) Ti source, and (b) 18-8 stainless steel, SUS304.

hatched box at 10^{-3} – 10^{-2} Pa indicates the range of ambient vacuum pressure. For Al source, its vapor pressure at the deposition temperature was lower than the ambient vacuum pressure. Therefore, Al was not deposited on all the substrates. On the other hand, for Cr, Ni and Ti sources, their vapor pressure values were higher than the ambient vacuum pressure, and their deposited layer was obtained for all the substrates. On the case of 18-8 austenitic stainless steel, SUS304, Ni's vapor pressure value at the

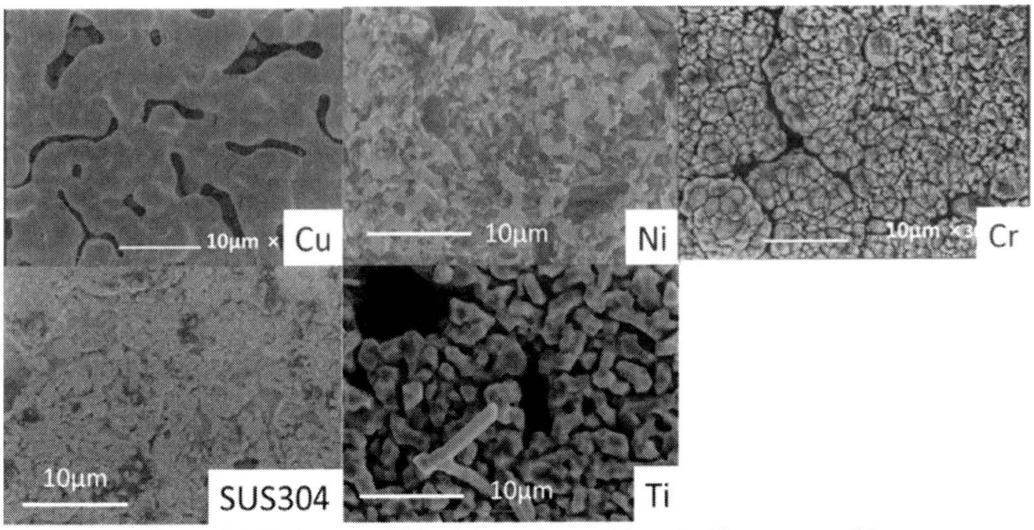

Figure 4. SEM images (SEI) of the surfaces of deposited layer on graphite.

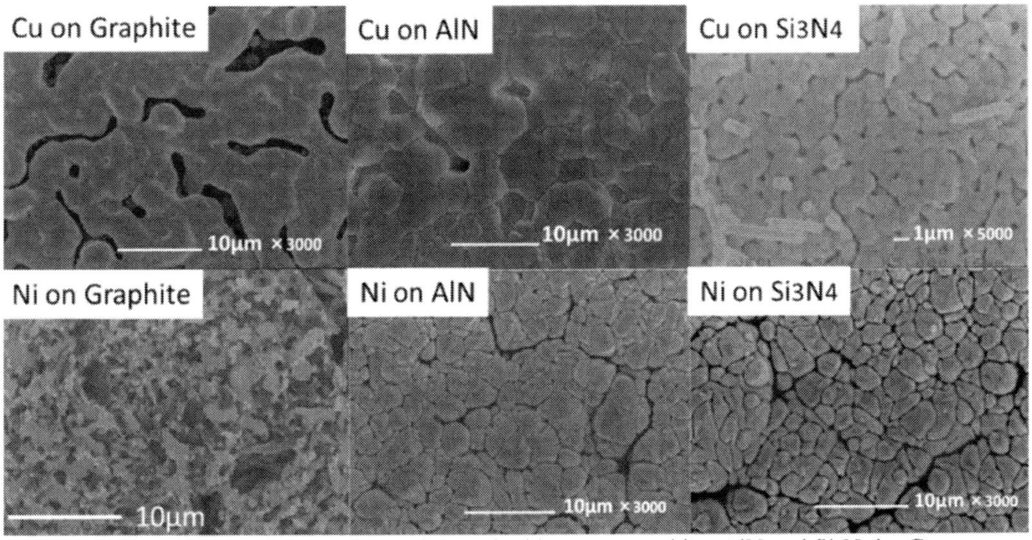

Figure 5. SEM images of the surfaces of deposited layer on graphite, AlN and Si_3N_4 by Cu source and Ni.

deposition temperature is an order of magnitude lower than Cr's value. So, the deposition layer by SUS304 might not contain Ni. As just described, because the deposition ability can be explained by the saturated vapor pressure, it can denote that the deposition mechanism of this method by induction heating utilizes the sublimation of metals.

This surface modification method using induction heating can be considered a type of PVD. A different point between this method and other PVD methods is that the evaporating method of source metal is the sublimation of metals. That means the source metals are not necessary to melt. Another different point is the distance between a source metal and a substrate is very small or none. According to these different points, this surface modification method using induction heating might have an advantage that the process could be conducted at relatively lower temperature with high deposition rate. The process temperature of this method might not be lower than that of sputter deposition, electron beam deposition, and laser beam deposition, but it will be lower than evaporative deposition by electrical

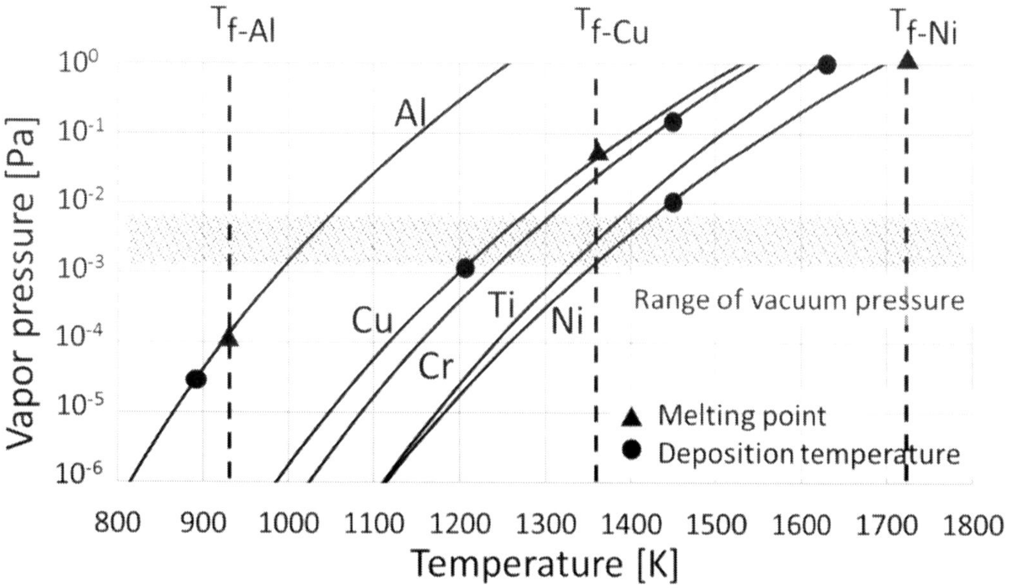

Figure 6. Saturated vapor pressure of source metals

resistive heating. The deposition rate of this surface modification method requires farther research, but it is expected to be high due to the narrow gap between a source metal and a substrate.

5. Conclusions

It was proposed that the new surface modification method for graphite and ceramics using the induction heating. The combinations of source metals and deposition substrates were examined and the applicability of this method was shown. And the following conclusions were obtained.

(1) The deposition did not depend on the target materials but the source materials.
(2) High vapor pressure of source metals makes deposition easier and it suggested this modification was conducted through the sublimation from the solid source material.
(3) Deposition films of Graphite/SUS304 and Graphite/Ti were detected carbide. Deposition films of Cu, Cr and Ni were pure.

References

[1] Komaki Y 2013 *JETI : Japan energy & technology intelligence*, **61** 65 *in Japanese*.
[2] Weltsch Z and Lovas A 2013 *Acta Physica Polonica A* **124** 78.

IOP Conf. Series: Materials Science and Engineering **61** (2014) 012026 doi:10.1088/1757-899X/61/1/012026

Interfacial structure of multi-layered thin-films produced by pulsed laser deposition for use in small-scale ceramic capacitors

Takao Araki[1], Takanori Hino[2] and Masahiro Ohara[3]

[1] Professor, Institute for International Relation, Ehime University, Matsuyama Japan
[2] Associate Professor, Department of Environmental Materials Engineering, Niihama National College of Technology, Niihama, Japan
[3] Professor, Department of Materials Science and Biotechnology, Graduate School of Science and Engineering, Ehime University, Matsuyama, Japan

E-mail: t-alucky@mb.pikara.ne.jp

Abstract. The aim of this study was to develop thin film capacitors with superior properties that could provide an alternative to materials currently used in conventional multi-layer ceramic capacitors fabricated by sintering. To this end, an artificial dielectric super lattice technique, incorporating pulsed laser deposition, was applied to improving the dielectric properties of thin film capacitors. This method permits the A-site atoms of a perovskite ABO_3 structure to be selected layer by layer at a nanoscopic scale; consequently, multi-layer $BaTiO_3$-$SrTiO_3$ thin films were produced on $Pt(111)/Ti/SiO_2/Si(100)$ and $SrTiO_3(111)$ substrates. Hetero-epitaxial grain growth was observed between $BaTiO_3$ and $SrTiO_3$, with the lattice mismatch between them introducing a compressive residual strain at the interface. The dielectric properties of these multi-layer thin-film capacitors were found to be superior to those of conventional solid-solution thin films once the thickness of the layers and the ratio of the two oxides were optimized.

1. Introduction

Barium titanate ($BaTiO_3$) is widely used as a dielectric material in a variety of everyday applications, including memory IC cards, chip capacitors and flash memory. Chip capacitors, otherwise known as multi-layer ceramic capacitors (MLCCs), are of particular importance as these are used in most compact electronic devices such as laptop computers, PDAs, and mobile phones. In the case of passive components, which include MLCCs, there is a high demand for an improvement in the dielectric constant and a reduction in the layer thickness in order to reduce their physical size. This is hindered, however, by the fact that a thickness of around 1 μm represents the technical limit for conventional powder metallurgy processes. The development of improved thin film technologies is therefore essential to the development of next-generation capacitors.

Although thinner dielectric layers can be easily produced by physical or chemical vapor deposition techniques, the dielectric characteristics of MLCCs formed in this way are usually not as good as those of bulk ceramics produced by powder metallurgy processes. These dielectric characteristics have been

Content from this work may be used under the terms of the Creative Commons Attribution 3.0 licence. Any further distribution of this work must maintain attribution to the author(s) and the title of the work, journal citation and DOI.
Published under licence by IOP Publishing Ltd

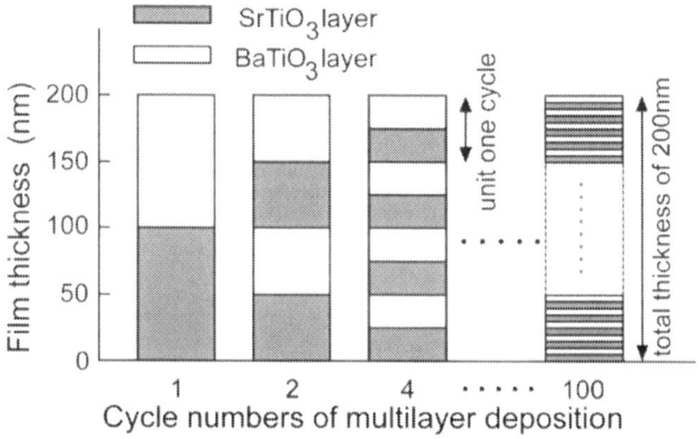

Figure 1. Cross-sectional illustration of a multi-layer $(BaTiO_3)_{1-x}(SrTiO_3)_x$ thin film.

somewhat improved through forming solid-solution films; but even so, equivalent properties are still far from being achieved.

The discovery that the strain induced by the formation of super lattice films improves their dielectric properties has prompted a considerable amount of research into this field [1-2]. Typically, these super lattice films are produced through layer-by-layer growth of artificial crystals at a nanoscopic scale [1-2]. In the present study, a pulsed laser deposition (PLD) technique was utilized to disintegrate bulk materials into atoms, which were then deposited on a suitable substrate to form super lattice films. In this way, multi-layer $BaTiO_3$-$SrTiO_3$ thin films were produced and their dielectric characteristics were evaluated.

2. Experimental procedure

The deposition targets were prepared by sintering compacts, which consisted of high purity $BaTiO_3$ (BTO onward) and $SrTiO_3$ (STO) powders. Single layer films of $Ba_{1-x}Sr_xTiO_3$ (BSTO) and multi-layer films of $(BaTiO_3)_{1-x}(SrTiO)_x$ (BTO-STO) were then fabricated by irradiating a KrF excimer laser onto the sintering targets, and ablating particles from the target onto the substrate [3-4]. The two types of single-crystal substrate were prepared as 1.0x1.0x0.5 mm^3 wafers of Pt(111)/Ti/SiO_2(100) by DC sputtering onto a 1%Nb-doped single-crystal $SrTiO_3$(111) substrate measuring 0.5x0.5x0.5 mm^3.

The BSTO and BTO-STO films were each deposited onto individual substrates, which were heated to a temperature of 973 K in a O_2-8%O_3 atmosphere under gas pressures of 1 and 8 Pa. The laser was irradiated onto the rotating target at a fluency of 3.0 J/cm^2 and a repeated pulse frequency of 2 Hz. BTO-STO super lattice films were made to different layer thicknesses, while still maintaining a total film thickness of 200 nm, as shown Figure 1. The thickness of each BTO and STO layer was determined by the number of layers deposited and by the overall proportion of STO, i.e., x in $(BaTiO_3)_{1-x}(SrTiO_3)_x$; with films being produced to x values of 0.25, 0.50 and 0.75. The thickness of each layer was controlled by the number of laser pulses. The $Ba_{1-x}Sr_xTiO_3$ single-layer thin films were produced using targets made from powders of BTO and STO mixed to ratios of 0.00, 0.25, 0.50, 0.75 and 1.00. The thickness of all films produced was assessed by surface roughness measurement and atomic force microscopy (AFM), with the cyclic layer thickness being estimated from the satellite peaks obtained by X-ray diffraction (XRD).

The crystal structure of the films was also examined by XRD, using CuKα radiation and a scan speed of 1°/min across a2θ angle range of 20° to 120°. Electrical measurements were conducted using a configuration of platinum/BTO film/STO substrate, the platinum electrodes being deposited by sputtering through a metal shadow mask with an area of 0.25 mm^2. The capacitance and dielectric loss tangent were measured by a LCR meter (HP4284A) at frequencies ranging from 100 Hz to 1 MHz and temperatures of between 250 to 400 K. Dielectric constants were calculated by the following equation:

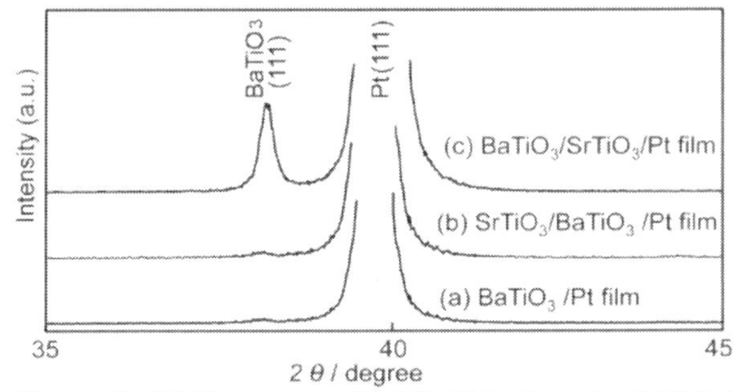

Figure 2. XRD patterns of (a) $BaTiO_3$ film, (b) $SrTiO_3$/$BaTiO_3$ film and (c) $BaTiO_3$/$SrTiO_3$ film deposited on Pt/Ti/SiO_2 substrates.

$$Cp=\varepsilon_0 \cdot \varepsilon_r \cdot S/t, \qquad (1)$$

where Cp is the capacitance, ε_r is the dielectric constant, S is the electrode area, t is the thickness of film and ε_0 is the dielectric constant in a vacuum (8.854×10^{-12} F/m).

3. Results and discussions

3.1 Effect of build-up sequence on crystal grain growth

In the case of both the 100 nm thick STO film formed on a 100 nm thick BTO layer on a Pt(111)/Ti/SiO_2/Si substrate, and the 200 nm thick BTO film produced directly on the same substrate, the XRD peak intensity of BTO(111) was extremely low, as shown in Figure 2. On the other hand, with the 100 nm thick BTO film on a 100 nm thick STO layer, this BTO(111) diffraction can be clearly observed; thus providing a strong indication of epitaxial BTO grain growth on a Pt(111)/Ti/SiO_2/Si substrate. It is considered that this epitaxial grain growth occurs due to a relaxation of the lattice mismatch between the BTO(111) and Pt(111) by the formation of the STO layer between them. The lattice spacing of Pt(111) in the [1-10] direction is 0.5548 nm, and that of BTO and STO are 0.5648 nm and 0.5522 nm, respectively. The lattice mismatch can therefore be calculated by Eq (2):

$$\Delta f = (\ell f - \ell s)/(\ell s), \qquad (2)$$

where ℓf is the lattice spacing of the film and ℓs is the lattice spacing of the substrate.

The lattice mismatches between Pt and BTO, and between Pt and STO, are calculated to be 1.80 and 0.47 %, respectively. Although the XRD peak of STO(111) normally observed at 2θ=39.9 is not evident in Figure 2, this is simply because the small lattice mismatch between STO(111) and Pt(111) means that it is obscured by the far more intense Pt(111) peak. The lattice mismatch between Pt(111) and BTO(111) in the [10-1] and [0-11] direction is 2.38 %, whereas the mismatch of the BTO (111)/Pt(111) interface is 2.28 % in the [1-10] direction and 2.86 % in the [10-1] and [0-11] direction. The mismatch of the STO(111)/Pt(111) interface is the smallest amongst all of the material combinations tested in this study. However, the hetero-joining of materials of the same crystal structure has been known to produce epitaxial crystal growth, even with a large lattice mismatch [5]. It is therefore considered that the lattice spacing and crystal orientation of the STO(111) formed on Pt(111) promotes the formation of BTO(111), which has a relatively large lattice mismatch with the Pt substrate. Consequently, subsequent experiments utilized STO(111) single crystals produced by STO deposition as the substrate, in order to more precisely control the crystal growth of the deposited films.

3.2 $Ba_{1-x}Sr_xTiO_3$ single layer thin film

With the single layer BSTO thin film deposited on a STO(111) substrate, only XRD peaks of BSTO(111) were observed; this clearly indicating epitaxial grain growth of the deposited layer from

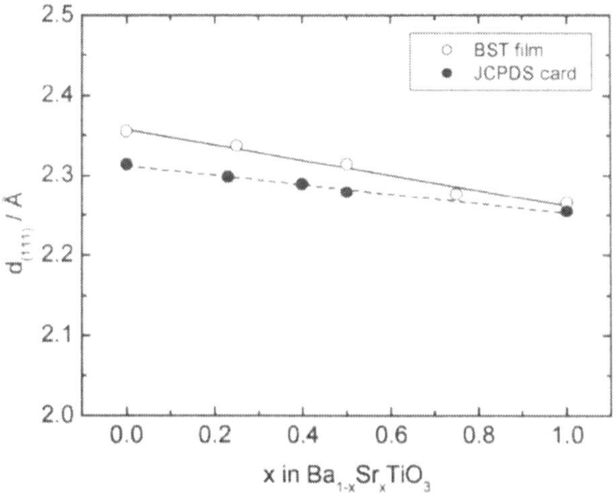

Figure 3. Temperature dependence of the dielectric constant and dielectric loss for (a) x=0.25, (b) x=0.50 and (c) x=0.75 in single layer $Ba_{1-x}Sr_xTiO_3$ films.

the substrate. The lattice spacing of the (111) plane (d(111) value), as calculated from the diffraction angle of the BSTO(111) films, exhibited a linear relationship with regards to composition (x), as shown in Figure 3. For comparison, the lattice spacing of the (111) plane estimated from JCPDS cards for BTO and STO using Vegard's law is also shown. From this, we can see that the difference between the two lines becomes larger with a reduction in x, indicating that the lattice mismatch between the BSTO film and STO substrate also becomes larger. However, even under such conditions, epitaxial crystal growth of BSTO from the STO(111) substrate is still maintained.

A compressive strain of about 0.5 GPa has previously been reported at the interface of a super lattice [6], which was found to lead to an increase in the dielectric constant. It is therefore expected that the use of a super lattice will show great potential for producing materials with unprecedented properties [7]. Since this compressive strain is introduced parallel to the (111) plane, it is increased when the crystal grows along the (100) plane. The fact that the BSTO crystal grew epitaxially, despite the large lattice mismatch, suggests that the lattice of the films is expanded by the compressive strain as a rhombohedral crystal in the <111> direction. To ascertain whether there is indeed any difference in the dielectric properties between the upper and lower electrodes, the electrical field was applied in a vertical direction relative to the substrate.

Figure 4 shows the temperature dependence of the dielectric constants of the BSTO single layer films. Of significance is the fact that the BSTO thin films do not appear to exhibit a maximum value around the phase transformation temperature of 393K, a phenomenon which is commonly seen in bulk BTO ceramics. This arises because the phase transition temperature of the BTO-STO ceramic system becomes lower with an increase in STO content, and thus the temperature at which a maximum occurs is reduced [8]. In practice, however, the temperature corresponding to a maximum dielectric constant in STO bulk ceramics has been reported to be as low as absolute zero [8-9]. The BSTO films produced in the present work show a very different behavior to their bulk phase with regards to the temperature dependence of their dielectric characteristics. For instance, in the case of the film with a composition of x=0.25, there is no change in the dielectric constant value of 320 with a change in temperature. This is attributed to the temperatures used in this experiment being well below the Curie temperature of BSTO, the point beyond which BSTO exhibits a characteristic positive temperature coefficient. The dielectric loss, tan δ, was found to increase at elevated temperatures, but was sufficiently low (~0.02) below room temperature. On the other hand, the frequency dependence of the dielectric constant did vary as the temperature increased. When the STO content of the film was increased, i.e. x=0.50, the

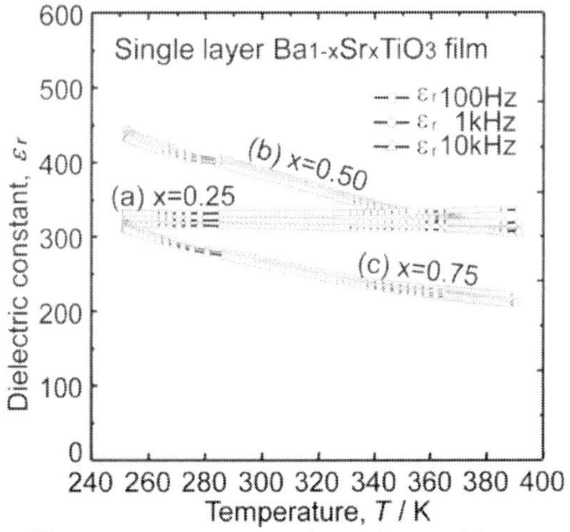

Figure 4. Temperature dependence of the dielectric constant and dielectric loss for (a) x=0.25, (b) x=0.50 and (c) x=0.75 in single layer $Ba_{1-x}Sr_xTiO_3$ film.

dielectric constant was found to increase with decreasing temperature. Moreover, for a given frequency, the dielectric constant of the x=0.75 film was smaller than that of the x=0.25 film. The dielectric constant of the bulk ceramic was greatest at x=0.60 and a temperature of about 210K [10] with the $Ba_{1-x}Sr_xTiO_3$ thin films exhibiting a similar tendency.

As shown in Figure 5, the dielectric constant at room temperature did moderately increase in relation to x to a maximum at x=0.50, but abruptly decreased at higher x values. Both the dielectric constant and dielectric loss were found to be dependent on frequency in the case of the STO thin-film, though this was not observed in the other thin-films. This same frequency dependence was also observed in the case of the STO thin film deposited on a STO(111) substrate, most likely due to the internal strain resulting from the small difference in d(111) shown in Figure 3. From this point of view, the fact that the BSTO thin film in which x=0.0 has the largest lattice mismatch means that it should have the highest dielectric constant; and yet the results obtained are quite contradictory. In the $Ba_{1-x}Sr_xTiO_3$ thin

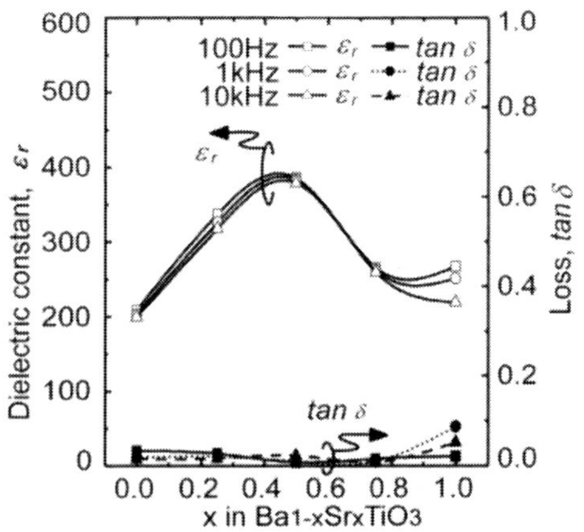

Figure 5. Dielectric constants and dielectric loss at room temperature for single layer $Ba_{1-x}Sr_xTiO_3$ films

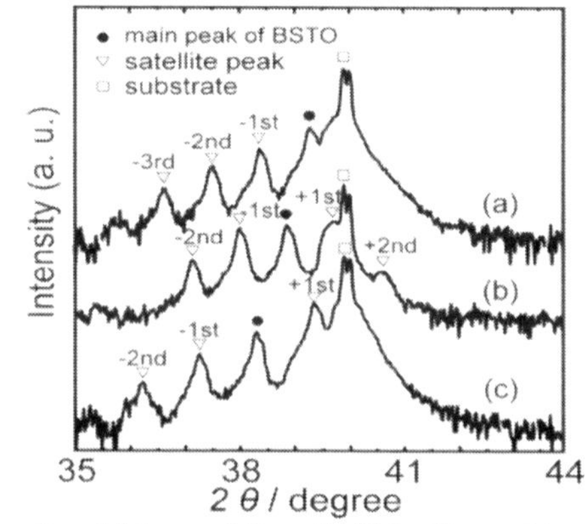

Figure 6. XRD patterns of multi-layered $Ba_{1-x}Sr_xTiO_3$ films with a stacking period of 20, the thickness of each layer being (a) $[BTO(2.5\ nm)/STO(7.5\ nm)]_{20\ cyclic\ layers}$, (b) $[BTO(5.0nm)/STO(5.0\ nm)]_{20\ cyclic\ layers}$ and (c) $[BTO(7.5\ nm)/STO(2.5\ nm)]_{20\ cyclic\ layers}$.

film, a variation in the temperature dependence of the dielectric constant was observed with different levels of STO addition. Furthermore, such temperature dependence was seen only when x was greater than 0.5, as shown in Figure 4.

The results presented in this section indicate that the effect of STO addition on the dielectric characteristics of single layer BSTO films is dependent on the amount of STO addition (i.e. x). Given that STO addition in BTO-STO ceramics is known to reduce the influence of the phase transformation temperature, the results shown in Figure 4 and Figure 5 are believed to incorporate the effects of the phase transformation temperature changes on the dielectric characteristics.

3.3 $(BaTiO_3)_{1-x}(SrTiO_3)_x$ multi-layer thin film

In the previous section, the factors influencing the dielectric characteristics of single layer films were found to be related to residual compressive strain at the material interface and the ratio of STO addition. This subsequently led to the idea that the use of multiple thin layers may have the potential to enhance this effect by increasing the number of hetero epitaxial interfaces. The PLD process used in this study can individually control the composition of each layer, and the number of artificial interfaces, simply by changing the thickness of each layer. Consequently, this section explores the possibility of improving the dielectric characteristics through the use of multi-layer $(BaTiO_3)_{1-x}$ $(SrTiO_3)_x$ thin films.

As shown in Figure. 6, the XRD patterns of $(BaTiO_3)_{1-x}(SrTiO_3)_x$ multi-layer thin-films with compositions of x=0.25, 0.50 and 0.75 have a clearly discernible BSTO(111) main peak, and a number of satellite peaks, when the number of layers exceeds 10. The presence these higher-order satellite peaks adjacent to the main peak indicates the formation of a coherent film with a periodically modulated structure, in which there is a single in-plane lattice parameter throughout the entire film thickness [11]. This BTO-STO super lattice consisted of layers with two distinctly different lattice constants, and hence, a compressive stress was induced in the BTO layers by the STO layers. Conversely, a tensile stress was induced in the STO layers by the BTO layers[12].

In the BTO-STO multi-layer thin films, the diffraction angle of the main BSTO(111) peak was shifted to a higher angle and approached the main peak of STO(111) with increasing STO content. It should be noted that this STO(111) peak is the same as that of the single-layer STO thin films. The layer thickness of the super lattice was subsequently calculated from the diffraction angles of the main BSTO(111) peak and its satellite peaks according to Eq. (3).

$$\Lambda=(m\lambda)/(2(\sin\theta_m - \sin\theta_0)) \quad (3)$$

From this, the cyclic layer thickness of the BTO-STO super lattice thin films were calculated to be (a) 9.0 nm, (b) 11.0 nm and (c) 10.7 nm, which were almost equivalent to the designed thickness of 10 nm.

Figure 7 shows the effect of the layer thickness on the dielectric constant and dielectric loss for multi-layer BSTO thin films with average compositions of x=0.25, x=0.50 and x=0.75. For reference, the values for single layer BTO and STO thin films are also shown in this figure. This clearly demonstrates that the dielectric constant for the multi-layer film in which x=0.25 is increased relative to a single layer film as a result of the multiple layers. Furthermore, the dielectric constant of the multi-layer film drastically increased when the cyclic thickness of STO and BTO was reduced below 7.5 nm and 2.5 nm, respectively; whilst a maximum value of 530 was obtained at BTO(3 nm)/STO (1nm). The dielectric loss, tan δ, remained at less than 0.02 regardless of the change in the average composition.

The dielectric constant at x=0.50 was observed to increase when the cyclic thickness was below BTO(10 nm)/STO(10 nm), whereas a maximum value of 560 was obtained at BTO(2 nm)/STO(2 nm).

Further reduction of the cyclic layer thickness reduced the dielectric constant; however, the dielectric constant obtained above BTO(5 nm)/STO(5 nm) was found to be smaller than that of a single thin film of x=0.5. Put simply, this means that the dielectric constant is reduced to a level comparable to that of a single layer film when the cyclic layer thickness is within a range of between10 and 100 nm.

Furthermore, although the dielectric constant of a single layer film in which x=0.50 is larger than if x=0.25, as shown in Figure 5; the maximum dielectric constant obtained with a multi-layer thin film of x=0.5 was almost equivalent to that of x=0.25, with composition therefore having no observable effect.

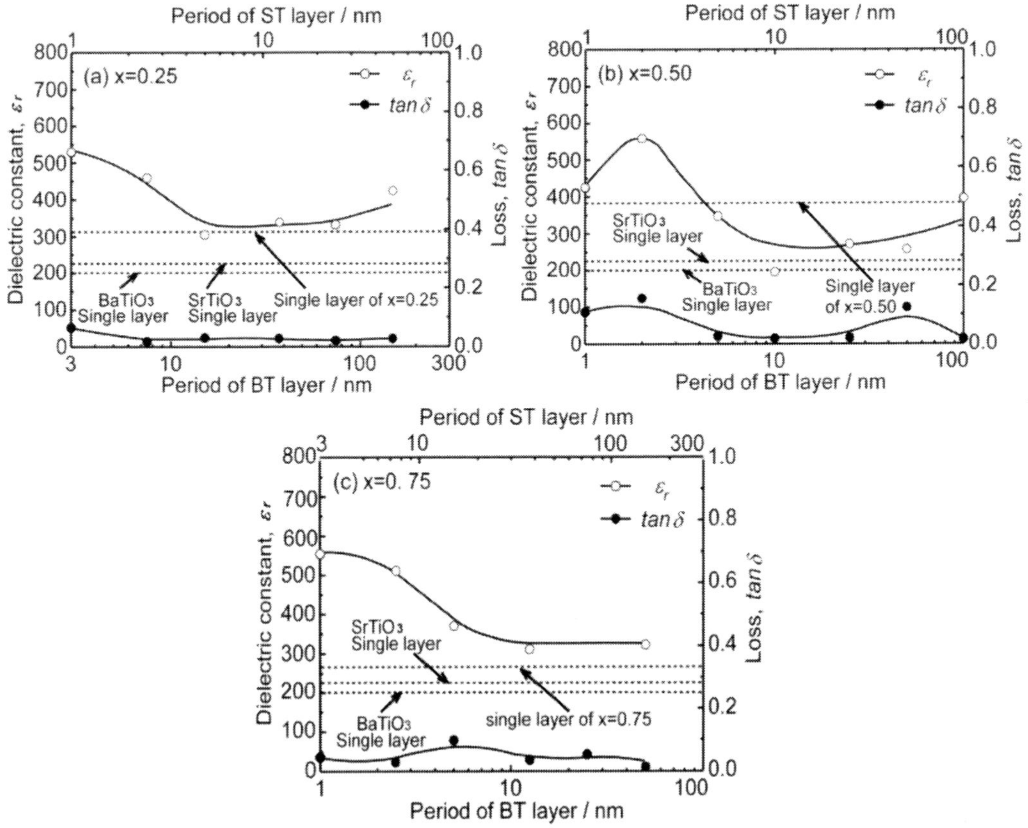

Figure 7. Dielectric constants and dielectric loss at a frequency of 10kHz at room temperature for (a) x=0.25, (b) x=0.50 and (c) x=0.75 for multi-layered $(BaTiO_3)_{1-x}(SrTiO_3)_x$ films.

Other factors with the potential to induce strain may have influenced these results, although the detailed mechanism of these is not well understood currently. Nevertheless, the effect of strain induced by the multiplication of layers has been said to be enhanced as the cyclic layer thickness decreases. It has also been reported that a maximal dielectric constant is obtained when the cyclic layer thickness is 0.8 nm [9] or 0.45 nm [10], with an orientation of (001) [9] and (111) [10], respectively. In both of these cases, a substrate temperature of about 873 K was.

In the present work, the maximal dielectric constants of 530 for x=0.25 and 560 for x=0.5 were almost equivalent to those value reported in the literature, but their corresponding cyclic layer thicknesses were notably different. The reason for this is considered to be the difference in the substrate temperature used, which at 973 K was 100 K higher than that used in the aforementioned studies. This increased temperature is likely to result in greater diffusion of elements across the interface of each layer, which would subsequently affect the dielectric constant.

The dielectric constant of the multi-layer films in which x=0.75 was larger than that of the single layer thin films in which x=0.75. Moreover, the dielectric constant drastically increased when the cyclic thickness was reduced below BTO (2.5 nm)/STO (7.5 nm); the maximum value of 550 being obtained with the lowest thickness, i.e. BTO (1 nm) /STO (3 nm).

4. Conclusions

Three types of layered (Ba,Sr)TiO_3 thin films were fabricated by pulsed laser deposition and their dielectric characteristics were examined in detail. The findings can be summarized as follows:

(1) In the case of a single layer $BaTiO_3$ (BTO) thin film deposited on $SrTiO_3$ (STO) film, a $BaTiO_3$ (111) diffraction peak was clearly observed. This indicates an epitaxial crystal relationship exists between $BaTiO_3$ films and Pt(111) substrates, which results from the relaxation of the lattice mismatch between them that is induced by the $SrTiO_3$ layer.

(2) The dielectric constant of $Ba_{1-x}Sr_xTiO_3$ single layer thin films is dependent on the STO composition, x. The largest dielectric constant was obtained when x=0.5; increasing at low temperatures when $x \geq 0.5$, but showing no temperature dependency when $x < 0.5$.

(3) Examination of multi-layer $(BaTiO_3)_{1-x}(SrTiO_3)_x$ thin films by XRD showed clear evidence of a main peak attributable to $(BaTiO_3)_{1-x}(SrTiO_3)_x(111)$, and additional satellite peaks, when the number of layers exceeded 10. The cyclic layer thickness of this super lattice was estimated from the diffraction angles to be approximately 10 nm, which corresponds well with the result expected from the conditions of pulsed laser deposition used.

(4) In the $(BaTiO_3)_{1-x}(SrTiO_3)_x$ multi-layer thin films, the dielectric constant was found to improve with an increasing number of cyclic layers. A $(BaTiO_3)_{1-x}(SrTiO_3)_x$ multi-layer thin film with a cyclic layer thickness of 2 nm, 50 layers and x=0.50, i.e. $[\mathrm{BTO(2\ nm)/\ STO\ (2\ nm)}]_{50\ \mathrm{cyclic\ layers}}$, x=0.50, exhibited the maximal dielectric constant of 560.

References

[1] Hino T *et al* 1999 *J. Soc. Mat. Sci.* **48** 777
[2] Hino T *et al* 2003 *Vacuum* **70** 47
[3] Tabata H, Tanaka H and Kawai T 1994 *Appl. Phys. Lett.* **65** 1970
[4] Koinuma H, Aiyer H N and Matsumoto Y 2000 *Sci. Tech. Adv. Mat.* **1** 1
[5] Yang G Z *et al* 1997 *Phy. Lett.* **14** 478
[6] Wiener-Avnear N 1994 *Appl. Phys. Lett.* **65** 1784
[7] Abe K and Komatsu S 1995 *J. Appl. Phy.* **77**, 6461
[8] Haeni J H *et al* 2004 *Nature* **430** 758
[9] Muller D A *et al* 2004 *Nature* **430** 657
[10] Koinuma H 2001 *The oxide electronics* (Tokyo: Baifukan) p137 [in Japanese]
[11] Hino T *et al* 2007 *J. Laser Micro/Nanoengineering* **2** 166
[12] Harigai T *et al* 2009 *Thin Solid Films* **609** 13

Effect of strain and deformation route on grain boundary characteristics and recrystallization behavior of aluminum

Tetsuo Sakai*, Hiroshi Utsunomiya and Yasuo Takahashi***

* Dept. of ECO Materials Processing, Joining and Welding Research Institute, Osaka University
11-1, Mihogaoka, Ibaraki, Osaka, 567-0047, Japan
** Division of Materials and Manufacturing Science, Osaka University
2-1 Yamada-oka, Suita, Osaka, 565-0871, Japan

E-mail: sakai@mat.eng.osaka-u.ac.jp

Abstract. The effect of strain and deformation route on the recrystallization behavior of aluminum sheets has been investigated using well lubricated cold rolling and continuous equal channel angular extrusion. Three different deformation routes in plane strain corresponding to (1) simple shear, (2) compression, and (3) the combination of simple shear and compression were performed on 1100 aluminum sheet. Fixed amounts of the equivalent strain of 1.28 and 1.06 were accumulated in each route. In case of the combined deformation route, the ratio of shear strain to the total equivalent strain was varied. The recrystallized grain size was finer if the combined deformation route was employed instead of the monotonic route under the same amount of equivalent strain at either strain level. The density of high angle grain boundaries that act as nucleation sites for recrystallization was higher in materials deformed by the combined route. The orientation imaging micrographs revealed that the change in deformation route is effective for introducing a larger number of new high angle grain boundaries with relatively low misorientation angle.

1. Introduction

It has been reported that the introduction of shear strain during the rolling of sheet materials promotes the refinement of recrystallized grains. A layer of extremely fine recrystallized grains appears beneath the surface of sheets that are hot rolled under high friction conditions [1-3]. In the accumulative roll-bonding process (ARB), rolling without lubrication accelerates grain refinement and strengthening [4,5]. Additional rolling after equal channel angular extrusion (ECAE) also enhances the grain refinement [6-8]. The promotion of grain refinement by a shear strain combined with compressive strain can be, in part, obviously attributed to the increase in equivalent strain. However, it has not been completely understood if the increased grain refinement is totally due to the increase in the amount of equivalent strain or, partly, to the change in strain path. The effect of the shear strain can probably not only be ascribed to the increase in the amount of equivalent strain. It should be considered that this additional strain activates slip systems others than those required by compressive deformation. The activation of multiple slip systems results in a complex deformation microstructure, which includes a large number of nucleation sites that may lead to finer recrystallized grains. The authors have revealed that the combination of compression and shear deformation is effective for grain refinement under a

Content from this work may be used under the terms of the Creative Commons Attribution 3.0 licence. Any further distribution of this work must maintain attribution to the author(s) and the title of the work, journal citation and DOI.
Published under licence by IOP Publishing Ltd

fixed amount of strain [9]. However, the effect of the deformation route on the deformation microstructure has not been observed in detail.

In the present study, the effects of the amount of equivalent strain and deformation route on the recrystallization behavior and grain boundary characteristics of 1100 aluminum sheets are investigated by introducing a fixed amount of equivalent strain along three different strain paths, namely simple shear only, compression only, and a combination of simple shear and compression. The ratio of the shear strain to the compressive strain was varied in the combined deformation route.

2. Experimental

Three deformation routes under plane strain condition, specifically (1) simple shear only, (2) compression only, and (3) combination of simple shear and compression, were employed to aluminum 1100-O sheets of commercial purity. Sheets being 2.0 mm thick, 19 mm wide, and 2 m long were processed by conventional rolling and/or conshearing at room temperature. A fixed amount of the equivalent (von Mises) strain was introduced in either path. Two levels of equivalent strain were introduced, namely ε_{eq}= 1.28 and ε_{eq}= 1.06. Experimental details of each strain path are described below.

The simple shear was introduced by continuous ECAE of the sheet materials, using the conshearing process [10,11]. The die angle θ for ECAE was 65°. In order to impose the shear deformation uniformly and cumulatively, not only the strip was turned upside down, but also the extruding direction was reversed after each pass. Conshearing was carried out up to eight passes to obtain the equivalent strain of 1.28. The mean inclination angle of the grains with respect to the rolling direction at mid-plane α' was determined from micrographs. The imposed shear strain γ and the equivalent strain ε_{eq} were estimated by the following equations:

$$\gamma = \tan(90^\circ - \alpha') \tag{1}$$

and

$$\varepsilon_{eq} = \frac{1}{\sqrt{3}}\gamma \ . \tag{2}$$

The amount of shear strain was varied by varying the pass number of conshearing.

Plane strain compression was introduced by conventional cold rolling with good lubrication. The sheets were rolled with $\phi = 130$ mm rolls at a roll peripheral speed of 2 m/min. The equivalent strain was calculated by the following equation:

$$\varepsilon_{eq} = -\frac{2}{\sqrt{3}}\ln(1-r) \ , \tag{3}$$

Table 1. Deformation routes and ratio of shear strain to total equivalent strain.

notation	equivalent strain	deformation route	strain	shear strain ratio
s1(1.28)	1.28	simple shear only	γ=2.21	1.00
s0.8(1.28)	1.28	simple shear +compression	γ=1.82, r=17.3%	0.83
s0.6(1.28)	1.28	simple shear +compression	γ=1.42, r=32.9%	0.64
s0.3(1.28)	1.28	simple shear +compression	γ=0.66, r=54.1%	0.30
s0(1.28)	1.28	compression only	r=67.0%	0.00
s1(1.06)	1.06	simple shear only	γ=1.82	1.00
s0.8(1.06)	1.06	simple shear +compression	γ=1.42, r=18.8%	0.77
s0.4(1.06)	1.06	simple shear +compression	γ=0.66, r=44.5%	0.36
s0(1.06)	1.06	compression only	r=60.1%	0.00

in which r is the reduction in thickness, defined as:

$$r = \frac{\Delta h}{h_0}, \tag{4}$$

where Δh is the change in thickness induced by rolling and h_0 is thickness before rolling.

The combined strain was introduced by conshearing, followed by conventional rolling. The equivalent strain of the combined path was calculated as follows:

$$\varepsilon_{eq} = \frac{1}{\sqrt{3}}\gamma - \frac{2}{\sqrt{3}}\ln(1-r). \tag{5}$$

The pass number of conshearing and the reduction in thickness by rolling were varied to obtain a fixed total amount of equivalent strain with various ratios of shear strain to the total equivalent strain. The different deformation routes conducted in the present study are listed in Table 1.

The sheets processed by the nine different routes were annealed at temperatures in the range of 423 K - 673 K for 1.2 ks in oil or salt bath. The Vickers hardness was measured on the longitudinal section of as-processed and annealed specimens. The mean intercept length of the recrystallized grains was taken as grain size. Field emission scanning electron microscope (FE-SEM) equipped with electron backscatter diffraction (EBSD) system was used to obtain orientation imaging micrographs (OIMs).

3. Results

The effect of the shear strain ratio on the softening behavior of the specimens deformed to the equivalent strain of 1.28 is shown in Figure 1. Among the as-processed materials, the rolled specimen s0(1.28) and s0.3(1.28) show higher hardness than the simply sheared sample s1(1.28) and s0.8(1.28). The sharp decrease in hardness caused by the recrystallization is found in all five specimens. The simply sheared specimen shows the highest recrystallization temperature.

The effect of the ratio of shear strain to the total equivalent strain on the recrystallized grain size is

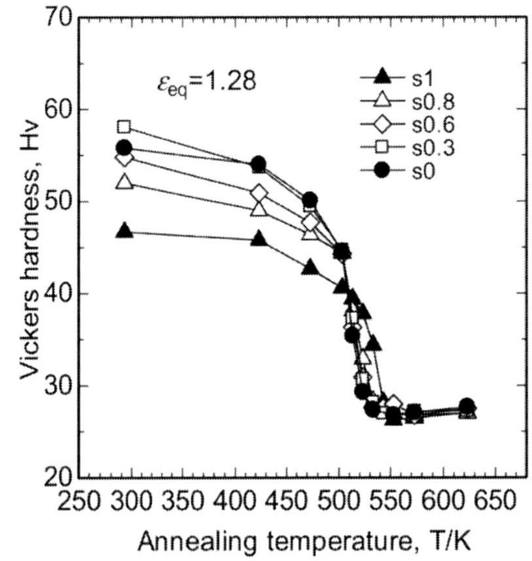

Figure 1. Effect of deformation route on softening behavior during annealing.

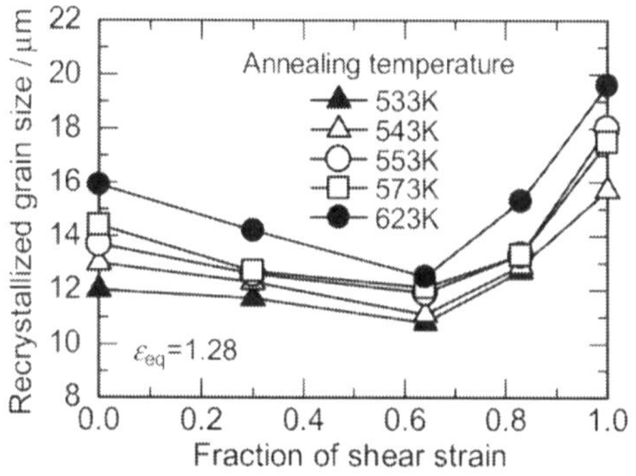

Figure 2. Effect of deformation route on the recrystallized grain size.

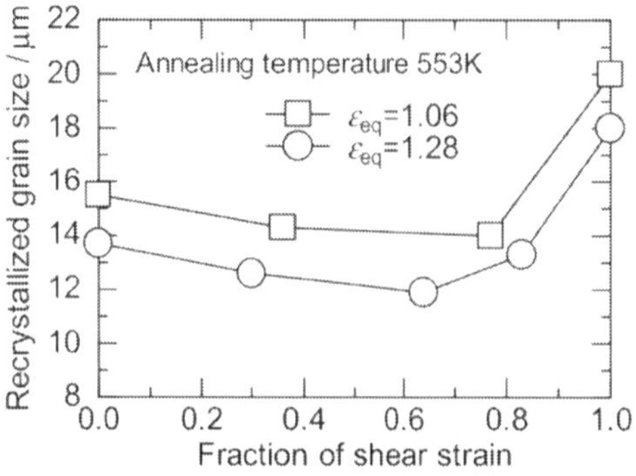

Figure 3. Effect of the total equivalent strain on the grain size in dependence of the ratio of shear strain to total strain.

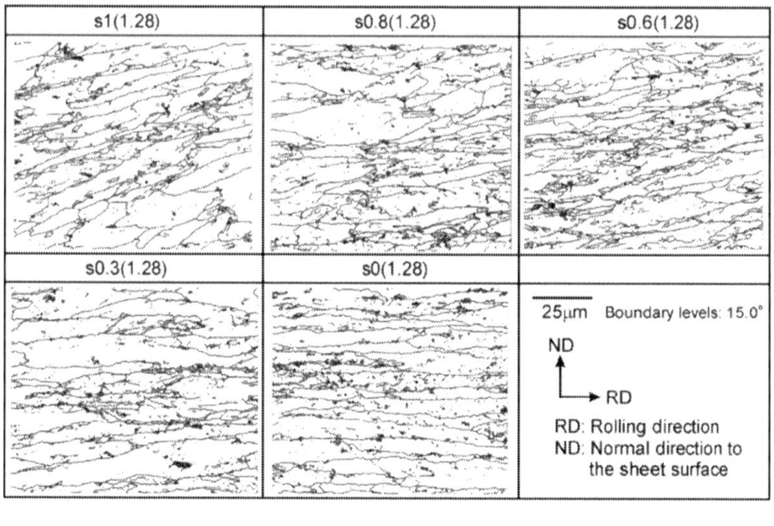

Figure 4. Effect of the ratio of shear strain to total strain on the arrangement of high angle grain boundaries in the specimens deformed to the equivalent strain of 1.28.

clearly shown in Figure 2 for the total equivalent strain of 1.28. The grain size depends on the ratio of the shear strain to the total equivalent strain at all annealing temperatures tested. Near the strain ratio of 0.6, the finest grain size was obtained. Corresponding to the hardness of the as-processed specimens, the simply sheared sample shows the largest grain size. The second largest grain size is observed in the rolled specimen s0(1.28). A monotonic deformation is less effective for grain refinement than the combined deformation route. This suggests that the combination of nearly equal shear and compressive strain is most effective for grain refinement. The grain size increases almost linearly with the increase in annealing temperature. It should be noted that these results are obtained by discontinuous static recrystallization because the applied strain is insufficient to induce continuous recrystallization that often occurs in severely deformed materials. Figure 3 shows the effect of the total equivalent strain on the recrystallized grain size. The grain size decreases with increasing equivalent strain. The combined deformation route is more effective for grain refinement, regardless of the amount of total strain.

4. Discussion

In the present study, the effect of the deformation route on the recrystallization behavior has been clarified by varying the ratio of shear strain to compressive strain under the condition of a constant total equivalent strain. The combined deformation route corresponding to a change in strain path promotes grain refinement. In order to discuss the mechanism, OIMs of the as-processed specimens were taken by FE-SEM equipped with an EBSD system. The OIMs of the as-processed materials are shown in Figure 4. The solid lines of the micrographs show boundaries with a misorientation angle larger than 15°, i.e., high angle grain boundaries (HAGBs). In the material that has been subject to simple shear (s1(1.28)), lamellar blocks surrounded by HAGBs are elongated in an inclined direction. The material after compression (s0(1.28)) has a lamellar structure elongated in the rolling direction (RD), but these grains are thinner than the sheared grains in the s1(1.28) specimen. Likewise, the materials resulting from the application of the combined deformation path (s0.8(1.28), s0.6(1.28), s0.3(1.28)) also have a lamellar structure of elongated grains. However, the density of HAGBs is higher and the grain size is smaller than in case of the materials processed by monotonic deformation. The HAGBs are likely to be formed by grain subdivision [12] due to the change in stable end orientations. The number of nucleation sites for recrystallization is larger in the material that is subject to the combined deformation path. The change in strain path may increase the stored energy and the density of HAGBs, thereby resulting in finer recrystallized grains.

Figure 5 shows the effect of the change in deformation route on the deformation microstructure. The specimens s1(1.28) and s0.8(1.28) are compared. The sample s1(1.28) was obtained by 8-pass ECAE, whereas s0.8(1.28) was obtained by 17% rolling after 6-pass ECAE. The same equivalent strain of 1.28 was introduced to each specimen. In comparison with the 6-pass ECAE processed specimen (s1(1.06)), the distance between high angle grain boundaries decreased by additional 2-pass ECAE in s1(1.28). The colors of the grains remain unchanged, implying that the grains reached stable end orientation already after simple shear at the equivalent strain of 1.06. However, by additional 17% rolling to 6-pass ECAE, the density of HAGBs slightly increases in the s0.8(1.28) specimen and the

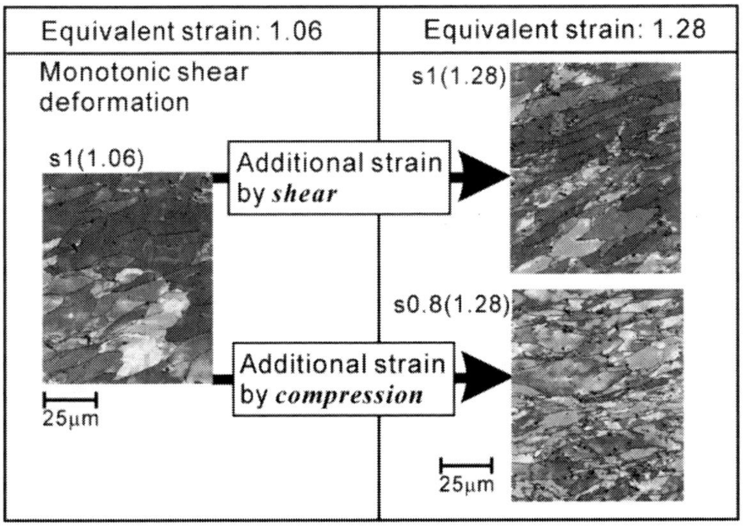

Figure 5. Effect of the mode of additional strain after monotonic shear deformation on the arrangement of high angle grain boundaries and grain orientations.

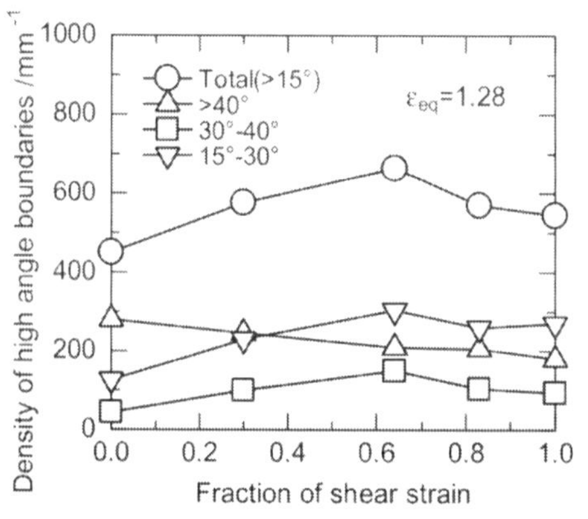

Figure 6. Effect of deformation route on the density of high angle grain boundaries.

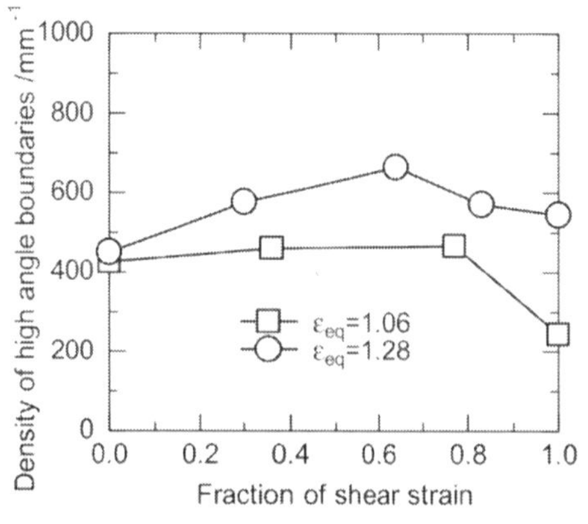

Figure 7. Effect of the total equivalent strain on the density of high angle grain boundaries as function of the shear strain fraction.

colors change more frequently than in s1(1.28), suggesting that the grain orientation changes more frequently in the specimen deformed by combined strain. The grains with stable end orientation after simple shear are rearranged by additional compression. Changing the deformation route clearly influences the deformation microstructure, grain orientation, and subsequent recrystallization behavior.

The effect of the ratio of shear strain to the total strain on the density of HAGBs of the specimens deformed to the equivalent strain of 1.28 is shown in Figure 6. The density of HAGBs was obtained by counting the number of intersections of a line of unit length with HAGBs [13]. In case of the specimen s0.6, the density of HAGBs is highest among the five investigated materials, corresponding to the finest recrystallized grain size. The HAGBs are grouped into three classes according to their misorientation angle. The density of HAGBs with misorientation angle higher than 40°, between 30° and 40°, and below 30° are shown in Figure 6. The density of HAGBs with misorientation angle smaller than 40° is higher if employing the combined deformation route, while that of HAGBs with misorientation angle being larger than 40° gradually decreases with increasing the ratio of shear strain. This suggests that the change in the deformation route promotes the increase of HAGBs with relatively low misorientation angle. HAGBs with misorientation angle of more than 40° seem to be difficult to be formed by grain subdivision. Figure 7 shows the effect of the total equivalent strain on the density of HAGBs in dependence of the fraction of shear strain. The effect of the shear strain fraction on the HAGBs density is more prominent in case of large strain deformation.

The change in deformation route may also affect the formation and arrangement of deformation bands. It is expected that more deformation bands would be formed by employing the combined deformation route. However, statistics of deformation bands were not analyzed in detail in the present study.

5. Summary

The dependency of the recrystallization behavior on the deformation route has been investigated under plane strain condition for 1100 aluminum sheets. Equivalent strain of either ε_{eq} = 1.28 or ε_{eq} = 1.06 was applied to the material along the three different strain paths (1) simple shear, (2) compression, and (3) the combination of simple shear and compression. The results are summarized as follows:

(1) The recrystallized grain size after combined deformation is smaller than in case of simple shear or compression, regardless of the total amount of equivalent strain.

(2) The finest recrystallized grain size was obtained at the ratio of 0.6 of the shear strain to the total equivalent strain, regardless of the total amount of equivalent strain.

(3) The grain refinement resulting from the combined deformation route is attributed to the increase in HAGBs.
(4) The change in deformation route promotes the increase of HAGBs with relatively low misorientation angle.

References

[1] Sakai T, Saito Y and Kato K 1987 *Trans. ISIJ* **27** 520
[2] Sakai T, Saito Y, Hirano K and Kato K 1988 *Trans. ISIJ* **28** 1028
[3] Sakai T, Saito Y, Matsuo M and Kawasaki K 1991 *ISIJ International* **31** 86
[4] Utsunomiya H , Tanda K, Saito Y, Sakai T and Tsuji N 1999 *J. Jpn. Soc. Tech. Plast.* **40** 1187
[5] Lee S H, Saito Y, Tsuji N, Utsunomiya H and Sakai T 2002 *Scripta Mat.* **46** 281
[6] Azushima A, Aoki K, Asada H and Kimura Y 2000 *CAMP-ISIJ* **13** 617
[7] Akamatsu H, Fujinami T, Horita Z and Langdon T G 2001 *Scripta Mat.* **44** 759
[8] Wei Q, Kecskes L, Jiao T, Hartwig K T, Ramesh K T and Ma E 2004 *Acta Mat.* **52** 1859
[9] Sakai T, Mori K and Utsunomiya H 2007 *Materials Science Forum* **550** 315.
[10] Saito Y, Utsunomiya H, Suzuki H and Sakai T 2000 *Scripta Mat.* **42** 1139.
[11] Utsunomiya H, Hatsuda K, Sakai T and Saito Y 2004 *Mat. Sci. Eng. A,* **372** 199.
[12] Hughes D A and Hansen N 1997 *Acta Mat.* **45** 3871
[13] DeHoff R T and Rhines F N 1968 *Quantitative Microscopy* (New York: McGraw-Hill)

Experimental and numerical study on magnetic pulse welding to improving the life time of one-turn flat coil

Tomokatsu Aizawa and Mehrdad Kashani

Tokyo Metropolitan College of Industrial Technology

1-10-40, Higashi-ohi, Shinagawa-ku, Tokyo, 140-0011, JAPAN

Email: kashani@asrl.org

Abstract. The objective of this investigation was to obtain an improved understanding of magnetic pulse welding processes and the mechanisms of electromagnetic forces applied to a one-turn flat coil. Results of simple numerical investigation and experiments show that the electromagnetic forces on the middle part of the coil can be canceled out by loading the welded workpieces on both sides of the coil and can prevent coil deformation and improve its lifetime.

1. Introduction

Magnetic pulse welding (MPW) is an excellent and fast method for achieving similar or dissimilar metal joints. MPW uses magnetic pressure to drive the primary metal against the target metal, sweeping away surface contaminants while forcing intimate metal-to-metal contact, thereby producing a solid-state weld. Several technical research papers have reported on MPW. For example, Tamaki and Kojima [1] and Shribman et al. [2] used conventional MPW with a solenoid coil for joining tubular parts and investigated its features. Aizawa et al. [3] studied the applicability and the potential of MPW for joining several different aluminum alloys and steel sheets using a one-turn flat coil, which is a new solution for application to automotive and vehicle lightweight structures. Recently, Aizawa et al. [4] proposed a new MPW device with a newly designed one-turn flat coil and developed it for flexible printed circuit board lap joints.

MPW has been theorized and tested for several decades; however, optimization of MPW systems and improvements to the lifetime of the coil may play a major role in future developments of MPW systems for use in a wide range of industrial applications. A field shaper can be used in conventional MPW with a solenoid coil. In this case, the magnetic pressure that has to be reacted by the solenoid coil is smaller than the pressure that acts onto the workpiece, thereby significantly increasing the service life of the solenoid coil compared with a direct acting coil. However, the electromagnetic force arrangement in MPW with a one-turn flat coil is complicated and depends on the loading positions of the workpieces. The goal of our work was to investigate the effect of loading two sets of workpieces on both sides of the coil to decrease the total electromagnetic forces on the middle part of the coil. The ratio of the electromagnetic forces acting on the workpieces was also calculated numerically for several coil sizes for two cases (1) loading one set workpieces on one side of the coil, (2) loading two

Content from this work may be used under the terms of the Creative Commons Attribution 3.0 licence. Any further distribution of this work must maintain attribution to the author(s) and the title of the work, journal citation and DOI.
Published under licence by IOP Publishing Ltd

sets of workpieces on both sides of the coil and the results are compared with experimental results in the present paper.

2. Numerical modelling

The principle of MPW using a one-turn flat coil has been reported in previous work [3, 4]. In this section, details of the electromagnetic forces acting on the workpieces and the middle part of the coil are described by a simple model. Figure 1 shows a typical view of the coil and current and return current path with horizontal electromagnetic forces ($\vec{g}$) acting on the central part of the coil. It can be observed that the horizontal forces between the central current path and the two return current paths are nearly the same and that they are able to cancel each other out. Vertical electromagnetic forces appear on the coil and on the surface of the workpieces when discharge starts with loaded metal sheets.

Figure 2 shows the arrangement of the electromagnetic forces for two cases: (a) loading workpieces on only one side of the one-turn flat coil and (b) loading workpieces on both sides of the one-turn flat coil. When a high current is applied to the coil, then a high magnetic field $\vec{B}$ is suddenly generated and penetrates the Al sheets. The eddy currents (current density $\vec{i}$) are produced because of this high magnetic field on the surface of the Al sheets (mainly on the base metal sheet). As a result, an electromagnetic force $\vec{f} = \vec{i} \times \vec{B}$ acts mainly on the base metal sheet, which accelerates away from the coil and collides rapidly with the target metal sheet. Here, $\vec{f}_{coil}$ is the force applied to the central part of the coil.

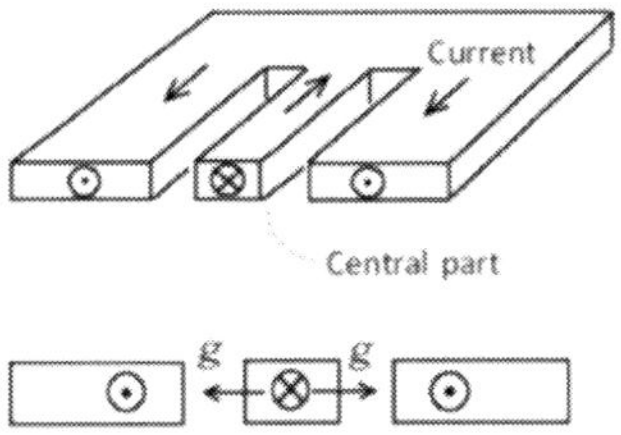

Figure 1. Typical view of one-turn flat coil with forces and current direction without workpieces.

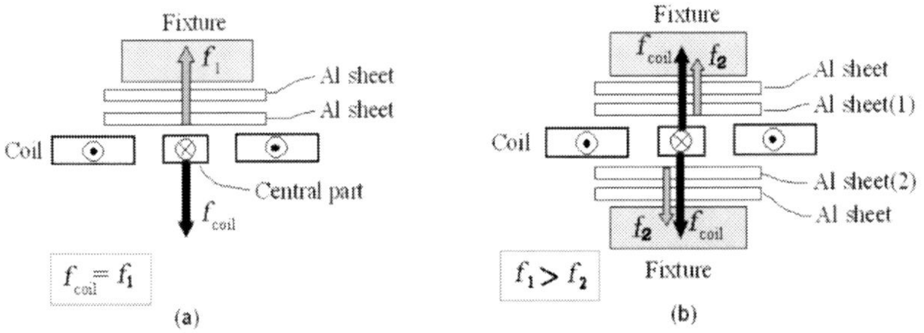

Figure 2. Arrangement of electromagnetic forces: $\vec{f}_{coil}$ acts on the central part of the coil, and $\vec{f}_1$ and $\vec{f}_2$ act on the Al sheets in two cases: (a) loading workpieces on only one side of the one-turn flat coil and (b) loading workpieces on both sides of the one-turn flat coil.

It is expected that the electromagnetic forces acting on the central part of the coil ($\vec{f}_{coil}$) will be divided into two parts of opposing directions when loading two sets of workpieces on the coil sides (figure 2(b)), and these two parts are able to cancel each other out vertically. However, when loading one set of workpieces on the coil, a strong force ($\vec{f}_{coil} = \vec{f}_1$) is applied to the coil and may damage and deform the coil structure. Conversely, $\vec{f}_2$ is applied to the base metals and plays a major role in welding process, decrease in the case of (b). Its magnitude is smaller than that of $\vec{f}_1$ for the same energy discharge.

3. Numerical calculation

Two forces, f_1 and f_2, were applied to the base metals (Al sheets) and play a major role in the welding process. They can be calculated by using a simple assumption for current density, as shown in figure 3. For calculation of the total magnetic field at point P in both cases (it can be assumed that point P is at a distance of $c/10$ from the base metal surface), we assume that the current and eddy current density are nearly same and they are following through the coil and base metal surface uniformly. The magnetic field at point P has three components. The first component is B_c, which is produced by the main current on the central part of the coil, and B_1 and B_2, which are produced by eddy currents at the top and bottom side of the base metal. The dimensions a, b and c represent the thickness, width, and insulator thickness, respectively, between the coil surface and base metal. The current density i [A/m] can be obtained by $i = \frac{I}{2(a+b)}$, where I is total discharge current. In this model, the eddy current is assumed to flow uniformly in the range of coil width (b) on the base metal surface and its density is equal to ki [A/m]. The coefficient k is a specified value between equations (1) and (2). These magnetic fields B_c, B_1, and B_2 can be given as follows:

$$B_c = \left\{\frac{\mu I}{2\pi(a+b)}\right\}\left[\tan^{-1}\left(\frac{b}{2c}\right) + \tan^{-1}\left(\frac{b}{2(a+c)}\right) + \frac{1}{2}\ln\left\{(a+b)^2 + \left(\frac{b}{2}\right)^2 \Big/ \left(c^2 + \left(\frac{b}{2}\right)^2\right)\right\}\right], \quad (1)$$

$$B_1 = k\left\{\frac{\mu I}{2\pi(a+b)}\right\}\left[\tan^{-1}\left(\frac{b}{2(c/10)}\right)\right], \quad (2)$$

and
$$B_2 = k\left\{\frac{\mu I}{2\pi(a+b)}\right\}\left[\tan^{-1}\left(\frac{b}{2(a+2c)}\right)\right]. \quad (3)$$

The value of "c" is smaller than "a" and "b" (c is insulator thickness in the order of 100 μm), and therefore, "c" is neglected in equations (1) and (3). We assume that $B_c = B_1$ when determining k. The ratio of $\vec{f}_2$ and $\vec{f}_1$ (force on base metal for each cases) will be given by

$$\frac{f_2}{f_1} \propto \frac{(2B_C - B_2)^2}{(2B_C)^2}. \quad (4)$$

The numerical values of k, $B_c(T)$, and f_2/f_1 for some coil dimensions are summarized in table 1. The results of f_2/f_1 for (a = 3, b = 5), (a = 5, b = 3), and (a = 8, b = 5) are compared with the maximum shearing strength values from the experimental results.

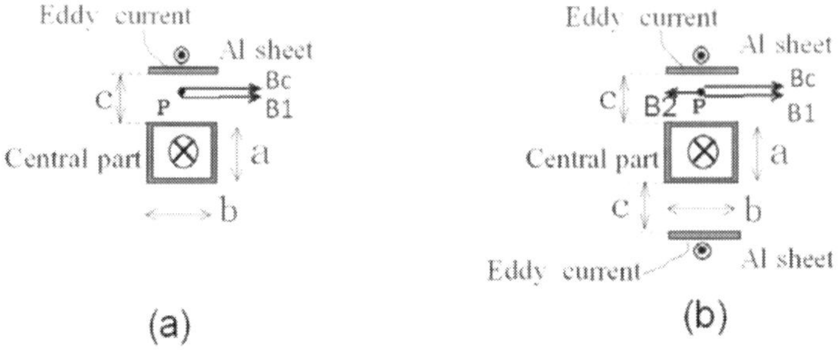

Figure 3. Distribution of current density in central part of coil and base metal for (a) workpieces loaded on one side of coil and (b) workpieces loaded on both sides of coil.

Table 1. Calculation results for ratio f_2/f_1 and $B_c(T)$.

Coil Size(mm) [C=0.1mm]	a=2 b=6	a=3 b=5	a=4 b=4	a=5 b=3	a=6 b=2	a=5 b=8	a=6 b=7	a=7 b=6	a=8 b=5	a=9 b=4
k	1.7	1.7	1.8	2.0	2.2	1.7	1.8	1.9	1.9	2.1
$B_c(T)$	10.1	10.0	10.5	11.5	12.9	6.2	6.4	6.6	7.0	7.6
f_2/f_1	0.49	0.62	0.74	0.83	0.90	0.63	0.70	0.76	0.82	0.87

4. Experimental procedure

A block diagram of the discharge system is shown in figure 4. It consists of a capacitor bank (C) and a spark gap switch (G) with an E-shaped one-turn flat coil. The capacitor bank is connected to the gap switch and the one-turn coil by a low inductance transmission line. Four capacitors with a total capacitance of 400 μF (4 × 100 μF) were charged to between 1.5 and 4.5 kV for obtaining a discharge energy of 0.6–4 kJ. The typical maximum discharge current for 1.4 kJ discharge was measured by a Rogowski coil to be about 180 kA. The one-turn flat coil was made of a Cr–Cu alloy, and its inductance was measured to be about 0.039 μH. The main discharge current passed through the middle part of the coil, and the thickness and width of the central part of the coil were changed between 3 mm and 5 mm to confirm the numerical estimation. Commercial pure aluminum A1050 and A6016 aluminum alloy sheets were prepared to carry out the welding for A1050/A1050 and A6016/A6016 combinations. All Al sheets were 100 mm long, 100 mm wide, and 1.0 mm thick. The joining surfaces between the Al sheets were cleaned by ethanol before starting the welding. Insulating sheets of 0.1–0.3 mm thickness were loaded between the coil surface and the Al sheets. The welding characteristics can be improved by fixing the initial gap (1 mm) between two Al sheets [3]. After welding, the welded sample was divided into 10 pieces, which were then used in the tensile shearing strength test.

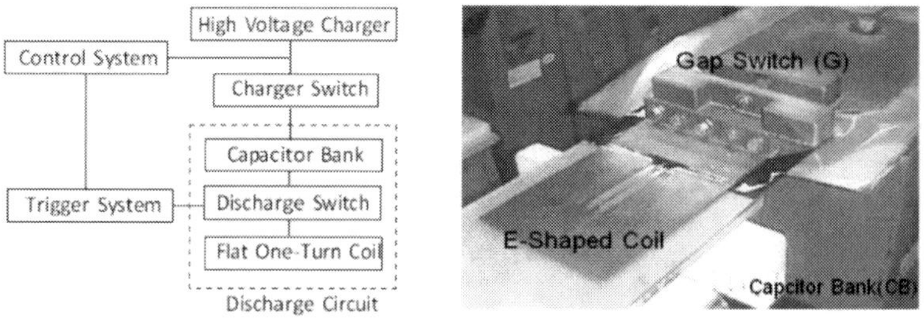

Figure 4. Block diagram of discharge system and appearance of discharge circuit.

5. Experimental results and discussion

5.1. Tensile shear test

Welded samples were investigated on a standard tensile shear testing machine at a test rate of 10 mm/min. Comparisons of shearing load for two-side and one-side welding were obtained by preparing several welded samples in various experiments with different coils and different dimensions of the middle part of the coil. The tensile shear test results for three coils of different sizes are summarized in figure 5. A1050 was used for preparing welded samples with lower energy discharges, and A6061 alloy was used for investigating higher energy discharges. The comparison shows that the average of maximum shearing strength is nearly the same in both cases, but the minimum discharge energy for achieving a good joint when loading samples on both sides of the coil was about 1.5 times higher than that in the case of one-side loading. If we assume that the ratio of f_2 and f_1 is directly related to the average of the maximum shearing strength of the interface joint in both cases, then the experimental results, shown in the graphs in figure 5, also confirm the numerical estimation, qualitatively.

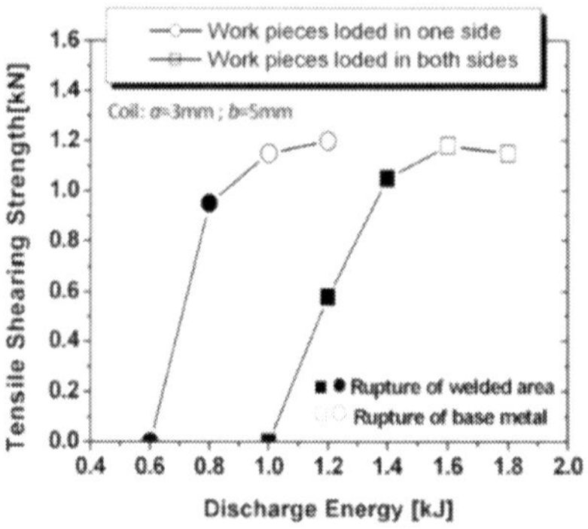

(a)

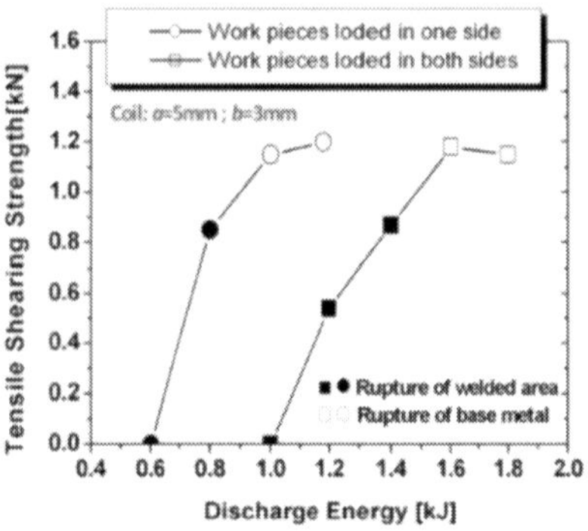

(b)

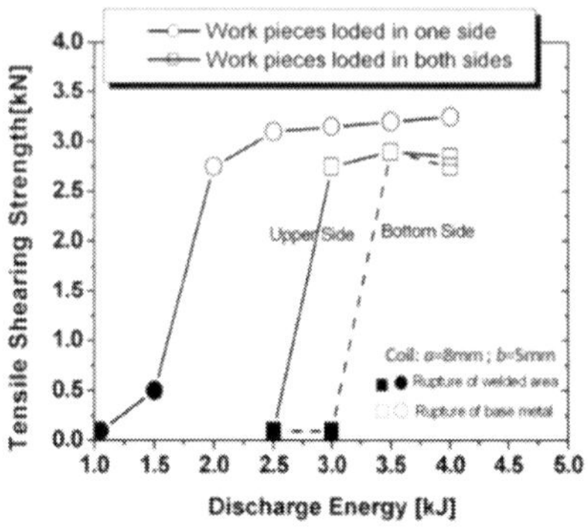

(c)

Figure 5. Comparison of tensile shearing strength versus bank energy for two-side and one-side loading cases. (a) (A1050 with 1 mm thickness); central part of coil size: 3 mm thick, 5 mm wide, (b) (A1050 with 1 mm thickness); central part of coil size: 5 mm thick, 3 mm wide and (c) (A6061 with 1 mm thickness) 8 mm thick, 5 mm wide.

5.2. Coil deformation

The deformation of the middle part of the coil was measured by a standard surface roughness measuring instrument (figure 6(a)). After each discharge, the deformation of the middle part of the coil was measured, and before the next measurement, the deformation was corrected by generating a discharge without any samples. The deformation of the middle part of the coil was low for when loading samples on both sides of the coil for low energy discharge (less than 2 kJ for A1050/A1050 combination). However, the deformation of the middle part of the coil was increased by increasing the discharge energy when loading samples on one side of the coil. The aluminum alloy A6061 sheet with thickness of 1 mm was used to further investigate coil deformation over a wide range (1 to 4 kJ) of bank energy discharges. The middle part of the coil was 8 mm thick and 5 mm wide in this experiment. Figure 6(b) shows a comparison of the deformation ratio of the middle part of the coil (ratio of deformation and coil thickness) versus the bank energy for two-side and one-side loading.

The deformation ratio was obtained by the average value measured at the three points along the middle part of the coil. The maximum deformation was about 0.02 mm in the case of two-side sample loading, and it was about 0.3 mm in the case of one-side sample loading. Reproducibility of the deformation values was poor, but the difference between the two cases remained considerable.

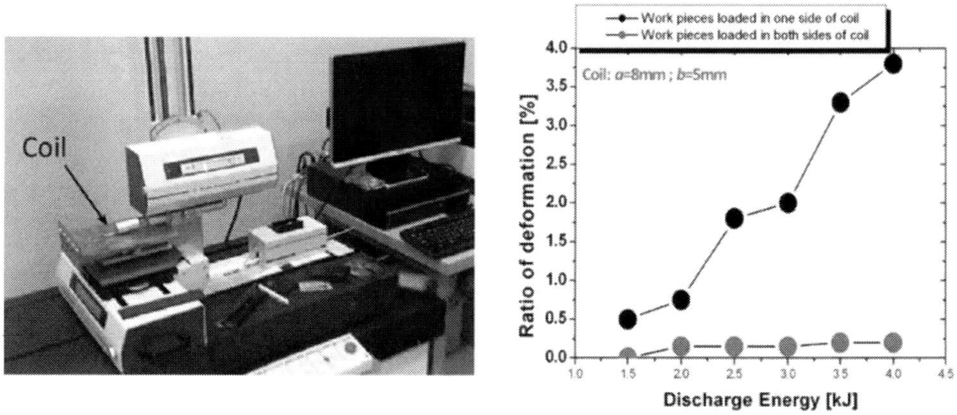

(a) (b)

Figure 6. (a) Standard surface roughness measuring instrument and (b) comparison of middle part of coil deformation ratio (ratio of deformation and coil thickness) versus bank energy for two-side and one-side loading cases (A6061/A6061; sample thickness 1 mm; coil: a = 8 mm; b = 5 mm).

6. Conclusions

In this work, the MPW processes and mechanisms of electromagnetic force were investigated for two types of loading, and the main results are summarized as follows. (1) The energy required for welding workpieces of the same materials loaded on both sides of the coil was approximately 1.5 times higher than that required for welding loaded workpieces on one side of the coil. (2) The energy required for welding one pair of workpieces on both sides of the coil was about 3/4 that required for welding the workpieces on one side of the coil. (3) The coil deformation in the case of loading on both sides of the coil was about 1/10 or less than that in the case of loading on one side of the coil. Therefore, loading the workpieces on both sides of the coil can improve the energy efficiency and the coil lifetime significantly.

Acknowledgments
The authors wish to express their thanks to Prof. K. Okagawa and Dr. M. Ishibashi of Tokyo Metropolitan Collage of Industrial Technology for cooperation on coil roughness measurements.

References
[1] Tamaki K, Kojima M. 1988 *Trans. Japan Welding Society* **19** (**1**) 53.
[2] Shribman V, Stern A, Livshitz Y, Gafri O, 2002 *Welding Journal* **18** 33.
[3] Aizawa T, Kashani M, Okagawa K, 2007 *American Welding Journal* **86** (**5**) 119.
[4] Aizawa T, Okagawa K, Kashani M: 2013 *Journal of Materials Processing Technology*, **213** (**7**) 1095.

Melting points and chemical bonding properties of 3d transition metal elements

Wataru Takahara

Division of Materials and Manufacturing Science, Graduate School of Engineering, Osaka University, 2-1 Yamada-oka, Suita, Osaka 565-0871, Japan

E-mail: takahara@mapse.eng.osaka-u.ac.jp

Abstract. The melting points of 3d transition metal elements show an unusual local minimal peak at manganese across Period 4 in the periodic table. The chemical bonding properties of scandium, titanium, vanadium, chromium, manganese, iron, cobalt, nickel and copper are investigated by the DV-Xα cluster method. The melting points are found to correlate with the bond overlap populations. The chemical bonding nature therefore appears to be the primary factor governing the melting points.

1. Introduction

In general, the melting points of transition metals are much higher than those of main-group metals. The outermost as well as inner shell electrons contribute to the bonding in transition metals. Across Period 4 in the periodic table, the melting points of 3d transition metal elements show a maximal peak around vanadium and chromium. Further, the melting points decrease from chromium to copper and zinc. However, the melting points show a local minimal peak at manganese. This behavior is not entirely clear although there have been attempts to explain the anomaly [1, 2].

In the present work, the chemical bonding properties of scandium, titanium, vanadium, chromium, manganese, iron, cobalt, nickel and copper are investigated by the DV-Xα cluster method to clarify the local minimal peak at manganese.

2. Computational method

DV-Xα molecular calculations were carried out using the SCAT code developed by Adachi et al. [3]. The calculations were based on the Hartree-Fock-Slater approximation with an exchange correlation between electrons based on the Slater local exchange-correlation potential V_{xc} given by

$$V_{XC} = -3\alpha\left[\frac{3}{8\pi}\rho(\vec{r})\right]^{1/3}, \quad (1)$$

where $\rho(\vec{r})$ is the local electron density at position $\vec{r}$ [2-4]. In this calculation, α is fixed at a constant value of 0.7, and the self-consistent charge approximation is employed. The molecular orbital wave function ϕ is expressed by a linear combination of atomic orbitals (LCAO) as

$$\phi_l = \sum_i C_{il}\ \chi_i, \quad (2)$$

where C_{il} is the coefficient and χ_i is a basis function. Numerical atomic orbitals are employed as the basis functions in the DV-Xα method.

Figure 1 shows the model clusters analyzed in the present work. Sc_{19} cluster is based on a hexagonal close-packed (hcp) crystal structure. Ti_{27}, V_{27}, Cr_{27}, Mn_{27} and Fe_{27} clusters are based on

Content from this work may be used under the terms of the Creative Commons Attribution 3.0 licence. Any further distribution of this work must maintain attribution to the author(s) and the title of the work, journal citation and DOI.
Published under licence by IOP Publishing Ltd

body-centered cubic (bcc) crystal structures. Co_{19}, Ni_{19} and Cu_{19} clusters are based on face-centered cubic (fcc) crystal structures. The atomic distances of these clusters are determined from the high temperature crystal structure data near the melting points [5, 6]. Table 1 lists the high temperature crystal structure data for cluster modeling. Sc_{19} cluster is based on the hcp crystal structure at 1773K, where lattice constants a = 0.3368nm and c = 0.5366nm [5]. Ti_{27}, V_{27}, Cr_{27}, Mn_{27} and Fe_{27} clusters are based on bcc crystal structures at 1670K, 1773K, 1773K, 1413K and 1781K, where lattice constants are 0.33338nm [5], 0.30850nm [5], 0.2936nm [5], 0.350nm [6] and 0.29396nm [5], respectively. Co_{19}, Ni_{19} and Cu_{19} clusters are based on fcc crystal structures at 1671K, 1473K and 1343K, where lattice constants are 0.36214nm [5], 0.35987nm [5] and 0.3692nm [5], respectively. DV-Xα molecular calculations are performed for these model clusters. The atomic positions are fixed in the calculations for all the cluster models. The atomic basis functions used in the present calculations are 1s-4p for Sc-Cu. No point groups are used to produce symmetry orbitals. Self-consistency within 0.001 electrons is obtained for the final orbital populations. Non-spin-polarized cases are considered in the molecular orbital calculations.

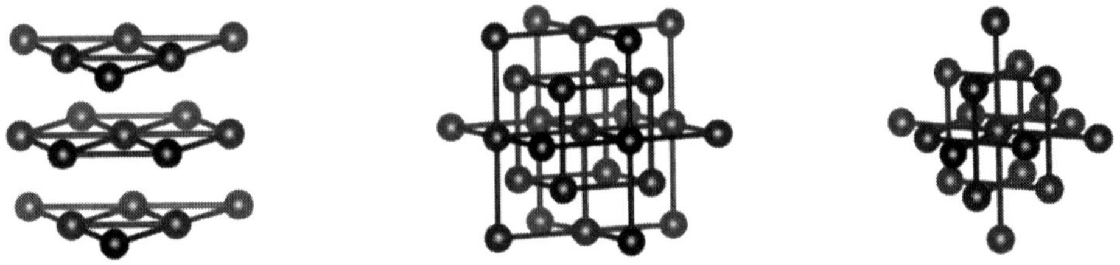

(a) Sc_{19} (hcp) (b) Ti_{27}, V_{27}, Cr_{27}, Mn_{27}, Fe_{27} (bcc) (c) Co_{19}, Ni_{19}, Cu_{19} (fcc)

Figure 1. Hcp, bcc, and fcc based cluster models.

Table 1. Hcp, bcc, and fcc based cluster models.

Atomic number	Element	Melting point, MP (K)	Reference temperature for modeling, RT (K)	Based crystal structure (high temperature phase)	Lattice constants (nm)	RT/MP
21	Sc	1812	1773	hcp	a=0.3368 c=0.5366	0.98
22	Ti	1939	1670	bcc	0.33338	0.86
23	V	2190	1773	bcc	0.30850	0.81
24	Cr	2130	1773	bcc	0.2936	0.83
25	Mn	1519	1413	bcc	0.350	0.93
26	Fe	1809	1781	bcc	0.29396	0.98
27	Co	1768	1671	fcc	0.36214	0.95
28	Ni	1728	1473	fcc	0.35987	0.85
29	Cu	1358	1343	fcc	0.3692	0.99

Mulliken population analysis [7] is employed for evaluating the bond overlap population [2-4, 8]. The bond overlap population, Q_{AB}, is defined as

$$Q_{AB} = \sum_{i \in A, j \in B} Q_{ij} \ , \tag{3}$$

where

$$Q_{ij} = \sum_l Q_{ij}^l \ , \tag{4}$$

$$Q_{ij}^l = f_l C_{il} C_{jl} S_{ij} \ , \tag{5}$$

$$S_{ij} = \int \chi_i^* \chi_j dv \ , \tag{6}$$

here S_{ij} is called the overlap integral in the LCAO molecular orbital method [3, 4, 8], f_l is the occupation number of the l th molecular orbital, and Q_{ij} is the overlap population between atomic orbitals i and j [4, 8]. The bond overlap population Q_{AB}, which is defined as the sum of Q_{ij} over all i belonging to atom A and all j belonging to atom B, is considered the measure of strength of covalent bonding between atoms A and B [4, 8].

3. Results and discussion

Figure 2 shows the bond overlap population between the central metal atom and neighboring metal atoms in the Sc_{19}, Ti_{27}, V_{27}, Cr_{27}, Mn_{27}, Fe_{27}, Co_{19}, Ni_{19}, and Cu_{19} clusters. The melting points of these elements are also plotted in the figure. The filled circles show the bond overlap populations and the open circles show the melting points. The bond overlap population was derived from the DV-Xα orbitals in the calculations considering only the Xα exchange correlation, without electron correlation. Although the finer details with respect to energy are rather sensitive to the way electron correlation is included, information concerning the wave functions is insensitive to electron correlation and such DV-Xα calculations can be considered reliable [8-12]. Both the bond overlap populations and the melting points show the local minimal peak at manganese. As the bond overlap population is the measure of the strength of covalent bonding, the local minimal peak of the bond overlap population at manganese indicates the relative weakening of the chemical bonding of manganese compared to the two-sided elements, chromium and iron.

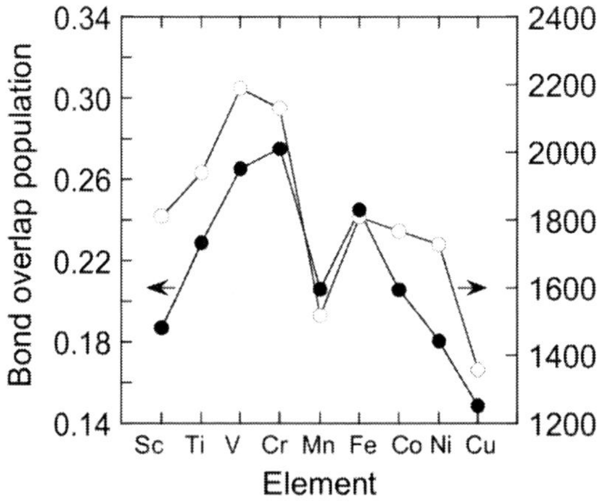

Figure 2. Bond overlap population between the central metal atom and neighboring metal atoms in the Sc_{19}, Ti_{27}, V_{27}, Cr_{27}, Mn_{27}, Fe_{27}, Co_{19}, Ni_{19}, and Cu_{19} clusters. Melting points of these elements are also plotted.

In the transition metals, the changes such as melting points are often associated with the number of d-electrons, n_d. In the simple Friedel model, the peak is at $n_d = 5$, for half-filled d-shell. However, this is not the case at manganese [1]. There have been attempts to explain the anomaly at manganese with account of the correlation effect [1]. In contrast, the present calculations consider only Xα exchange correlation, without electron correlation. Further, non-spin-polarized cases are considered in the present calculations. Although the unusual behavior of manganese is often discussed in connection with the magnetic properties, the chemical bonding nature appears to be the primary factor governing the melting points.

4. Conclusions

Chemical bonding properties of 3d transition metal elements, scandium, titanium, vanadium, chromium, manganese, iron, cobalt, nickel and copper, are investigated by the DV-Xα cluster method. The bond overlap populations show a local minimal peak at manganese. The calculations consider only Xα exchange correlation, without electron correlation, and only non-spin-polarized cases. The chemical bonding nature appears to be the primary factor governing the anomaly of melting point at manganese.

Acknowledgments

The computation was partially carried out using the computer facilities at Cybermedia Center, Osaka University and at Research Institute for Information Technology, Kyushu University.

References

[1] Irkhin V Yu and Irkhin Yu P 2007 *Electronic structure, correlation effects and properties of d- and f-metals and their compounds* (Cambridge International Science Publishing)
[2] Adachi H 1997 *Adv. Quant. Chem.* **29** 49
[3] Adachi H, Tsukada M and Satoko C 1978 *J. Phys. Soc. Jpn.* **45** 875
[4] Adachi H 1997 *Mater. Trans., JIM* **38** 485
[5] Villars P and Calvert L D 1991 *Pearson's Handbook of Crystallographic Data for Intermetallic Phases* Second Edition (Materials Park, OH: ASM International)
[6] Asada T and Terakura K 1993 *Phys. Rev. B* **47** 15992
[7] Mulliken R S 1955 *J. Chem. Phys.* **23** 1833
[8] Adachi H 2006 *Hartree-Fock-Slater Method for Materials Science, The DV-Xα Method for Design and Characterization of Materials*, ed H Adachi, T Mukoyama and J Kawai (Springer-Verlag Berlin Heidelberg) p 3
[9] Morinaga M, Murata Y and Yukawa H 2006 *Hartree-Fock-Slater Method for Materials Science, The DV-Xα Method for Design and Characterization of Materials*, ed H Adachi, T Mukoyama and J Kawai (Springer-Verlag Berlin Heidelberg) p 23
[10] Morinaga M, Nasu S, Adachi H, Saito J and Yukawa N 1991 *J. Phys.: Condens. matter* **3** 6817
[11] Takahara W, Maeda H and Minami F 2003 *Adv. Quant. Chem.* **42** 275
[12] Takahara W 2012 *J. Phys.: Conf. Ser.* **379** 012040

Density distribution of point-contact current

Kazuki Wakabayashi[1], Pasomphone Hemthavy[1], Shigeki Saito[2], and Kunio Takahashi[1]

[1] Department of International Development Engineering, Tokyo Institute of Technology, I4-11, 2-12-1 O-okayama, Meguro-ku, Tokyo 152-8552, JAPAN
[2] Department of Mechanical and Aerospace Engineering, Tokyo Institute of Technology, I1-47, 2-12-1 O-okayama, Meguro-ku, Tokyo 152-8552 JAPAN

E-mail: wakabayashi.k.af@m.titech.ac.jp

Abstract. The density distribution of point-contact current is theoretically investigated. Finite current is introduced to a semi-infinite body from the outside. The following assumptions are made: the contact area is circular, magnetic effects are ignored, the contact area is equipotential, the potential equals zero at an infinite distance from the contact area, and the electric permittivity and resistivity are constant. The density distribution of point-contact current for a non-steady state is obtained in an integral form and the derivation process is discussed. The results show that the current density distribution for the non-steady state agrees with that for the steady state. Further, the results suggest that the analytical solution of the density distribution of point-contact current for the steady state is applicable to thermal analysis as well.

1. Introduction

The density distribution of point-contact current is significant for the design of various processes and devices such as resistance welding and mechanical relays. Melting by current input has been investigated considering the heat input only from the surface [1, 2]. However, heat is also generated inside a body because there exists a current density distribution. It is important to evaluate the effects of the current density distribution. Greenwood [3] investigated the steady-state heat generation inside a body; in this state, the heat generation at each point of the body equals the heat loss by diffusion at that point. Kubono [4] investigated heat generation for a non-steady state. In both these researches [3, 4], the same current density distribution was employed; however, the method for obtaining this distribution was not clarified. In the present research, the current distribution for the simplest boundary conditions is theoretically derived.

2. Theory

Figure 1 shows the simplest model that has an axially symmetric semi-infinite conductive body. The dark part represents the contact area. V [V] is the potential at the contact area; a [m] is the contact radius; and I [A] is the amount of current introduced in the body. The following assumptions are made here: (1) magnetic effects are ignored; (2) the contact area is equipotential V; (3) potential equals zero at an infinite distance from the contact area; (4) electric permittivity ε [N/V^2] is constant; and (5) resistivity ρ [Ω m] is constant.

Content from this work may be used under the terms of the Creative Commons Attribution 3.0 licence. Any further distribution of this work must maintain attribution to the author(s) and the title of the work, journal citation and DOI.
Published under licence by IOP Publishing Ltd

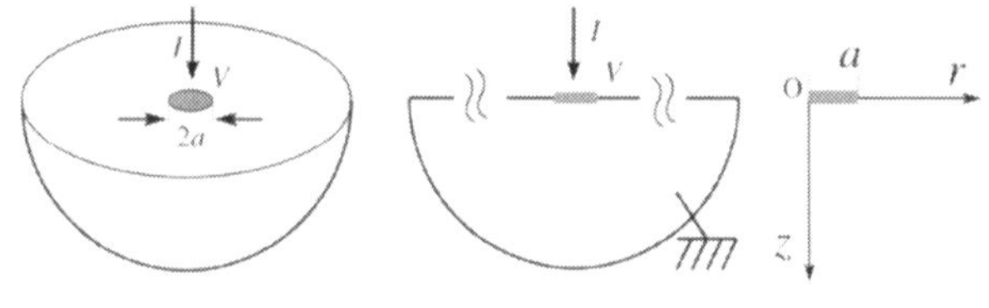

Figure 1. Model: Current through a semi-infinite bulk with voltage applied across contact area.

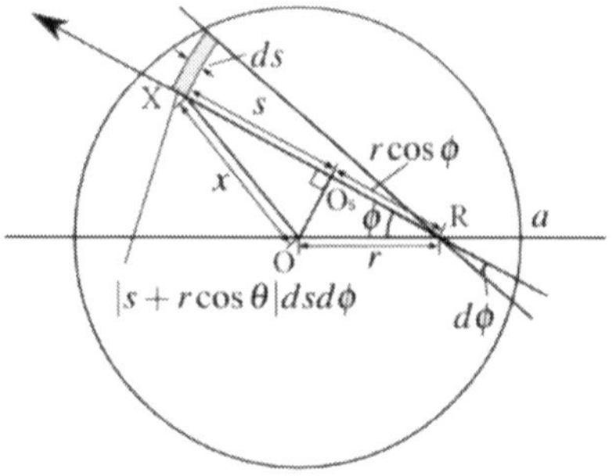

Figure 2. Integration path for deriving potential distribution in contact area.

The basic equations for this model are Gauss' law without charge div$\boldsymbol{E}(\boldsymbol{r}) = 0$ and the relationship between potential and electric field $\boldsymbol{E}(\boldsymbol{r}) = -\mathrm{grad}\, V(\boldsymbol{r})$. By combining these relations, the Poisson equation is obtained as follows:

$$\Delta V(\boldsymbol{r}) = 0 \tag{1}$$

The potential distribution must satisfy the Poisson equation and boundary conditions described in the assumptions. In order to obtain the density distribution for point-contact current, an imaginary charge distribution in the contact area is assumed. The charge distribution is represented by equation (2).

$$c(r) = \frac{4V\varepsilon}{\pi} \frac{1}{\sqrt{a^2 - r^2}} \tag{2}$$

An integral pass is introduced as shown in Figure 2. $r, x,$ and s are the length of OR, OX, and O_sX, respectively. x is expressed by r and s as $x = \sqrt{s^2 + r^2 \sin^2\phi}$. ϕ is the angle between OR and RX. The charge in the black part is $c(x)|s + r\cos\phi|dsd\phi$ because the length of RX is $|s + r\cos\phi|$. The potential change at R due to the charge is $c(x)|s + r\cos\phi|dsd\phi(4\pi\varepsilon|s + r\cos\phi|)^{-1}$. The potential at R is obtained by integrating the potential change over the contact area where the range of s and ϕ are $[-\sqrt{a^2 - r^2\sin^2\phi}, \sqrt{a^2 - r^2\sin^2\phi}]$ and $[0, \pi]$, respectively. The potential is obtained as follows:

$$V(r) = \int_0^{\pi} \int_{\sqrt{a^2 - r^2\sin^2\phi}}^{-\sqrt{a^2 - r^2\sin^2\phi}} \frac{c(x)|s + r\cos\phi|dsd\phi}{4\pi\varepsilon|s + r\cos\phi|}$$
$$= V$$

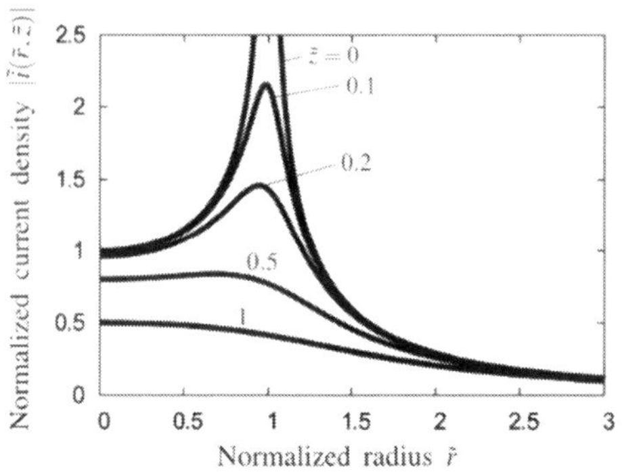

Figure 3. Current density distribution inside bulk.

It means that the potential based on equation (2) in the contact area satisfies the boundary condition for this model. It is clear that the potential based on equation (2) at an infinite distance from the origin is equal to zero.

In addition, the electric field based on equation (2) satisfies Gauss' law inside the semi-infinite body since the electric field based on Coulomb' law satisfies Gauss' law. Therefore, the potential distribution based on equation (2) satisfies equation (1). Therefore, the potential distribution based on equation (2) is the solution for the given system owing to the uniqueness of the Poisson equation. The electric field distribution is obtained on the basis of equation (2) because the electric field distribution has a one-to-one relationship with the potential distribution. Then, the current density distribution is obtained by Ohm's law given by equation (3).

$$\boldsymbol{i}(\boldsymbol{r}) = \frac{1}{\rho}\boldsymbol{E}(\boldsymbol{r}) \tag{3}$$

The current density distribution can be obtained as follows in a normalized vector form, on the basis of the steps described above. This equation is originally derived.

$$\tilde{\boldsymbol{\imath}}(\tilde{r},\tilde{z}) = \frac{1}{\pi}\int_0^{\pi}\int_0^1 \frac{s}{\sqrt{1-s^2}}\frac{1}{(\tilde{r}^2+\tilde{z}^2+s^2-2\tilde{r}s\cos\theta)^{3/2}}\begin{pmatrix}\tilde{r}-s\cos\theta\\ \tilde{z}\end{pmatrix}dsd\theta\,, \tag{4}$$

where $\tilde{r} := r/a\,, \tilde{z} := z/a, \tilde{\imath} := i/(2V/\pi\rho a)$.

3. Results and discussion

Figure 3 shows the results of equation (4). The current density distribution $|\tilde{\boldsymbol{\imath}}(\tilde{r},\tilde{z})|$ has a singular point at $(\tilde{r},\tilde{z}) = (1,0)$, which is the edge of the contact area. $|\tilde{\boldsymbol{\imath}}(\tilde{r},\tilde{z})|$ is close to $|\tilde{\boldsymbol{\imath}}(\tilde{r},\tilde{z})| = 1/(\tilde{r}^2+\tilde{z}^2)$ and is sufficiently far from the origin (0,0). In some special cases, the analytic solutions of $|\tilde{\boldsymbol{\imath}}(\tilde{r},\tilde{z})|$ are obtained as follows:

$$|\tilde{\boldsymbol{\imath}}(\tilde{r},0)| = \begin{cases}\dfrac{1}{\sqrt{1-\tilde{r}^2}} & (\tilde{r}\le 1)\\ \dfrac{1}{\tilde{r}\sqrt{\tilde{r}^2-1}} & (\tilde{r}>1)\end{cases} \tag{5}$$

$$|\tilde{\boldsymbol{\imath}}(0,\tilde{z})| = \frac{1}{1+\tilde{z}^2} \tag{6}$$

Equation (4) agrees with the current density distribution suggested in literature [3]

$$|\boldsymbol{i}(\xi,\eta)| = \frac{I}{2\pi a^2[(1+\xi^2)(\eta^2+\xi^2)]^{1/2}}, \tag{7}$$

where $\tilde{r} = [(1+\xi^2)(1-\eta^2)]^{1/2}$, $\tilde{z} = \xi\eta$.

This agreement suggests that the current density distribution obtained under steady-state conditions is applicable for the non-steady state under the assumptions introduced in this research. As shown in Figure 3, the current density distribution has a singular point at the edge of the contact area. In general, if the heat input has an infinite value at some point, the temperature has an infinite value at that point. It is suggested that since the current density distribution is obtained under simple assumptions and temperature should have a finite value, it is necessary to treat the singular point appropriately for thermal analysis.

4. Conclusion

The density distribution of point-contact current is theoretically investigated under simple assumptions. The density distribution is obtained in an integral form and shows great agreement with the current density distribution suggested by Greenwood [3]. In particular, for special cases, the two distributions shows analytical agreement for $\tilde{z} = 0$ and $\tilde{r} = 0$. This result suggests the applicability of the current density distribution obtained by Greenwood for the thermal analysis of a system.

References

[1] Bor-Jeng W and Nannaji S 1993 *IEEE Trans. on Com. Hybr. and Manu Tech* **16** 456
[2] Omori T 1965 *J.Inst.Elect. Engnr.Jpn* **85-5** 111
[3] J A Greenwood and J B P Williamson 1958 *Proc. R. Soc.,A, Mathe and Phys* **246** 13
[4] Kubono T 1985 *Trans.Inst. Elect.Engnr.Jpn* **104** 47

IOP Conf. Series: Materials Science and Engineering **61** (2014) 012031 doi:10.1088/1757-899X/61/1/012031

Nickel-titanium-based contact for n-type silicon carbide to combine high ohmic conductivity and mechanical properties

Masakatsu Maeda[1], Masatoshi Sano[2] and Yasuo Takahashi[1]

[1] Joining and Welding Research Institute, Osaka University, Mihogaoka, Ibaraki, Osaka 567-0047, Japan

[2] Graduate School of Engineering, Osaka University, Suita, Osaka 565-0871, Japan

E-mail: maeda@casi.osaka-u.ac.jp

Abstract. This paper reports a method to improve mechanical properties retaining the electrical properties of Ni-based ohmic contact material for n-type 4H-SiC. Ni/Ti bilayered films varying only in the thickness of the Ti layer were deposited on SiC substrates and annealed at 1273 K for a very short time in vacuum. The interfacial structures were analyzed by X-ray diffraction. The electrical and mechanical properties were measured by DC conduction test and constant-load scratch test, respectively. An appropriate thickness of the Ti layer on Ni improves the mechanical properties retaining the electrical properties by forming TiC instead of the free carbon.

1. Introduction

Silicon carbide (SiC) is a wide-bandgap compound semiconductor expected in application to heavy-duty power electronic devices [1]. Replacement of conventional Si-based power electronic devices with SiC-based ones can provide significant improvement in energy efficiency, service life, and total mass of the power electronic system. To realize the devices, however, several problems have to be overcome. One of them is to form a reliable and low-resistance ohmic electrode on SiC [1-3].

On n-type SiC, the ohmic contact is commonly formed by deposition of Ni on SiC and subsequent annealing at temperatures higher than 1173 K [2-5]. The interfacial reaction during the heat treatment forms NiSi adjacent to the SiC. However, NiSi is not the only phase produced by the reaction. The most fatal product of the reaction is the free carbon, consisting of graphite and/or amorphous carbon. The free carbon deteriorates the mechanical properties of the contact. To suppress the formation of the free carbon, an element X that has a high affinity with carbon should be added in the deposited layer to form X-carbide instead of the free carbon. A candidate for element X is Ti. Ti is a strong carbide former [6]. Furthermore, TiC has a low work function [7], which is favorable for an ohmic contact on n-type SiC. Although several papers reported Ni-Ti contacts on SiC [8, 9], the mechanical properties are rarely discussed [9].

The present study reports the mechanical properties of Ni-Ti contacts on n-type 4H-SiC in relation with the electrical properties. TiC contact on SiC is also examined.

2. Control of interfacial reaction among SiC, Ni, and Ti

The interfacial reaction between SiC and Ni is considered based on the Ni-Si-C ternary phase diagram. Figure 1 shows the isotherm at 1173 K [10]. SiC and Ni cannot coexist in equilibrium. Therefore, the interfacial reaction proceeds during annealing to establish local equilibrium. Since Ni does not form

Content from this work may be used under the terms of the Creative Commons Attribution 3.0 licence. Any further distribution of this work must maintain attribution to the author(s) and the title of the work, journal citation and DOI.
Published under licence by IOP Publishing Ltd

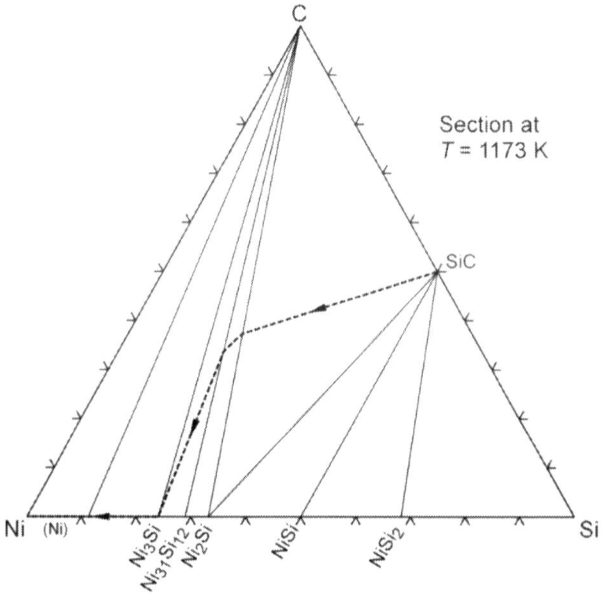

Figure 1. Isothermal section of the Ni-Si-C ternary phase diagram at 1173 K [10]. The broken line with arrows indicates the diffusion path reported in the literature [14].

any stable carbides nor dissolve carbon, the interfacial reaction produces Ni-Si compounds and the free carbon. Therefore, the formation of the free carbon is inevitable when only Ni reacts with SiC. The addition of Ti overcomes this problem. Formation of TiC proceeds spontaneously, since the formation significantly reduces the chemical potential of carbon in the system. However, it has to be noted that Ti can also react with Si.

Naka et al. reported that the reaction between SiC and pure Ti forms $Ti_5Si_3C_x$ and Ti_3SiC_2 at the interface [11]. To avoid the formation of these products, the chemical potential of Ti at the reaction interface has to be lowered to a sufficient level. The chemical potential can be reduced by forming compound phases and/or by diluting Ti at the reaction interface. Thermodynamic calculations suggest that the formation of NiTi and Ni_3Ti intermetallic compounds lowers down the chemical potential of Ti to an appropriate level [12, 13]. On the other hand, it is effective to dilute Ti by placing a Ti-source at a portion distant from the reaction interface, suppressing the diffusion flux of Ti through the phase adjacent to the SiC. The Ni layer should be thick for this purpose, since the thickness of the Ni layer determines the chemical potential gradient and diffusion flux of Ti. Consequently, the Ti layer was formed on the Ni layer in the present study in order to make Ti form Ni-Ti intermetallic compounds and to supply Ti to the interface by diffusion of Ti through the intermetallic compound. In the present study, the thickness was set to 100 nm as a pilot case.

The phase diagram shown in Figure 1 presents another concern. Fukai et al. reported that the interfacial phase sequence of a SiC/Ni diffusion pair becomes SiC/(Ni_2Si + C)/($Ni_{31}Si_{12}$ + C)/Ni_3Si/Ni by annealing at 1273 K for 3600 s [14]. The sequence corresponds to the diffusion path indicated on Figure 1 with a broken line. The diffusion path indicates that NiSi cannot be formed while Ni exists as the end-member of the diffusion pair. Two methods to overcome this problem are considered. One is to change the annealing temperature at which SiC, NiSi, and C coexist in equilibrium. However, the three-phase equilibrium cannot be achieved at temperatures below the eutectic of Ni and Ni_3Si at 1416 K [15]. The other method to form NiSi adjacent to SiC is to make Ni_2Si react with SiC by reducing the chemical potential of C. The addition of Ti in the film reduces the chemical potential significantly by forming TiC instead of the free carbon. Figure 2 shows the original and modified diffusion paths on the Ni-Si-C ternary chemical potential diagram at 1173 K [12, 16]. It is clearly understood that the diffusion path reported in the literature goes through a high chemical potential of C. Due to this, NiSi is kept away from formation. By preventing the formation of the free carbon, the diffusion pair is

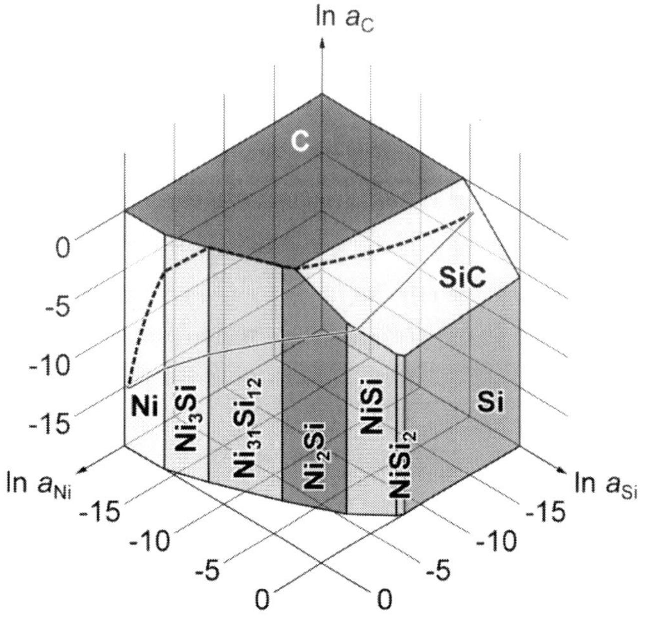

Figure 2. Isothermal section of the Ni-Si-C ternary chemical potential diagram at 1173 K. The broken line indicates the diffusion path shown in Figure 1, whereas the cored solid line depicts the modified diffusion path by addition of Ti.

allowed to take a direct path with monotonous chemical potential gradients of all constituent elements. In addition, the formation of a Ni-Ti intermetallic compound reduces the chemical potential of Ni at one of the end-members, which magnifies the driving force to establish a diffusion path through a low chemical potential of C.

Therefore, the key point of the contact formation process proposed in the present study is to control the interfacial reaction behavior of Ti, i.e., to facilitate TiC formation in order to eliminate the free carbon, preventing the formation of other Ti-Si-C byproducts.

3. Experimental procedure

SiC substrates used in the present study were N-doped n-type 4H-SiC substrates cut to a square shape of 5.0 mm on each side from a 360-μm-thick wafer of which surface was inclined 8° from the (0001) crystallographic plane toward $[11\bar{2}0]$. The inclination is commonly given to SiC single-crystal wafers for controlling the polytype of epitaxial layers stacked on the substrate to form a semiconductor device. Although the surface is different from the exact crystallographic plane, it is conventionally treated the same as the (0001) plane, based on the fact that most of the surface consists of the (0001) facet. Only the (0001) Si-face of the SiC substrates was used. Ni and Ti were deposited consecutively on the substrates by radio-frequency magnetron sputtering. The thickness of the Ni layer was kept constant at 100 nm, whereas that of the Ti layer was varied from 0 to 640 nm. The samples were then heated to 1273 K in a vacuum of 3×10^{-3} Pa and immediately cooled down after reaching the target temperature (hereinafter described as the annealing time of 0 s). For comparison, a TiC contact was formed on some SiC substrates by deposition of a C/Ti bilayer and subsequent annealing at 1273 K for 0 s in vacuum. Since this process can form TiC on the substrate without an interfacial reaction between the contact film and the substrate [17], Ti-Si-C byproducts were not formed at the interface.

The interfacial structures were analyzed by X-ray diffraction (XRD). The electrical properties were measured by direct-current conduction test. The mechanical properties were evaluated by constant-load scratch test. A Rockwell-C indenter was plunged into the contact layer with a force of 5.0 N and swept along the surface of the layer. The depth of the trench formed by the scratch was measured by atomic force microscopy at a position 1.0 mm distant from the start point of the scratch test. The reciprocal value of the trench depth was used as the strength index of the contact layer.

4. Results and discussion

Figure 3 shows the XRD patterns of the samples with various thicknesses of Ti layers after annealing. The initial thicknesses of the Ti layers for the patterns (a), (b), (c), and (d) are 0, 16, 80, and 640 nm, respectively. The pattern (a) consists of the peaks of SiC and Ni_2Si. The peaks of NiSi and the free carbon do not appear in the pattern. Failure of detection of the NiSi layer indicates that the layer is very thin, which agrees with previous reports [4, 5]. On the other hand, the free carbon will not appear if it is amorphous. The pattern (b) also consists of the peaks of SiC and Ni_2Si, which is quite similar to the pattern (a). The initial thicknesses of the Ni and Ti layers of this sample were 100 and 16 nm, respectively, which corresponds approximately to the composition of Ni-10 mol%Ti. Therefore, the peaks of NiSi, TiC, and Ni_3Ti are expected to appear, in addition to the peaks appearing in the pattern. It is likely that the Ti layer is consumed by the interlayer reaction before the Ni adjacent to the SiC changes to Ni_3Ti. SiC reacts with this unchanged Ni. In this way, the XRD pattern of the sample is almost the same as that of pattern (a).

In the pattern (c) of Figure 3, a peak of TiC appears at the diffraction angle of 41.88°, indicating that a considerable amount of TiC is successfully formed at the interface and thus the formation of the free carbon is suppressed. Even in this pattern, however, the peaks of Ni-Ti intermetallic compounds are not found. It is hardly expected that a film of which total thickness is 180 nm is completely consumed by the reaction with SiC. Further analysis is needed to clarify the interfacial phenomena during annealing and the resultant structure.

The interfacial reaction fails to form Ni-Si compounds and TiC when the Ti layer is very thick. In the pattern (d) of Figure 3, Ti_5Si_3 and Ti_3SiC_2 are detected instead of TiC. Furthermore, the film after annealing is easily peeled off the substrate, even with careful handling. Therefore, the electrical and mechanical properties of this sample could not be measured.

The results shown in Figure 3 suggest that there is an optimum thickness of the Ti layer and/or an optimum thickness ratio between the Ti and Ni layers to facilitate the TiC formation. Among the samples shown in Figure 3, the sample with a 100-nm-thick Ni layer and an 80-nm-thick Ti layer seems to be the closest to the optimum condition. The composition of the film corresponds approximately to Ni-33 mol%Ti. The Ni-Ti binary phase diagram suggests that Ni_3Ti and NiTi are stable at this composition [15], i.e., the results agree with the thermodynamic consideration described in the previous section.

Figure 4 shows the electrical current–voltage profiles of the SiC/Ni, SiC/Ni/Ti (16 nm), SiC/Ni/Ti (80 nm), and SiC/C/Ti samples after annealing at 1273 K for 0 s. The plots of each sample lie on a

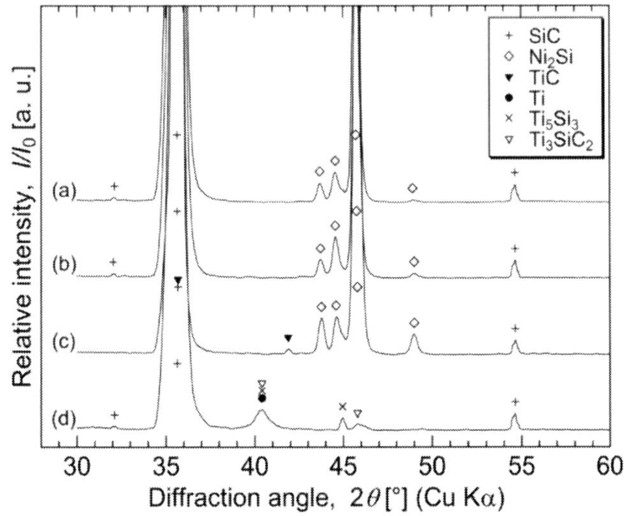

Figure 3. XRD patterns of the SiC/Ni/Ti samples after annealing at 1273 K for 0 s in vacuum. The initial thicknesses of the Ti layers are (a) 0, (b) 16, (c) 80, and (d) 640 nm.

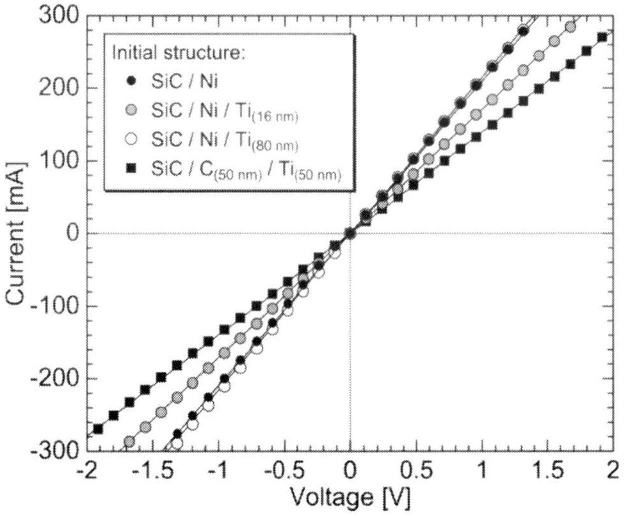

Figure 4. Current-voltage profiles of the samples after annealing at 1273 K for 0 s.

first-order least-square line in good agreement, indicating that all samples show ohmic conduction. Among the samples, the highest electrical conductance is obtained with the SiC/Ni/Ti (80 nm) and SiC/Ni samples. On the other hand, the lowest electrical conductance is obtained with the SiC/C/Ti samples. Although it is proven that TiC is an ohmic contact former for n-type SiC, NiSi contacts exhibit higher electrical conductance. Nevertheless, TiC is an indispensable phase for improving both the electrical and mechanical properties of the contact interface by eliminating the free carbon phase.

The electrical conductance of SiC/Ni/Ti (16 nm) appears lower than that of the SiC/Ni samples. Since the phase formed adjacent to SiC is considered to be the same as that of the SiC/Ni interface, the degradation of the conductance is attributed to the formation of a Ni-Ti intermetallic compound at the Ni/Ti interface. It should be noted that Ni-Ti intermetallic compounds are also formed at the interface of the SiC/Ni/Ti (80 nm) sample. It is considered that the degradation of the conductance in the film is compensated for by reducing the contact resistance, i.e., by replacing the non-ohmic SiC/free-carbon interface with an ohmic SiC/TiC interface.

Figure 5 shows the mechanical indices of the samples against their electrical conductance. The highest mechanical index is obtained with the SiC/C/Ti samples. This result evinces the successful formation of TiC at the interface, i.e., it is likely that the high hardness of TiC made the trench formed

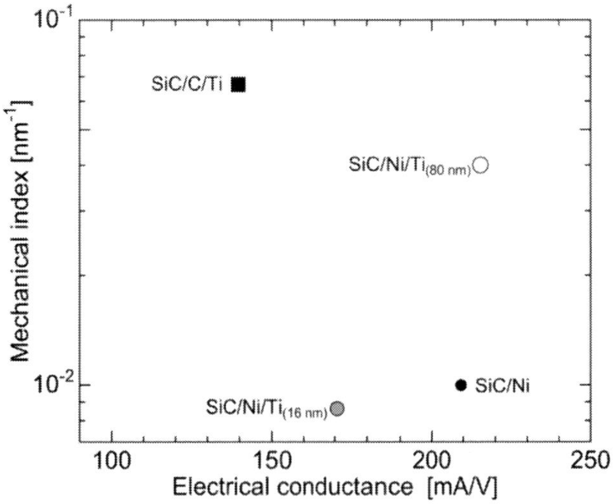

Figure 5. Relation between the electrical properties and mechanical properties of the samples prepared in the present study.

by scratching shallow. In addition, the adhesion between the contact electrode and the substrate is strong and ductile enough to endure the scratch test. The SiC/Ni/Ti (80 nm) sample also endures the scratch test. Although its mechanical index appears lower than those of the SiC/C/Ti samples, it achieves high electrical conductance and mechanical strength simultaneously. On the other hand, the mechanical indices of the SiC/Ni and SiC/Ni/Ti (16 nm) samples are almost one order of magnitude lower than those of the other two samples. The contact films on these samples are completely removed by scratching, i.e., the depths of their trenches are the same as the thicknesses of the contact films. Therefore, the fact that the mechanical index of the SiC/Ni/Ti (16 nm) sample is lower than that of SiC/Ni has no meaning, but the contact film of SiC/Ni/Ti (16 nm) is thicker than that of SiC/Ni. However, it should be noted that an inappropriate combination of thicknesses of the Ti and Ni layers deteriorates both the mechanical and electrical properties. The interfacial phenomena during the formation process and the resultant performance of the contact film are very sensitive to the thicknesses of the Ti and Ni layers.

5. Conclusions

The present paper proposed and demonstrated the improvement of the mechanical properties of a Ni-based ohmic contact for n-type SiC by the addition of a Ti layer. The following points became clear.

(1) SiC/Ni samples achieve low-resistance ohmic contact after annealing at 1273 K for a short time. However, the mechanical properties are significantly deteriorated by the annealing.

(2) SiC/C/Ti samples become ohmic after annealing at 1273 K for a short time. The contact films perform high mechanical properties. However, the electrical conductance is lower than that of SiC/Ni contacts after annealing.

(3) SiC/Ni/Ti samples also become ohmic after annealing at 1273 K for a short time. Appropriate thicknesses of the Ti and Ni layers significantly improve the mechanical properties, retaining high electrical conductivity. On the other hand, inappropriate setting of the thicknesses deteriorates both the mechanical and the electrical properties.

References

[1] Madsen L D 2001 *J. Electron. Mater.* **30** 1353
[2] Porter L M and Davis R F 1995 *Mater. Sci. Eng. B* **34** 83
[3] Pécz B 2001 *Appl. Surf. Sci.* **184** 287
[4] Tanimoto S, Miyabe M, Shiiyama T, Suzuki T, Yamaguchi H, Nakashima S and Okumura H 2011 *Mater. Sci. Forum* **679-680** 465
[5] Han S Y, Shin J-Y, Lee B-T and Lee J-L 2002 *J. Vac. Sci. Technol. B* **20** 1496
[6] Arvieu C, Manaud J P, Quenisset J M 2004 *J. Alloy. Compound.* **368** 116
[7] Zaima S, Shibata Y, Adachi H, Oshima C, Otani S, Aono M and Ishizawa Y 1985 *Surf. Sci.* **157** 380
[8] Park J H and Holloway P H 2005 *J. Vac. Sci. Technol. B* **23** 486
[9] Ohyanagi T, Onose Y and Watanabe A 2008 *J. Vac. Sci. Technol. B* **26** 1359
[10] 1997 *Handbook of Ternary Alloy Phase Diagrams CD-ROM* ed P Villars, A Prince and H Okamoto (Materials Park, Ohio: ASM International)
[11] Naka M, Feng J C and Schuster J C 1997 *Metall. Mater. Trans. A* **28** 1385
[12] Kubaschewski O, Alcock C B and Spencer P J 1993 *Materials Thermochemistry* 6th edition (Oxford, UK: Pergamon) p. 300
[13] Du Y, He C, Schuster J C, Liu S and Xu H 2006 *Z. Metallkd.* **97** 543
[14] Fukai T, Naka M, Shibayanagi T and Schuster J C 1997 *Trans. Join. Weld. Res. Inst.* **26-2** 27
[15] 1996 *Binary Alloy Phase Diagrams, 2nd edition plus updates on CD-ROM* ed T B Massalski, H Okamoto, P R Subramanian and L Kacprzak (Materials Park, Ohio: ASM International)
[16] Du Y and Schuster J C 1999 *Metall. Mater. Trans. A* **30A** 2409
[17] Maeda M and Takahashi Y 2013 *Int. J. Nanotechnol.* **10** 89

Nucleation and growth of Ti_3SiC_2 on SiC by interfacial reaction

Koji Ideguchi[1], Masakatsu Maeda[2] and Yasuo Takahashi[2]

[1] Graduate School of Engineering, Osaka University, 2-1 Yamadaoka, Suita, Osaka 565-0871, Japan
[2] Joining and Welding Research Institute, Osaka University, 11-1 Mihogaoka, Ibaraki, Osaka 567-0047, Japan

E-mail: maeda@casi.osaka-u.ac.jp

Abstract. In this study, we investigated the nucleation and growth of Ti_3SiC_2 by an interfacial reaction between a SiC substrate and a Ti/Al bilayered film at various temperatures. The specimens were prepared by depositing Ti/Al on a 4H-SiC substrate and subsequently annealing the substrate in a vacuum. The interfacial structures were analyzed by X-ray diffraction analysis and transmission electron microscopy. Ti_3SiC_2 nucleated even when annealed at a low temperature of 973 K. The Ti_3SiC_2 grains formed at 973 K were isolated and nonepitaxial with respect to the SiC substrate. By increasing the annealing temperature to 1273 K, the Ti_3SiC_2 grains became epitaxial with the SiC substrate, spreading preferentially on it, and coalesced with each other, forming a layer without grain boundaries.

1. Introduction

Silicon carbide (SiC) is a semiconductor compound that is suitable for next-generation power electronic devices [1]. However, to fully exploit the potential of SiC, it is essential that the ohmic contacts used are of low resistance [2, 3]. The Schottky barrier at the SiC/metal interface has to be lowered and/or thinned to achieve ohmic contacts. However, the deep valence band edge of SiC, which is approximately 6.5 eV under vacuum [4, 5], makes it difficult to form ohmic contacts on p-type SiC. When using pure metals, the formation of high Schottky barriers at the interfaces with p-type SiC cannot be prevented. One solution to this problem is to form a thin and uniform layer of Ti_3SiC_2 at the interface [6-8].

Ti_3SiC_2 is commonly formed by an interfacial reaction between SiC and Al_3Ti, starting from the multilayered structure of SiC/Ti/Al [6-8]. It is important to form Ti_3SiC_2 as a thin and uniform layer for manufacturing reliable SiC devices. The interfacial structure reported in the literature suggests that the Ti_3SiC_2 phase shows a distinct epitaxial relation with SiC [8] and grows predominantly perpendicular to its [0001] axis. Owing to this behavior, the growth tip of Ti_3SiC_2 grains can be separated from the SiC substrate, because the SiC surface is generally tilted with respect to the exact (0001) plane for polytype control [9]. On the other hand, the formation of Ti_3SiC_2 involves processing at the high temperature of 1273 K, which is not compatible with the manufacturing of electronic devices. A few studies have reported techniques of lowering the annealing temperature [10]. However, the nucleation and growth behaviors of Ti_3SiC_2 are not well understood. In particular, the behavior in the early stage, that is, while the individual grains of Ti_3SiC_2 are growing and forming a continuous layer on the SiC substrate, is important. This is because the crystallographic relation between Ti_3SiC_2

Content from this work may be used under the terms of the Creative Commons Attribution 3.0 licence. Any further distribution of this work must maintain attribution to the author(s) and the title of the work, journal citation and DOI.
Published under licence by IOP Publishing Ltd

and SiC, whether it is epitaxial or not, is defined at this stage. Understanding the behaviors at this stage is indispensable for controlling the interfacial reaction between Ti_3SiC_2 and SiC, as well as the structure of the resulting layer.

In the present study, the growth behavior of Ti_3SiC_2 was investigated, with the emphasis being on the temperature dependence of the crystallographic relation between Ti_3SiC_2 and SiC.

2. Experimental procedure

A single-crystal wafer of p-type 4H-SiC, which was tilted by 8° toward the $[11\bar{2}0]$ direction from the (0001) plane, was cut into 4-mm squares to be used as substrates. The thickness and specific sheet resistance of these SiC substrates were 370 μm and 1.368 Ω cm, respectively. Only the Si-terminated surface was used. The surfaces of the substrates were cleaned ultrasonically using acetone. A Ti (50 nm)/Al (600 nm) bilayered film was deposited on the substrates by radio frequency magnetron sputtering using high-purity targets of Ti and Al. Ar ion sputtering was performed before depositing the films. The specimens were annealed at a pressure of 1.3×10^{-3} Pa, while the temperature was varied between 973 and 1273 K using a resistance-heating vacuum furnace.

The interfacial structures of the specimens were analyzed by X-ray diffraction (XRD) analysis and transmission electron microscopy (TEM).

3. Results and discussion

Figure 1 shows the XRD patterns of the specimens annealed at 973, 1073, 1173, and 1273 K for 600 s. Figure 1(a) is the pattern of the specimen annealed at 973 K. Peaks attributable to Ti_3SiC_2 do not appear in the pattern, while these peaks appear in the patterns of the specimens annealed at temperatures of 1073 K and higher. Figure 1(b) is the pattern of the specimen annealed at 1073 K. Three peaks, at 39.44°, 40.52°, and 41.48°, which correspond to the $(10\bar{1}4)$, (0008), and $(10\bar{1}5)$ planes of Ti_3SiC_2, respectively, can be seen; the peaks have similar intensities. The appearance of these peaks indicates that Ti_3SiC_2 is formed by annealing at 1073 K for 600 s and that it has grown to an amount detectable by XRD analysis. However, the Ti_3SiC_2 layer formed at this temperature has a polycrystalline structure consisting of randomly oriented grains, that is, the epitaxial relation between Ti_3SiC_2 and SiC is not known. This is in contrast to the results reported by Tsukimoto et al. [8]. By increasing the annealing temperature to 1173 K, the relative intensities of the peaks corresponding to the $(10\bar{1}4)$ and $(10\bar{1}5)$ planes of Ti_3SiC_2 decrease, whereas the intensity of the peak attributable to

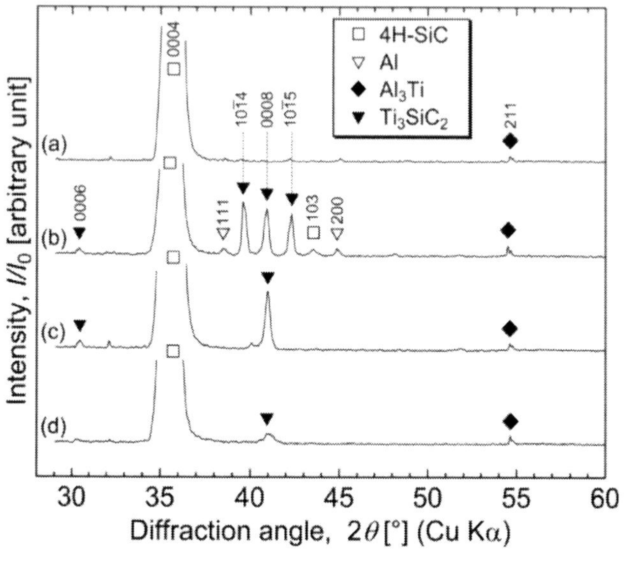

Figure 1. XRD patterns of the specimens annealed at various temperatures for 600 s. (a) 973 K, (b) 1073 K, (c) 1173 K, and (d) 1273 K.

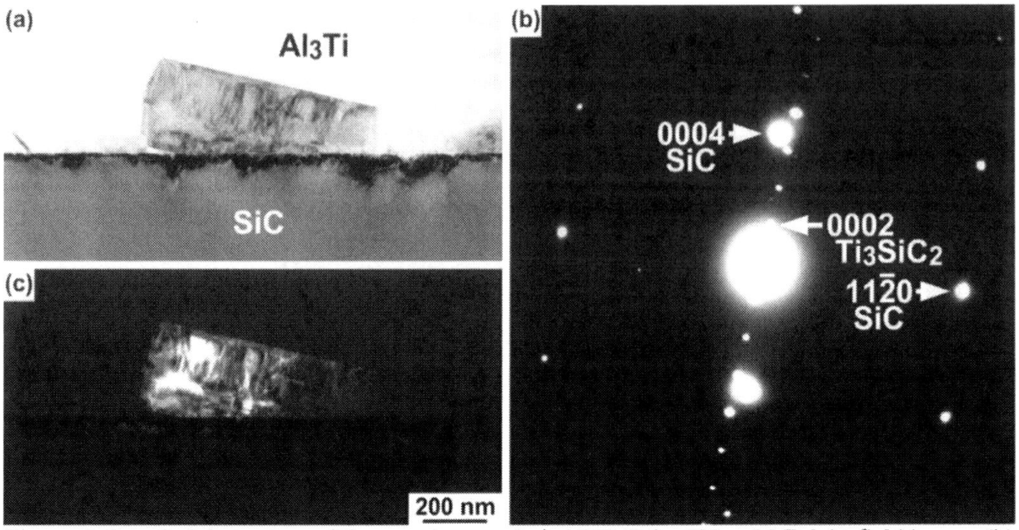

Figure 2. Interfacial structure of the sample annealed at 973 K for 600 s. (a) Bright-field image, (b) selected area electron diffraction pattern corresponding to the area shown in (a), and (c) dark-field image of the same area as (a), obtained using the Ti_3SiC_2 0004 diffraction.

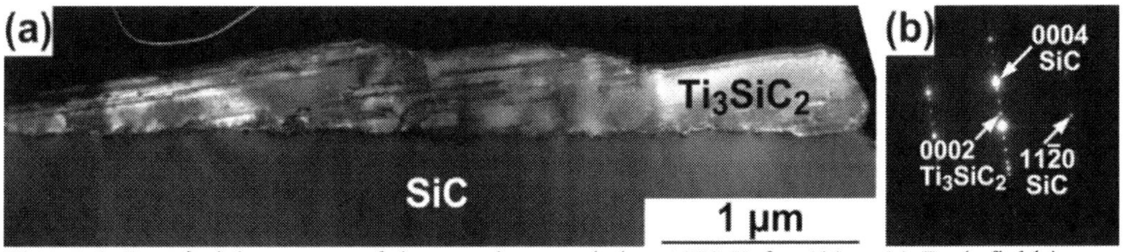

Figure 3. Interfacial structure of the sample annealed at 1273 K for 600 s. (a) Dark-field image obtained using the Ti_3SiC_2 0004 diffraction and (b) the corresponding selected area electron diffraction pattern.

the (0008) plane remains unchanged, as shown in Figure 1(c). Furthermore, only the peak of the (0008) plane is seen in the pattern of the specimen annealed at 1273 K, as can be confirmed from Figure 1(d). The differences in the changes in the relative intensities of the peaks indicate that Ti_3SiC_2 grains that have their (0001) plane parallel to the (0001) plane of SiC grow preferentially during annealing at temperatures of 1173 K and higher. This crystallographic relation corresponds to an epitaxial Ti_3SiC_2/SiC interface [8]. Therefore, there is a threshold temperature above which Ti_3SiC_2 grains that are epitaxial with the SiC substrate grow preferentially.

Figure 2 shows the interfacial structure of the specimen annealed at 973 K for 600 s. In the bright-field image (BFI) of the interface shown in Figure 2(a), a clear-faceted grain can be seen adjacent to the SiC substrate. The electron diffraction pattern (EDP) corresponding to the image, shown in Figure 2(b), suggests the existence of Ti_3SiC_2. The pattern also indicates that the (0001) plane of Ti_3SiC_2 is tilted to 2.2° with respect to that of SiC. An epitaxial relation between Ti_3SiC_2 and SiC was not observed for the specimens annealed at 973 K. The dark-field image (DFI) obtained using the diffraction pattern of the Ti_3SiC_2 layer shown in Figure 2(c) confirms that the grain observed in Figure 2(a) is of Ti_3SiC_2. It must be noted that peaks related to Ti_3SiC_2 do not appear in the XRD pattern of the specimen annealed under the same conditions (Figure 1(a)). This may be due to the small size and low number of Ti_3SiC_2 grains existing at the interface. It was difficult to investigate Ti_3SiC_2 grain growth in the sample annealed at 973 K, but the *I-V* (current-voltage) properties improved with an increase in the annealing time, t, with the characteristics becoming ohmic in nature at $t = 3600$s. It was revealed that Ti_3SiC_2 nuclei could grow to a small degree at 973 K or the interface between Ti_3SiC_2 grains and SiC might be changed to improve the I-V properties. It was, however, found that Ti_3SiC_2 seeds nucleated even at 973K, as shown in Figure 2.

The Ti_3SiC_2 grain in Figure 2(a) appears isolated. Some parts of the SiC surface are not covered by Ti_3SiC_2 grains. These facts indicate that the nucleation frequency of Ti_3SiC_2 is low. Furthermore, the position of the Ti_3SiC_2/SiC interface appears to be the same as the edge of the SiC substrate in the other parts that are not covered by Ti_3SiC_2 grains. Although the formation of Ti_3SiC_2 consumes Si and C from SiC and Ti from Al_3Ti, Ti_3SiC_2 grains are formed without penetrating the SiC substrate. On the basis of this fact, the Si and C for the nucleation of Ti_3SiC_2 are probably not supplied directly by the SiC substrate, but are obtained by diffusion through Al_3Ti.

Figure 3 shows the interfacial structure after annealing at 1273 K for 600 s. It can be seen from Figure 3(a) that the Ti_3SiC_2 grains have grown larger. No grain boundary is observed in the layer, that is, the layer consists of a single crystal. The grains are coalesced with each other, forming a continuous layer [8]. This is probably because the difference in the crystal orientations of the Ti_3SiC_2 grains is small. The Ti_3SiC_2 layer contains a number of stacking faults. On the other hand, the Ti_3SiC_2/SiC interface becomes rough, being different from that shown in Figure 2(a). The roughened interface implies that the Si and C for the growth of the Ti_3SiC_2 grains were supplied directly by the SiC substrate. It is assumed that the roughening occurs owing to the local strain field in the vicinity of the defects and/or dopant atoms in the SiC substrate. The electron diffraction pattern corresponding to this area, which is shown in Figure 3(b), indicates that the interface between Ti_3SiC_2 and SiC exhibits an epitaxial relation when annealing is performed at 1273 K.

These results suggest that the Ti_3SiC_2 grains easily establish an epitaxial relation with SiC when annealed at 1273 K, whereas they fail to be epitaxial when it is performed at 973 K. However, it is still unclear whether only the Ti_3SiC_2 grains epitaxial with SiC nucleate at 1273 K or whether nonepitaxial Ti_3SiC_2 grains are eliminated during the growth stage once they are formed. If the former mechanism holds true, then the temperature during the major part of the annealing process can be lowered by separating the Ti_3SiC_2 grain growth process from the Ti_3SiC_2 nucleation step.

To investigate further the nucleation behavior of Ti_3SiC_2 at 1273 K, another heat treatment was implemented. First, the specimens were heated to 1273 K. Immediately after this temperature had been reached, the specimens were cooled and held at 973 K for 600 s. Figure 4 shows the XRD pattern of

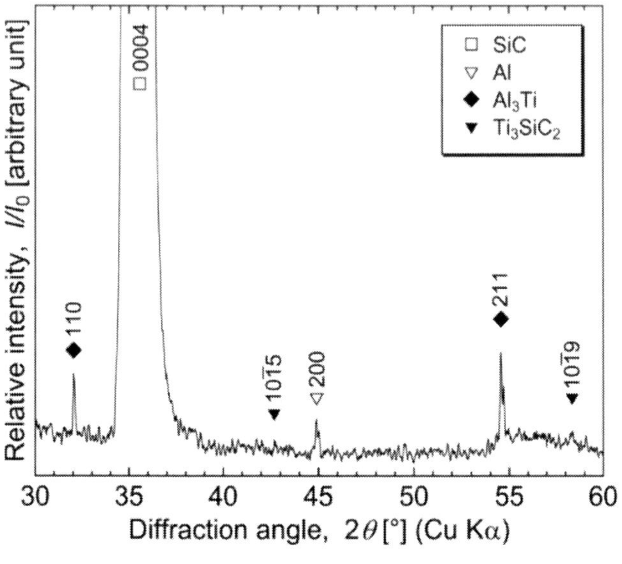

Figure 4. XRD pattern of the specimen annealed at 973 K for 600 s after reaching the maximum temperature of 1273 K for a very short time.

Table 1. Length of the *a*-axis and linear thermal expansion coefficients of 4H-SiC and Ti_3SiC_2.

Substance	Length of *a*-axis, a / nm	Linear thermal expansion coefficient, α / K^{-1}
4H-SiC (hexagonal)	0.3081 (ref. [11])	5.1×10^{-6} (ref. [13])
Ti_3SiC_2 (hexagonal)	0.3062 (ref. [12])	9.1×10^{-6} (ref. [13])

one specimen after annealing. Two weak peaks related to Ti_3SiC_2 appear in the pattern. In addition, when the specimens were annealed at 1073K for 0 s, clear peaks similar to the ones seen in Figure 1 (b) were observed; however, when the samples were annealed at 1273 K for 0 s, only weak peaks appeared. The reason for this phenomenon is not known. Both of these weak peaks were related to Ti_3SiC_2 that was nonepitaxial with SiC. It is suggested that randomly oriented (i.e., both epitaxial and nonepitaxial) Ti_3SiC_2 nuclei are formed even at 1273 K. However, only the epitaxial nuclei survive owing to the preferential growth at 1273 K.

The Ti_3SiC_2/SiC interface fails to be epitaxial when the specimens are annealed at 973 K. It is likely that the interfacial energy is not low at the low temperature, because of the *a*-axis lattice mismatch between SiC and Ti_3SiC_2. This lattice mismatch has to be compensated for by the elastic straining of the lattices of Ti_3SiC_2 and SiC, in order for an epitaxial interface to form. If the interfacial energy of the interface becomes higher than that of a randomly oriented interface because of the elastic straining, the interface will not be epitaxial. The lattice mismatch (Δa) can be expressed as

$$\Delta a = a_{\mathrm{SiC}}\left(1+\alpha_{\mathrm{SiC}}\Delta T\right) - a_{\mathrm{Ti_3SiC_2}}\left(1+\alpha_{\mathrm{Ti_3SiC_2}}\Delta T\right), \quad (1)$$

where a_i, α_i, and ΔT are the length of the hexagonal *a*-axis of the substance i (i = SiC or Ti_3SiC_2), the coefficient of linear thermal expansion of the substance i, and the temperature change from 293 K, respectively. The values of a_i and α_i are listed in Table 1. Although the actual coefficients of linear thermal expansion are anisotropic and temperature dependent, they are assumed to be constant in the temperature range considered in the present study. When the annealing temperature is increased, the lattice mismatch decreases and becomes zero at 1856 K. At 1173 K and 1273 K, the values of Δa are 8.3×10^{-4} and 7.1×10^{-4} nm, respectively. In addition, the elastic constants of SiC and Ti_3SiC_2 become lower at higher temperatures [14, 15]. Thus, the interfacial energy, which increases because of the lattice mismatch, decreased at higher temperatures. The present study revealed that the temperature for transitioning from a nonepitaxial interface to an epitaxial one lies between 973 and 1173 K.

4. Conclusions

The early stage of Ti_3SiC_2 formation at the interface between a SiC substrate and a Ti/Al bilayer film during annealing at temperatures ranging from 973 to 1273 K was analyzed through XRD analysis and TEM. The main conclusions of the study were the following.

(1) At every annealing temperature investigated in the present study, Ti_3SiC_2 nucleates adjacent to SiC. The Ti_3SiC_2 nuclei appear dispersed and isolated, indicating that the nucleation frequency is low. The orientation of each Ti_3SiC_2 nucleus is random.

(2) During annealing at temperatures of 1173 K and higher, randomly oriented Ti_3SiC_2 nuclei grow individually and form a polycrystalline layer. On the other hand, the nuclei epitaxial with the SiC substrate grow preferentially at temperatures higher than 1173 K. These epitaxial grains coalesce with each other and form a single-crystal layer.

References

[1] Krishna S, Scott R S and Baliga B J 1989 *IEEE Trans. Electron Devices* **36** 1811
[2] Porter L M and Davis R F 1995 *Mater. Sci. Eng. B* **34** 83
[3] Madsen L D 2001 *J. Electron. Mater.* **30** 1353
[4] van Haeringen W, Bobbert P A and Backes W H 1997 *Phys. Stat. Sol. B* **202** 63

[5] Davydov S Y 2007 *Semicond.* **41** 696
[6] Crofton J, Beyer L, Williams J R, Luckowski E D, Mohney S E and Delucca J M 1997 *Sol.-Stat. Electron.* **41** 1725
[7] Johnson B J and Capano M A 2004 *J. Appl. Phys.* **95** 5616
[8] Tsukimoto S, Nitta K, Sakai T, Moriyama M and Murakami M 2004 *J. Electron. Mater.* **33** 460
[9] Matsunami H 2004 *Jpn. J. Appl. Phys.* **43** 6835
[10] Tsukimoto S, Sakai T and Murakami M 2004 *J. Appl. Phys.* **96** 4976
[11] Powder Diffraction Files (Newtown Square, PA: ICDD) 29-1127
[12] Powder Diffraction Files (Newtown Square, PA: ICDD) 40-1132
[13] Pierson H O 1996 *Handbook of Refractory Carbides and Nitrides* (Amsterdam: Elsevier)
[14] Li Z and Bradt R C 1988 *Int. J. High Technol. Ceram.* **4** 1
[15] Radovic M, Barsoum M W, Ganguly A, Zhen T, Finkel P, Kalidindi S R and Lara-Curzio E 2006 *Acta Mater.* **54** 2757

Effect of crystal orientation on ohmic contact formation for n-type gallium nitride

Kota Kimura[1], Aiman bin Mohd Halil[1], Masakatsu Maeda[2] and Yasuo Takahashi[2]

[1] Graduate School of Engineering, Osaka University, 2-1 Yamadaoka, Suita, Osaka 565-0871, Japan
[2] Joining and Welding Research Institute, Osaka University, 11-1 Mihogaoka, Ibaraki, Osaka 567-0047, Japan

E-mail: maeda@casi.osaka-u.ac.jp

Abstract. The present paper describes the influence of surface orientation of n-type GaN on the electrical properties and the interfacial reaction between GaN and Ti during annealing. Although the contact formed on (0001) Ga-face of GaN performs the highest electrical conductance in the as-deposited state, the conductance deteriorates significantly by annealing even at a low temperature of 773 K. A considerable amount of Ti-Ga intermetallic compounds is formed at the deteriorated interfaces, to which the deterioration is attributed.

1. Introduction

Gallium nitride (GaN) is commonly applied in light-emitting and laser diodes [1]. GaN has also a breakthrough potential for high-frequency-operated power electronic devices [2]. These applications require low-resistance ohmic contacts at interfaces between the GaN and the metallic circuits. To form low-resistance ohmic contacts on GaN, it is necessary to reduce the height and/or width of the Schottky barrier. In the case of n-type GaN, the work function of the contact material has to be shallower than the conduction-band edge of GaN, which is 4.11 eV below the vacuum level [3]. TiN is known as one of those contact materials because it has a sufficiently shallow work function of 3.74 eV [3]. On the other hand, to decrease the thickness of the Schottky barrier, it is effective to increase the carrier density in the GaN just under the contact material, which can be accomplished by implantation of dopants. The formation of TiN by interfacial reaction between GaN and Ti generates nitrogen vacancies (V_N) in GaN, which work as donors at an energy level close to the conduction-band edge of GaN [4]. Therefore, it is preferable to form TiN on GaN by interfacial reaction between GaN and multilayered metallic film containing a Ti layer [5-8]. Furthermore, Maeda et al. have reported that V_N formed in the sub-interface of GaN play the dominant role in developing ohmic properties, instead of the formation of TiN adjacent to GaN [9].

The (0001) Ga-face of GaN has been used in most of the papers reporting ohmic contact formation. This is due to the restriction in the growth of GaN single-crystals, which are generally epitaxially grown on sapphire substrates by metal-organic chemical vapor deposition. Recently, some techniques to produce free-standing bulk GaN single-crystals have been developed [10, 11]. Although they are still not the mainstream of mass production, the quality and size of the single-crystals will be rapidly improved. Thus, GaN single-crystals with other surface orientations will soon be available. On the other hand, it has not been clarified whether or not the (0001) plane is the best orientation for

Content from this work may be used under the terms of the Creative Commons Attribution 3.0 licence. Any further distribution of this work must maintain attribution to the author(s) and the title of the work, journal citation and DOI.
Published under licence by IOP Publishing Ltd

achieving low-resistance conduction in the sheet and at the contact interface. Therefore, the electrical properties of the contacts on some representative surface orientations of GaN have to be compared.

The present study compares the electrical properties of GaN/Ti contacts formed on four different surface orientations of GaN: (0001) Ga-face, $(000\bar{1})$ N-face, $(10\bar{1}0)$, and $(11\bar{2}0)$.

2. Experimental procedure

Four kinds of free-standing n-type GaN substrates with surface orientations of (0001) Ga-face, $(000\bar{1})$ N-face, $(10\bar{1}0)$, and $(11\bar{2}0)$, respectively, were used. Hereinafter, the substrates are denoted by their surface orientations. A Ti layer was deposited on two separate areas with a 1.0 mm space between them on each substrate by radio-frequency (RF) magnetron sputtering. The substrates were fixed in the deposition chamber using 1.0-mm-wide Al ribbon masks. After evacuation of the chamber to 4.0×10^{-5} Pa, high-purity Ar of 8.0 Pa was introduced as the discharging gas. At first, the surfaces of the target and the substrates were sputter-cleaned consecutively. The RF power for sputter-cleaning of both the target and the substrates was 200 W. The sputter-cleaning time for the target was set at 600 s, whereas that for the substrates was set at 300 s. Finally, a layer of Ti was deposited under the RF power of 200 W and a deposition time of 600 s. The samples were annealed in a vacuum of 1.2×10^{-3} Pa. The annealing temperature was varied from 773 to 973 K. The samples were immediately cooled down in a vacuum furnace after reaching the target temperature, i.e., the samples were kept at the annealing temperature for a very short time (hereinafter the time is described as 0 s). Some of the samples were annealed at 973 K for 300 s in order to investigate the interfacial reaction behavior between GaN and Ti.

The interfacial structures were analyzed by X-ray diffraction (XRD) and transmission electron microscopy (TEM). Direct-current (DC) electrical conductivity was measured at 273 K by probing between the two Ti contacts separated by a distance of 1.0 mm formed on each sample.

3. Results and discussion

Figure 1 shows the DC current-voltage (*I*-*V*) profiles of as-deposited GaN/Ti contacts formed on the (0001), $(000\bar{1})$, $(10\bar{1}0)$, and $(11\bar{2}0)$ surfaces of GaN. Two important features appear in the figure. One is that the profiles of all the contacts show a linear *I*-*V* relation, indicating that all contacts are ohmic in the as-deposited state. Therefore, the electrical conductance of each sample is obtained as a single value from the gradient of each line. The *I*-*V* relation remains linear after annealing. The other feature is that the conductance changes depending on the surface orientation of GaN. The highest

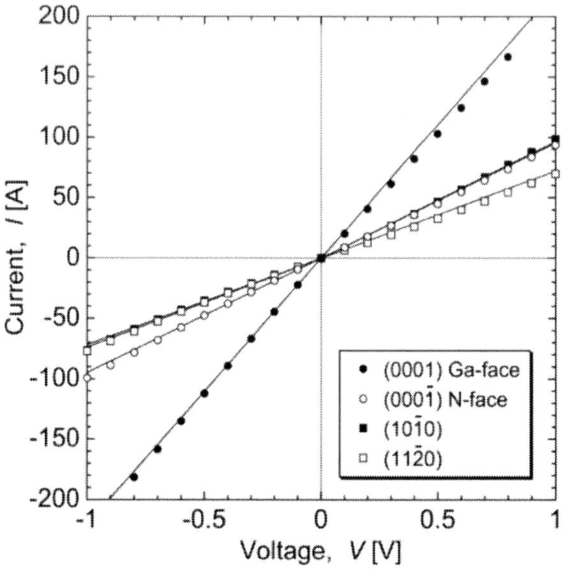

Figure 1. Current-voltage profiles of as-deposited GaN/Ti contacts formed on (0001), $(000\bar{1})$, $(10\bar{1}0)$, and $(11\bar{2}0)$ surfaces of GaN.

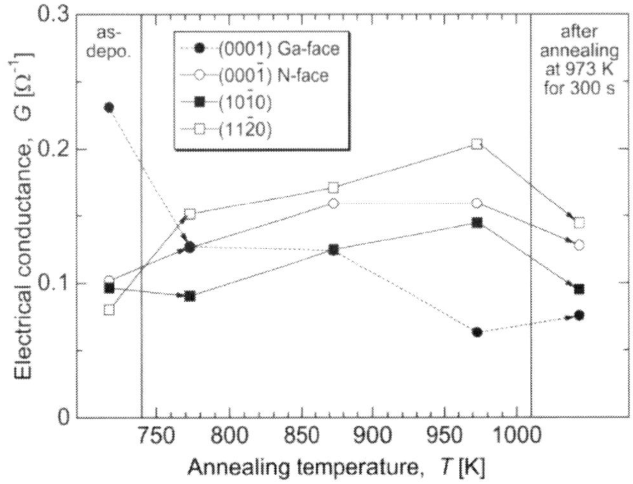

Figure 2. Changes in electrical conductance after annealing under various conditions.

conductance is obtained with the contact on (0001). On the other hand, the conductance of the contacts formed on the other three surface orientations of GaN appears similar, as shown in the left part of Figure 2. The conductance of the samples changes depending on the annealing temperature and time. The change behavior of the contacts on (0001) appears different from that of the contacts on all other surface orientations. Although the highest conductance is obtained with the contacts on (0001) in the as-deposited state, the conductance deteriorates significantly by annealing even at a low temperature of 773 K for a very short time, and it becomes worse by annealing at higher temperatures. On the other hand, the conductance of the contacts on all other surface orientations of GaN is improved at first by annealing, and the improvement becomes larger by annealing at a higher temperature. However, the conductance of these samples decreases upon extending the annealing time to 300 s at 973 K.

It is likely that the conductance is improved owing to the formation of V_N in the GaN sub-interface by interfacial reaction between GaN and Ti [9]. On the other hand, the deterioration could be caused by excessive reactions that form byproduct phases such as Ti-Ga compounds [12]. Thus, the deterioration of the conductance of the contacts is considered to be due to the transition from the interfacial reaction forming V_N in GaN to that forming Ti-Ga compounds at the interface. The transition starts earlier and at a lower temperature in the contacts formed on (0001) than in those formed on other surface orientations.

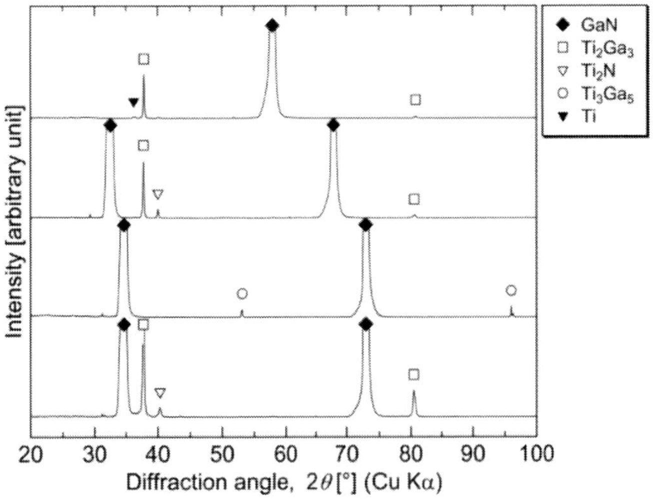

Figure 3. XRD patterns of the contact samples formed on (a) (11$\bar{2}$0), (b) (10$\bar{1}$0), (c) (000$\bar{1}$), and (d) (0001) after annealing at 973 K for 300 s.

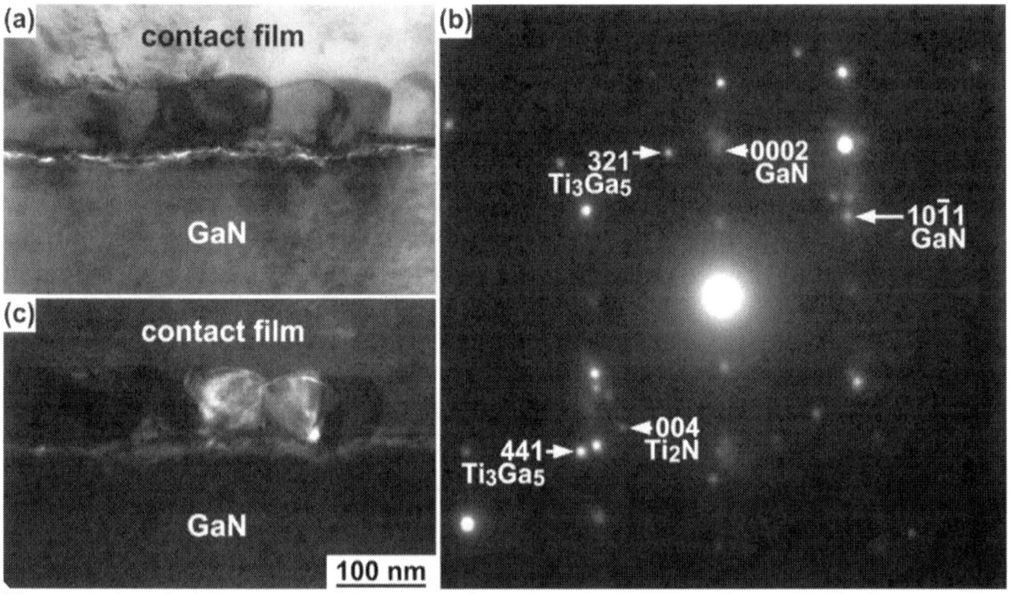

Figure 4. Interfacial structure of the contact formed on the (0001) surface of GaN after annealing at 973 K for 300 s. (a) Bright field image of the interface, (b) selected-area electron diffraction pattern corresponding to the area shown in (a), and (c) dark field image of the same area as (a) using Ti_3Ga_5 321 diffraction.

Figure 3 shows the XRD patterns of the contact samples after annealing at 973 K for 300 s. In the patterns of the samples formed on the (0001), ($10\bar{1}0$), and ($11\bar{2}0$) surfaces of GaN, peaks of GaN, Ti_2N, and Ti_2Ga_3 are identified, whereas GaN and Ti_3Ga_5 appear in the pattern of the sample formed on ($000\bar{1}$). Particularly, the peaks of Ti_2Ga_3 appearing in the pattern of the contact on (0001) are very strong, indicating that a considerable amount of Ti_2Ga_3 is formed at the interface. These results confirm the formation of Ti-Ga compounds at the interfaces with deteriorated electrical conductance. In addition, the reaction forming Ti-Ga compounds on the (0001) surface proceeds faster than that on the other surfaces of GaN.

Figure 4 shows a cross-sectional structure of the contact interface formed on the (0001) surface by annealing at 973 K for 300 s. In the bright-field image shown in Figure 4(a), a 60-nm-thick layer is formed adjacent to GaN by the interfacial reaction. The layer has a polycrystalline structure which consists of equiaxial grains. The electron diffraction pattern (EDP) corresponding to this area is shown in Figure 4(b) with some indices of diffraction spots. The EDP consists of a [$1\bar{2}10$] zone axis net pattern of GaN [13] and randomly oriented spots from Ti_2N [14] and Ti_3Ga_5 [15]. The dark-field image of the same area using the Ti_3Ga_5 (321) diffraction is shown in Figure 4(c). Some grains in the layer adjacent to GaN appear bright, proving that the layer is Ti_3Ga_5. This result suggests that the interfacial reaction forming V_N in GaN changes to that forming Ti-Ga compounds, i.e., from a donor-forming to a donor-eliminating reaction by excessive annealing. Therefore, the interfacial reaction has to be controlled to prevent the latter reaction in order to obtain a favorable ohmic interface.

4. Conclusions

Ti-based contacts were formed on four representative crystal surfaces of n-type GaN by deposition of Ti and subsequent annealing under various conditions to investigate the differences in the electrical properties and interfacial structure among the contacts on the four surfaces of GaN. The following points became clear:

(1) The behavior of electrical conductance by annealing appears different between the contacts on the (0001) Ga-face and those on the other three surface orientations of GaN. The contacts on the

(0001) Ga-face exhibit the highest conductance in the as-deposited state. However, the conductance deteriorates by annealing even at a low temperature of 773 K for a very short time. On the other hand, the conductance of the contacts on the other three surfaces improves at first and then deteriorates with annealing.

(2) The deterioration of the electrical conductance of the contacts is attributed to the formation of Ti-Ga intermetallic compounds adjacent to GaN.

(3) The surface orientation of GaN affects the rate and the incubation time of the interfacial reaction to form Ti-Ga compounds. Consequently, the annealing condition for the contact formation has to be determined by taking the anisotropy of the reaction rate into account.

References

[1] Akasaki I 2007 *J. Cryst. Growth* **300** 2
[2] Asif Khan M, Chen Q, Shur M S, Dermott B T, Higgins J A, Burm J, Schaff W J and Eastman L F 1997 *Sol.-Stat. Electron.* **41** 1555
[3] Mohammad S N 2004 *J. Appl. Phys.* **95** 7940
[4] Look D C, Reynolds D C, Hemsky J W, Sizelove J R, Jones R L and Molnar R J 1997 *Phys. Rev. Lett.* **79** 2273
[5] Luther B P, Mohney S E and Jackson T N 1998 *Semicond. Sci. Techonol.* **13** 1322
[6] Fan Z, Mohammad S N, Kim W, Aktas Ö, Botchkarev A E and Morkoç H 1996 Appl. Phys. Lett. **68** 1672
[7] Smith L L, Davis R F, Liu R J, Kim M J, and Carpenter R W 1999 *J. Mater. Res.* **14** 1032
[8] Maeda M and Takahashi Y 2013 *Int. J. Nanotechnol.* **10** 89
[9] Maeda M, Yamasaki T, and Takahashi Y, 2012 *J. Phys. Conf. Ser.* **379** 012020
[10] Motoki M, Okahisa T, Nakahata S, Matsumoto N, Kimura H, Kasai H, Takemoto K, Uematsu K, Ueno M, Kumagai Y, Koukitu A and Seki H 2002 *J. Cryst. Growth* **237-239** 912
[11] Mori Y, Imade M, Murakami K, Takazawa H, Imabayashi H, Todoroki Y, Kitamoto K, Maruyama M, Yoshimura M, Kitaoka Y and Sasaki T 2012 *J. Cryst. Growth* **350** 72
[12] Kimura K, Maeda M, and Takahashi Y 2013 *Mater. Trans.* **54** 895
[13] Powder Diffraction Files (Newtown Square, PA: ICDD) 50-792
[14] Powder Diffraction Files (Newtown Square, PA: ICDD) 23-1455
[15] Powder Diffraction Files (Newtown Square, PA: ICDD) 42-811

IOP Conf. Series: Materials Science and Engineering **61** (2014) 012034 doi:10.1088/1757-899X/61/1/012034

Effect of Ti_3SiC_2 formation on p-type GaN by vacuum annealing on the contact properties

Aiman bin Mohd Halil[1], Masakatsu Maeda[2] and Yasuo Takahashi[2]

[1] Graduate School of Engineering, Osaka University, 2-1 Yamadaoka, Suita, Osaka 565-0871, Japan

[2] Joining and Welding Research Institute, Osaka University, 11-1 Mihogaoka, Ibaraki, Osaka 567-0047, Japan

Email: aiman@casi.osaka-u.ac.jp

Abstract. In the present study, after the formation of Ti_3SiC_2 on p-type GaN by depositing Ti-Si-C ternary film with a composition stoichiometrically close to Ti_3SiC_2 and subsequent annealing at temperatures of 973 K and 1073 K (lower than the annealing temperature for a contact between p-type SiC and Ti_3SiC_2), the resulting contact properties were analysed by X-ray diffraction, a direct-current conduction test, and a Hall-effect measurement test. The X-ray diffraction results reveal that the Ti_3SiC_2 phase is successfully formed after the annealing. The direct-current conduction test shows that ohmic-like contacts are achieved after the formation of Ti_3SiC_2. However, the Hall-effect measurement test reveals that the dominant carrier type of the specimens is inverted from p-type to n-type even after the annealing at 973 K. The N vacancy formation during the annealing is likely the cause of this change. The contact properties of the annealed specimens are discussed because it is difficult to achieve ohmic contact formation between n-type GaN and Ti_3SiC_2.

1. Introduction

Because gallium nitride (GaN) can operate at higher temperatures, frequencies, voltages, and power densities than silicon, it is one of the most promising candidates as an alternative material for next-generation power electronic devices [1]. However, in spite of the significant development in the single-crystal growth and processing technologies for light-emitting diodes (LED) and laser diodes (LDs), the realization of GaN-based power electronic devices is still being held back by some major problems.

One of the main problem is the difficulties in forming an ohmic contact to p-type GaN. In the search for appropriate contact materials for p-type GaN, many options have been considered based on p-type GaAs contact materials such as Ni, Au, Pd and Pt [2-3]. This is due to the fact that these contact materials have been empirically proven to form good electrical properties with p-type GaAs and have been used for a wide variety of devices [4]. Jang et al. suggested that with a Au/Ni/Ru/Ag/Ni multilayered contact on p-type GaN, low resistance and thermally stable ohmic characteristics are achieved [2]. In another paper, Chen et al. suggested that a low-resistance ohmic contact is achieved with the formation of NiO and a specific microstructure by annealing Au/Ni/p-type GaN in an oxidative ambient [3]. Despite the many p-type GaN ohmic contact solutions that have been reported, most of the solutions are difficult to reproduce, even if the procedures are exactly followed. This

Content from this work may be used under the terms of the Creative Commons Attribution 3.0 licence. Any further distribution of this work must maintain attribution to the author(s) and the title of the work, journal citation and DOI.
Published under licence by IOP Publishing Ltd

suggests that the reported procedures are not solely responsible for the observed formations of the favoured contact interface structure.

On the other hand, a different approach to finding an appropriate contact material for p-type GaN has been used in the present study. GaN shows a higher degree of similarity with SiC than with GaAs in both crystallographic and electronic structures, as shown in Table 1 [5-8]. These are the most crucial factors which directly affect the formed contact structure and electrical properties. Thus, there is a high possibility that SiC contact materials are more suitable than GaAs contact materials for the formation of an ohmic contact to GaN.

The present study demonstrates this idea by forming Ti_3SiC_2 contacts on p-type GaN and examining the properties of the contacts. It has been reported that good ohmic contact has been achieved with p-type SiC and Ti_3SiC_2 contacts by annealing at 1273 K [9-14]. However, numerous investigations of p-type GaN have suggested that the elevated temperature during the contact formation can cause formation of N vacancies, which consequently reduce the hole concentration of the p-type GaN [15]. These vacancies are known to act as n-type dopant atoms with a donor level very close to the conduction band edge of n-type GaN [16]. Fujimoto et al. reported that Hall-effect measurements revealed that an n-type conductive layer was formed near the p-GaN surfaces by annealing at 1073 K in vacuum [17]. In this study, the formation of Ti_3SiC_2 is attempted by annealing at 973 K and 1073 K. Although there is possibility that the hole concentration of the p-type GaN is reduced by annealing at these temperatures, the investigation of the contact properties of p-type GaN and Ti_3SiC_2 is an important step to understand the practicality of this contact structure.

In the present study, Ti_3SiC_2 is formed on p-type GaN by 1) depositing a Ti-Si-C ternary film with a composition stoichiometrically equivalent to Ti_3SiC_2 on p-type GaN and 2) subsequent annealing. By forming Ti_3SiC_2 from the ternary film, no reaction is expected to occur between the deposited film and GaN, which would suppress the formation of N vacancies and lead to the preservation of the dominant carrier type of p-type GaN.

The direct-current conduction test and Hall-effect measurement is needed to examine the effect of the formation of Ti_3SiC_2 by annealing at 973 K and 1073 K. With the preservation of p-type carriers (holes) as the dominant carrier type, formation of an ohmic-like contact between p-type GaN and Ti_3SiC_2 is expected. However, with the dominant carrier type inverted from p-type to n-type, formation of a non-ohmic contact between the n-type GaN and Ti_3SiC_2 is expected.

In the present study, the effects of Ti_3SiC_2 formation on p-type GaN by annealing at 973K and 1073 K and the relation of these effect to the contact properties are discussed.

Table 1. Properties of GaN, SiC and GaAs.

Properties	GaN	4H-SiC	GaAs
Crystal system	Hexagonal	Hexagonal	Cubic
Lattice parameter, a / nm	0.3189	0.3073	0.5653
c / nm	0.5186	1.0053	-
Bandgap, E_g / eV	3.39	3.26	1.42
Electron affinity, χ / eV	4.11	3.20	4.07

2. Experimental procedure

The substrates used in the present study were 2.0-μm-thick p-type GaN epitaxially grown on a 330-μm-thick sapphire (0001) wafer with a 2.3-μm-thick undoped GaN buffer layer. The surface orientation and carrier density of the substrate were (0001) Ga-face and 3×10^{17} cm^{-3}, respectively. The wafer was cut into 4.0-mm-square substrates. Before the deposition process, the substrates were cleaned with acetone while applying ultrasonic vibration. Then, the substrates were fixed in a radio-frequency magnetron sputter deposition apparatus using 2.0-mm-wide Al masking ribbons to form 1.0-mm-square contacts at the corners of the substrates, as shown in figure 1. Before the sputter deposition, the surfaces of the Ti-Si-C alloy target and the substrates were sputter-cleaned to remove the native oxide layer. The sputter deposition of Ti-Si-C ternary films on the substrates was performed immediately after the sputter cleaning. Both the sputter cleaning and sputter deposition were performed under 0.8 Pa of 99.9999% high-purity Ar under a radio-frequency power of 200 W. The target used in the sputter deposition process was a Ti-Si-C ternary alloy sintered disk with a composition of Ti-19mol%Si-17mol%C. Sputter deposition for 600 s with this target results in the successfully formation of a 200-nm-thick film with a composition of Ti-17mol%Si-33mol%C on all p-type GaN substrates. The composition of the ternary films was close to the stoichiometric composition of Ti_3SiC_2. Some of the deposited samples were subjected to annealing at 973 K or 1073 K in a vacuum pressure of 1.3×10^{-3} Pa. The specimens were cooled down immediately after reaching the targeted temperature. The heating rate and cooling rate between 600 K and 1073 K are approximately 0.2 K/s and 0.7 K/s, respectively.

The contact structure and electrical properties of the specimens were analysed by X-ray diffraction (XRD) and a direct-current (DC) conduction test at room temperature. The electrical conduction profiles of the specimens were measured between two adjacent contacts. Finally, the dominant carrier type of the annealed specimens was analysed by a Hall-effect measurement test using the van der Pauw method at room temperature.

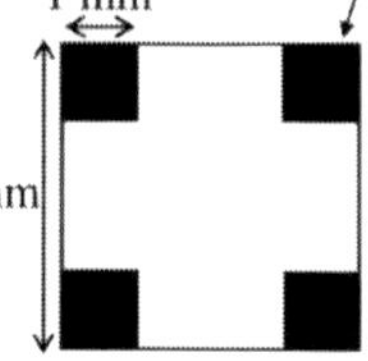

Figure 1. Prepared specimens.

3. Results and discussion

Figure 2 shows the XRD patterns of the specimens in the as-deposited state and after the annealing at 973 K and 1073 K. In the XRD pattern of the as-deposited specimen shown in figure 2(a), a weak peak from Ti appears in addition to the strong peaks, which are identified as being from GaN and sapphire. Other than that, no peak related to Ti-Si-C phases was found. This analysis of as-deposited specimens indicates that the deposited film is in an amorphous state. The XRD for the annealed specimens appear different from that of the as-deposited specimen, as shown in figure 2(b) and (c). Peaks corresponding to various crystallographic planes of Ti_3SiC_2 were identified. As a result of reducing the annealing temperature to 973 K from 1073 K, the residual Ti phase appeared, and fewer peaks corresponding to crystallographic planes of Ti_3SiC_2 were identified. The structural change of each of the two annealed specimens indicates that the Ti_3SiC_2 has a randomly oriented polycrystalline structure for each of the specimens. This proves that the Ti_3SiC_2 phase can be formed on GaN by deposition of the Ti-Si-C ternary film stoichiometrically close to Ti_3SiC_2 and subsequent annealing at 973 K or 1073 K.

It has been suggested that p-type SiC/Ti/Al needs to be annealed up to 1273 K for 120 s or longer to form Ti_3SiC_2 phase on SiC [14]. Hence, the Ti_3SiC_2 formation method applied in the present study has the advantages of lowering the annealing temperature and shortening the holding time. The lowering of the annealing temperature was achieved due to the fact that the deposited film is in a thermodynamically unstable state, resulting in the crystallization process that forms Ti_3SiC_2 phase being triggered at a lower temperature.

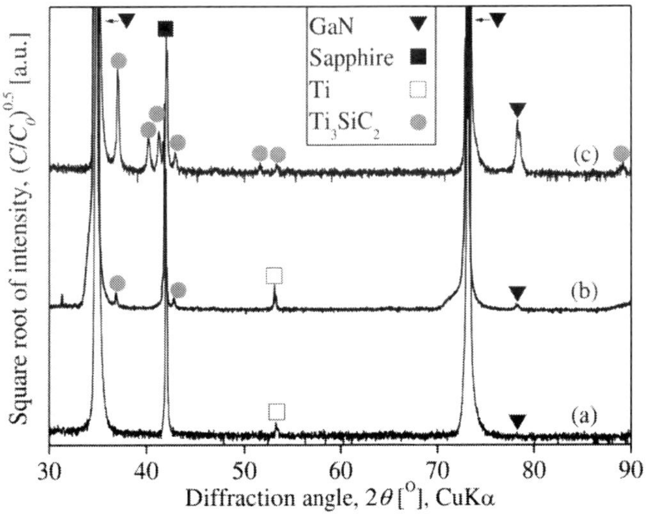

Figure 2. XRD patterns of the specimens. (a) As-deposited, (b) annealed at 973 K and (c) annealed at 1073 K.

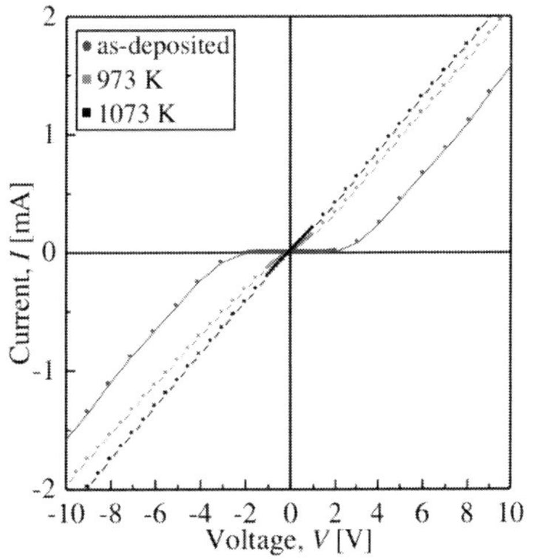

Figure 3. Electrical conduction profiles of the specimens: as-deposited, annealed at 973 K and annealed at 1073 K.

Figure 3 shows the changes in the electrical conduction profiles caused by annealing the specimens at 973 K or 1073 K. The curve of the as-deposited specimen shows a non-linear relation between voltage and current, indicating that the formed contact is non-ohmic. On the other hand, near-linear relations between voltage and current are observed for both of the annealed specimens, indicating that ohmic-like contacts have been successfully formed by the formation of Ti_3SiC_2 during annealing.

To investigate the dominant carrier type of the two annealed specimens, the Hall-effect measurement test has been performed at room temperature. The results revealed that the dominant carrier type of both specimens are n-type, i.e., the dominant carrier-type of the specimens changes from p-type to n-type as a result of the annealing at 973 K and 1073 K. The factor that is likely to be the cause of this change is the increasing of N vacancies in the p-type GaN subsurface resulting from the out-diffusion of N atoms during the annealing. These results indicate that the observed non-ohmic contact of the as-deposited specimen is formed between the p-type GaN and the deposited Ti-S-C ternary film. On the other hand, the observed ohmic-like contacts of the annealed specimens are formed between the n-type GaN (inverted from p-type GaN by annealing) and phases formed within the film during annealing. In the previous study, transmission electron microscopy revealed that other than the Ti_3SiC_2 phase, Ti_5Si_3 and $TiSi_2$ phases were also formed during the annealing [18]. Due to the high work function of Ti_3SiC_2 (5.07 eV), it is most likely that the observed ohmic-like contacts of the annealed specimens are formed between the n-type GaN and these byproducts. The work functions of Ti_5Si_3 and $TiSi_2$, which are respectively 3.71 eV and 4.10 eV, are lower compared to that of Ti_3SiC_2. In another study, Maeda et al. reported that with an enhanced interfacial reaction and the formation of N vacancies, ohmic contacts can be formed between n-type GaN and Ti even without annealing [19]. Thus, it is also possible that the observed ohmic-like contact of the annealed specimens is formed between n-type GaN and the residual Ti phase. Additionally, Mohammed reported that n-type GaN ohmic contacts can also be achieved by the formation of the TiN phase through the reaction between GaN and Ti film during the annealing [16]. Although the TiN phase was not found in the XRD patterns in this study, there is a possibility that the observed ohmic-like contacts of the annealed specimens are between n-type GaN and TiN.

To form a contact between p-type GaN and Ti_3SiC_2, a lower annealing temperature to form Ti_3SiC_2 is needed so that the dominant carrier type of the p-type GaN can be retained. For this purpose, an unconventional Ti_3SiC_2 formation method such as multilayer deposition by magnetron sputtering can be adopted [20].

4. Conclusions

A Ti-Si-C ternary film with a composition stoichiometrically close to Ti_3SiC_2 has been deposited on p-type GaN. By annealing the film at temperatures of 973 K and 1073 K, the Ti_3SiC_2 phase has been successfully formed. The direct-current conduction test shows that ohmic-like contacts have been achieved after the formation of Ti_3SiC_2. However, the Hall-effect measurement test reveals that the dominant carrier type of the specimens has changed from p-type to n-type even after the annealing at 973 K. The formation of N vacancies during the annealing is likely the cause of this change. The ohmic-like contacts are likely formed between n-type GaN and the byproducts of Ti_3SiC_2 formation (Ti_5Si_3 and $TiSi_2$), the residual Ti phase and/or the TiN phase. To form contacts between p-type GaN and Ti_3SiC_2, the annealing temperature to form Ti_3SiC_2 needs to be lowered so that the dominant carrier type of the p-type GaN can be retained.

References

[1] Madsen L D 2001 *J. Electron. Mater.* **30** 1353
[2] Jang H W and Lee J L 2004 *Appl. Phys. Lett.* **85** 4421
[3] Chen L-C, Ho K-K, Jong C-H, Chiu C C, Shih K-K, Chen F-R, Kai J J and Chang L 2000 *Appl. Phys. Lett.* **76** 3703
[4] Baca A G, Ren F, Zolper J C, Briggs R D and Pearton S J 1997 *Thin Solid Films* **308-309** 599
[5] Davydov S Y 2007 *Semicond.* **41** 696
[6] Sumakeris J, Sitar Z, Ailey-Trent K S, More K L and Davis R F 1993 *Thin Solid Films* **225** 244
[7] Gao M, Tsukimoto S, Goss S H, Tumakha S P, Onishi T, Murakami M and Brillson L J 2007 *J. Electon. Mater.* **36** 277
[8] Okumura H, Ohta K, Ando K, Rühle W W, Nagatomo T and Yoshida S 1997 *Solid State Electron.* **41** 201
[9] Crofton J, Beyer L, Williams J R, Luckowski E D, Mohney S E and Delucca J M 1997 *Solid State Electron.* **41** 1725
[10] Johnson B J and Capano M A 2004 *J. Appl. Phys.* **95** 5616
[11] Tsukimoto S, Nitta K, Sakai T, Moriyama M and Murakami M 2004 *J. Electron. Mater.* **33** 460
[12] Veisz B and Pécz B 2004 *Appl. Surf. Sci.* **233** 360
[13] Tsukimoto S, Ito K, Wang Z, Saito M, Ikuhara Y and Murakami M 2009 *Mater. Trans.* **50** 1071
[14] Maeda M and Takahashi Y 2013 *Int J. Nanotechnology* **10** 89
[15] Myers S M and Wright A F 2006 *J. Appl. Phys.* **99** 113506
[16] Mohammad S N 2004 *J. Appl. Phys.* **95** 7940
[17] Fujimoto I, Asamizu H, Shimada M, Moriyama M, Shibata N and Murakami M 2003 *J. Electron. Mater.* **32** 957

[18] Aiman bin Mohd Halil, Maeda M and Takahashi Y 2013 *Mater. Trans.* **54** No.6 890
[19] Maeda M, Yamasaki T and Takahashi Y 2012 *Trans. JWRI* **41** No. 1 45
[20] Vishnyakov V, Lu J, Eklund P, Hultman L and Colligon J 2013 *Vacuum* **93** 56

Bondability of Mg_2Si element to Ni electrode using Al for thermoelectric modules

Tomotake Tohei[1], Shinichi Fujiwara[1], Takahiro Jinushi[2] and Zenzo Ishijima[2]

[1]Yokohama Research Laboratory, Hitachi. Ltd., 292, Yoshida-cho, Totsuka, Yokohama, Kanagawa, 244-0817, Japan

[2]Hitachi Chemical Co., Ltd., Hitachi Powdered Metals, 2-1, Minoridai 5-chome, Matsudo, Chiba, 270-2295, Japan

E-mail: tomotake.tohei.yu@hitachi.com

Abstract. The purpose of this study has been to develop a low cost bonding technique for thermoelectric Mg_2Si/Si-Ge modules that provides reliable bonding. Aluminum was chosen as an alternative material to conventional silver alloy braze because of its cost advantage and bondability. The shear strength of an aluminum joint between a Mg_2Si element and nickel electrode was 19 MPa. The generation capacity of a prototype Mg_2Si/Si-Ge twin couple module was about 20% higher than that of a conventional Si-Ge/Si-Ge twin couple module at 923 K (ΔT = 620 K).

1. Introduction

The use of renewable energy resources such as sunlight, water flow and wind is expected to provide an alternative to fossil fuels with the aim of reducing CO_2 emissions [1-3]. Power generation from heat recovery, where electricity is generated from the recovery of exhausted heat, has also advanced for the same reason. Thermoelectric generation is one such technology for heat energy recovery. Thermoelectric modules can directly convert heat energy into electrical energy [4]. However, these modules have only been used in special applications, such as satellite power supply [5], because of their high cost and low conversion rate to electricity.

A thermoelectric generation system offers many advantages, including ease of maintenance, because of its simple structure and lack of moving parts. Therefore, many researchers have developed thermoelectric elements with higher conversion efficiencies using super lattice structures, nanocrystals, and new structure modules [6-14].

A thermoelectric module can improve the mileage of a vehicle by utilizing its exhaust heat. However, applying a thermoelectric generator in vehicles presents many challenges, such as the need to reduce heat stress in the bonding region between the thermoelectric element and electrodes, to develop a refractory structure for the thermoelectric module, and to develop a low cost and durable thermoelectric generator that can accommodate a wide range of operating temperatures.

We have previously reported that a Si-Ge thermoelectric element and electrode can be bonded with high reliability using low cost aluminum [15,16]. In this study, we have performed fundamental research focused on an aluminum diffusion bonding technique for mid- to high-temperature (about 573–873 K) thermoelectric modules, specifically the use of aluminum for bonding a Mg_2Si

Content from this work may be used under the terms of the Creative Commons Attribution 3.0 licence. Any further distribution of this work must maintain attribution to the author(s) and the title of the work, journal citation and DOI.
Published under licence by IOP Publishing Ltd

thermoelectric element to a nickel electrode. A prototype Mg_2Si/Si-Ge twin couple module was then assembled to evaluate the power generation of this thermoelectric module.

2. Experimental details

2.1. Bonding conditions

Table 1 shows the bonding conditions between the thermoelectric element and electrodes. Aluminum foil of 99 % purity was selected as the bonding material. Mg_2Si was chosen as the thermoelectric element for this research because it has a better conversion rate than Si-Ge from about 573 K to 873 K. Nickel was selected as the electrode material because its coefficient of thermal expansion (CTE) is 13.4 ppm/K, which is close to that for Mg_2Si (15.5 ppm/K). No. 1 and 2 were chosen as low pressure bonding (6.1×10^3 Pa), and No. 3 was chosen at high pressure bonding (3.0×10^7 Pa). In addition, with No. 2, the Mg_2Si surface was polished before bonding. Further, the Mg_2Si element is close to the aluminum foil under high pressure (3.0×10^7 Pa). The eutectic reaction of aluminum, magnesium and silicon is lower than the melting point of pure aluminum. Consequently, the Mg_2Si element and nickel electrode can be joined at temperatures below the melting point of pure aluminum due to interface interdiffusion between Mg_2Si and aluminum. In addition, aluminum and nickel produce a Al-Ni intermetallic compound with a high melting point of over 1127 K. These Al-Ni intermetallic compound layers show great promise as a high-melting-point joint between the Mg_2Si element and nickel electrode after bonding. Therefore, bonding temperature of test No. 3 was 893 K lower than the melting point of aluminum.

Figure 1 provides a schematic view of the bonding process and also shows the specimen size. The surface of each material was cleaned with acetone. After cleaning, the aluminum foil was then placed between the nickel electrode and Mg_2Si element, with the bonding process being performed in the chamber under an atmosphere of pure nitrogen. The bonding pressure was set to either 6.1×10^3 Pa or 3.0×10^7 Pa, and the temperature of the hotplate was set to 953 or 893 K for 60 s.

2.2. Shear strength test

Figure 2 presents a schematic view of the shear test for the Mg_2Si element/aluminum joint/nickel electrode thermoelectric structure. The nickel electrode was held on a shear tester stage, and the test was executed with a shear speed of 50 μm/s and shear height of 200 μm. The available loading range was under 1000 N. Strength was then calculated by dividing the measured shear strength by 3.7×3.7 mm^2.

2.3. Metallurgical observation

Characterization of the Mg_2Si to nickel joints made with aluminum was conducted by field emission-scanning electron microscopy (FE-SEM). Cross-sections of the bonding interface were prepared after the joint sample was first polished with polishing papers and diamond paste. FE-SEM observations and quantitative analysis by energy-dispersive X-ray spectroscopy (EDX) were then performed.

Table 1. Bonding conditions between the thermoelectric element and electrodes.

	Jointing material			Device						
No.	Material	Thickness (μm/s)	Elect rode	Material	Surface condition	Atmos phere	Ramp temp.Rate (K/s)	Max. Temp. (K)	Holding time(s)	Pressure (Pa)
1					As is			953		6.1×10^3
2	Al-foil	25	Ni	Mg_2Si	Polishing	N_2	1		60	
3					As is			893		3.0×10^7

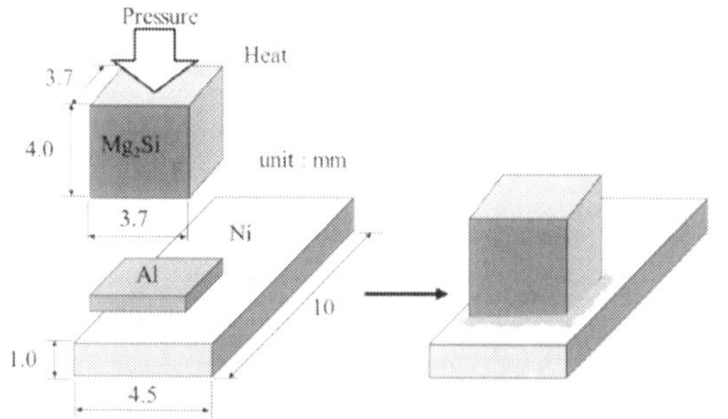

Figure 1. Schematic view of the bonding process to create a joint between the Mg_2Si element and nickel electrode with aluminum foil.

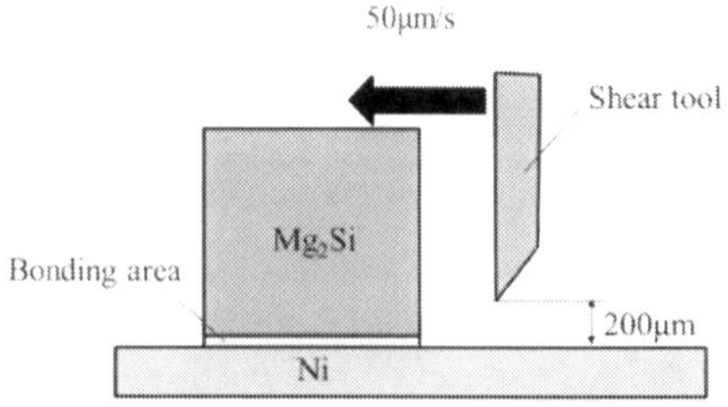

Figure 2. Schematic view of the shear test for the aluminum joint between the Mg_2Si element and nickel electrode.

3. Results and discussion

3.1. Bondability of the Mg_2Si element and nickel electrode with aluminum.

Figure 3 shows the results of the shear tests under each set of bonding conditions. The test number corresponds to bonding conditions 1, 2 or 3 in Table 1. From this, it is evident that No. 1 produced the lowest strength, whereas No. 3 produced the greatest strength. In the case of the latter, the shear strength of the joint between the Mg_2Si element and nickel electrode was 19 MPa.

Figure 4 shows cross-sectional views of the bonding interface between the Mg_2Si device and nickel electrode before and after the shear tests. The chart numbers correspond to the test numbers in Figure 3. Typically, an oxide film is present on the bonding surface, thus reducing its wettability and bondability [17]. However, a good joint without a discernible gap was confirmed under a bonding pressure of 6.1×10^{-3} Pa in the bonding of a Si-Ge element and aluminum foil without flux [15,16]. As shown in Figure 4, a gap exists in No. 1 at the interface between the Mg_2Si element and aluminum even before the shear test. In addition, the observation of a flat Mg_2Si surface indicates that Mg_2Si does seem not to react with molten aluminum. This is probably caused by the reduced wettability resulting from the magnesium oxide layer on top of the Mg_2Si element. Also, from the cross sectional view of Figure 4 No. 2 (Surface treated Mg_2Si element), a wavy interface is evident between the Mg_2Si element surface and aluminum joint. This can be inferred to mean that the Mg_2Si was partially dissolved by molten aluminum. However, a horizontal streaky gap exists in the aluminum joint from the cross sectional view before shear testing. This means that it is the newly formed surface of the Mg_2Si that reacts with molten aluminum, as the partial oxide film on the Mg_2Si surface is broken by the bonding pressure and the flow of molten aluminum. On the other hand, the oxide film remained as part of the aluminum joint, at a site distant from the Mg_2Si surface. Considering these results, the streaky gap in the aluminum joint was most likely formed by this residual oxide film.

Even with a surface-treated Mg_2Si element, the Mg_2Si surface is re-oxidized well before the bonding process between Mg_2Si and aluminum. As shown in Figure 4 No. 2 and Figure 5 (b), the Mg_2Si element/aluminum joint/nickel electrode structure is broken in the Mg_2Si element at the partially destroyed oxide film. Nevertheless, the joint between the Mg_2Si element and nickel electrode separates easily at the streaky gap part of the aluminum joint.

In contrast, in the case of No. 3, a streaky gap was not observed in the cross sectional view of the joint between the Mg_2Si element and nickel electrode. This is probably a result of the oxide film and molten aluminum being discharged at the same time in the direction of the edge by the high bonding pressure (3.0×10^7Pa), thus preventing a streaky gap from being generated. In addition, observation of the cross sectional view after shear test showed that the Mg_2Si element/nickel electrode bonding structure created using aluminum was broken in the Mg_2Si element. Thus, the joint between Mg_2Si element and nickel electrode of No. 3 has a high strength without a gap.

Furthermore, the No. 3 joint formed layers on an Al-Ni intermetallic compound (IMC) that included magnesium and silicon. It therefore seems reasonable to assume that the Mg_2Si elements were partially dissolved by the molten aluminum.

3.2. Power generation with a prototype Si-Ge/Mg_2Si hybrid module

A prototype Si-Ge/Mg_2Si twin couple module was bonded based on the results of the shear tests. Specifically, No. 3 of Table 1 was chosen as a reliable bonding condition. A Si-Ge element was used as the P-type semiconductor, and an N-type Mg_2Si element was used as the other leg. However, given that the CTE of Si-Ge (3.5 ppm/K) is very different to that of Mg_2Si (15.5 ppm/K), we prepared a molybdenum/nickel hybrid electrode to enable CTE matching between the thermoelectric element and electrode. Figure 6 shows an exterior view of the prototype Si-Ge/Mg_2Si twin couple module. The bondability of a Si-Ge element and molybdenum electrode using aluminum is discussed elsewhere [15,16].

Figure 7 shows the output power generation of the prototype Si-Ge/Mg_2Si twin couple module with the upper electrodes at 923 K and the lower electrodes at 303 K ($\Delta T = 620$ K). A power generation of 1 W/cm^2 was successfully achieved without incurring any damage to the module. This is 20% higher than the power generation of a Si-Ge/Si-Ge twin couple module, and thus a Si-Ge/Mg_2Si hybrid module clearly provides a significant improvement at mid- to high-temperature ranges.

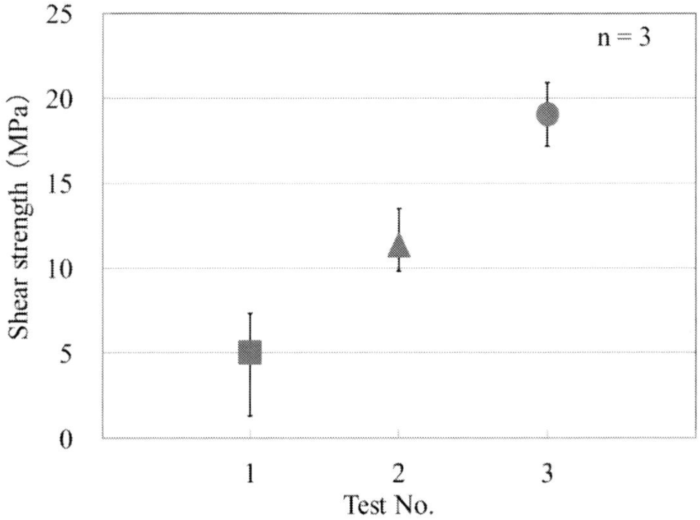

Figure 3. Results of shear strength tests between the Mg_2Si element and nickel electrode ('n' is the number of samples for each test condition).

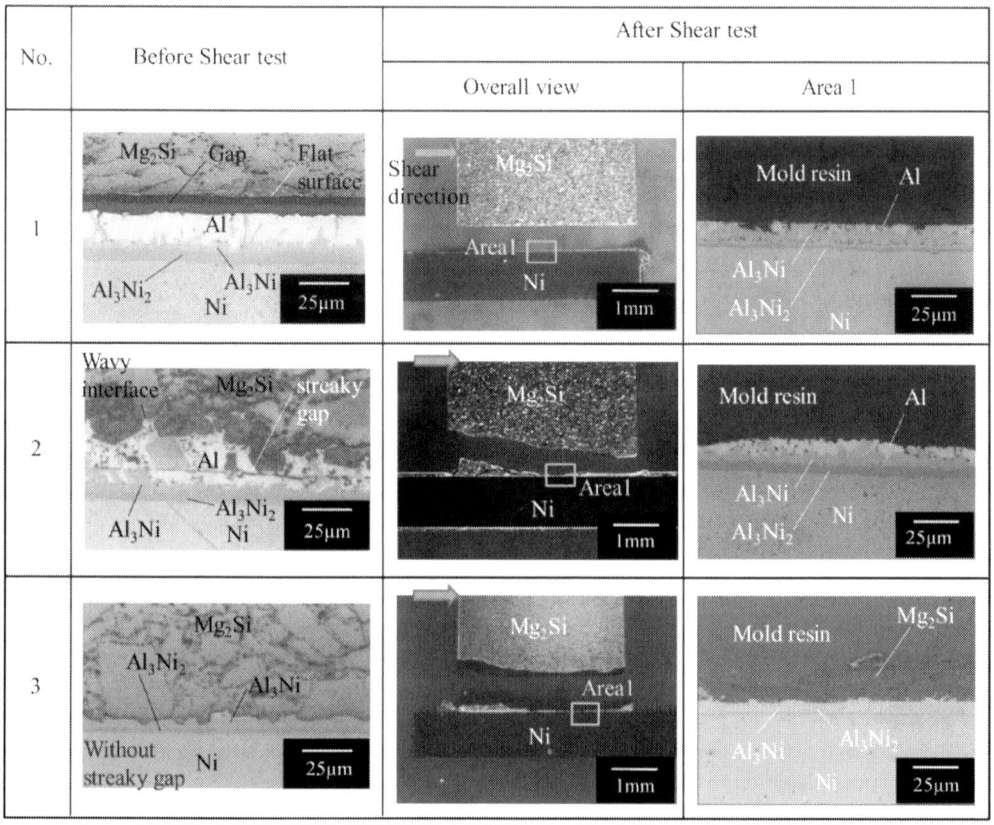

Figure 4. Cross-sectional views of the bonding interface between the Mg_2Si device and Ni electrode.

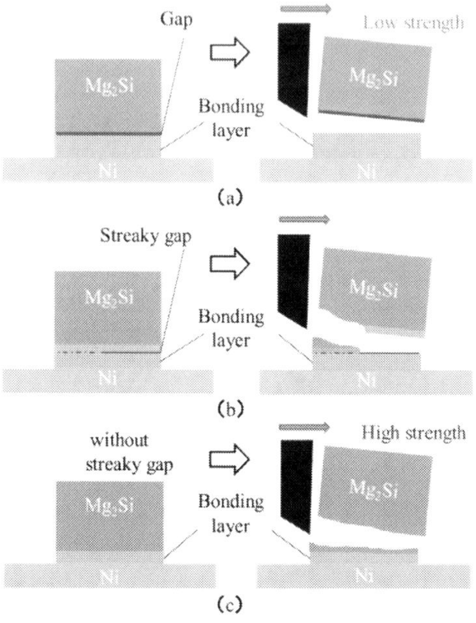

Figure 5. Comparison of fracture mode (a) No polishing of the Mg_2Si surface and low pressure (6.1×10^3Pa) bonding, (b) Polishing of the Mg_2Si surface and low pressure (6.1×10^3Pa) bonding, (c) No polishing of the Mg_2Si surface and high pressure (3.0×10^7Pa) bonding.

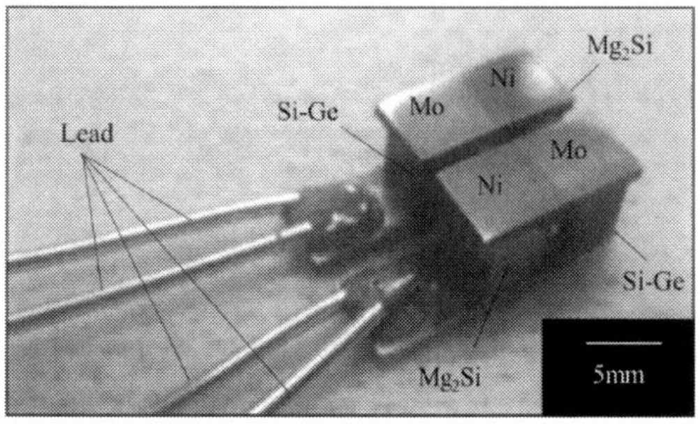

Figure 6. Exterior view of a prototype Si-Ge/Mg_2Si twin-couple module.

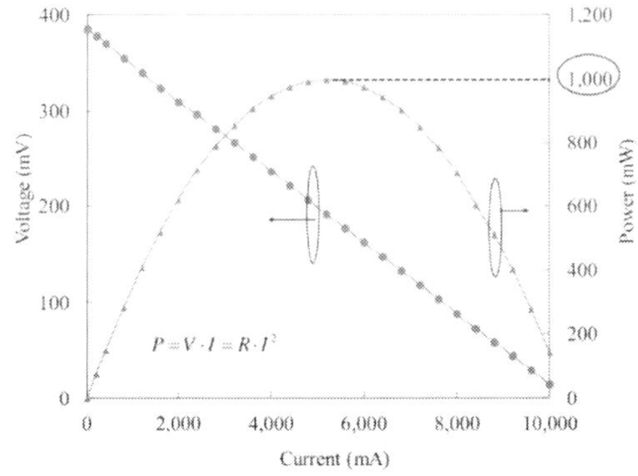

Figure 7. Power generation of the prototype module (upper and lower electrode ΔT = 620 K).

4. Conclusions

Aluminum foil was used as an alternative bonding material for a Mg_2Si element and nickel electrode because of its low cost and high bonding ability. The shear strength of the joint obtained was found to be sufficiently reliable when high pressure bonding was performed. In addition, a highly refractory joint was formed due to the formation of a high-melting-point Al-Ni layer that included magnesium and silicon. A prototype Si-Ge/Mg_2Si twin-couple module was subsequently bonded based on the results of the shear tests. Our main findings can be summarized as follows:

- The shear strength of the aluminum joint between the Mg_2Si element and Ni electrode under a bonding pressure of 3.0×10^7 Pa was 19 MPa, which is well above the target value of 10 MPa.
- The generating capacity of the prototype Si-Ge/Mg_2Si twin couple module was 1 W/cm^2 at ΔT = 620 K (high temperature side: 923 K), or 20% higher than that of a Si-Ge/Si-Ge twin couple module.

References

[1] Arvizu D and Balaya P 2011 *IPCC Special Report on Renewable Energy Sources and Climate Change Mitigation* 33
[2] Hayami H, Nakamura M and Yoshioka K 2005 *IEEE Trans.* on Systems Man and Cybernetics Part C Applications and Reviews. **35** 391
[3] Tagare D 2011 *Electricity Power Generation*: The Changing Dimensions: 173
[4] Rowe D 2012 *Thermoelectrics and its energy harvesting* 1st edition, (CRC press) 1-2
[5] Anderson D J, Sankovic J, Witt D and Abelson R D 2007 *Proc. of aerospace conference (montana, USA, 3-10 March 2007)* 1
[6] Venkatasubramanian R, Siivola E, Colpitts T and Oquinn B 2001 *Nature* **413** 597
[7] Poudel B, Hao Q, Ma Y, Lan Y C, Minnich A, Yu B and Yan X 2008 *Science* **320** 634
[8] Miyazaki K 2008 *OYO BUTURI* **78** 527
[9] Tanaka S, Takashiri M and Miyazaki K 2010 *Jpn. Journal of Thermophysical Properties* **24** 94
[10] Kaibe H, Aoyama I, Mukoujima M, Kanda T, Fujimoto S, Kurosawa T, Ishimabushi H, Ishida K, Rauscher L, Hata Y and Sano S 2005 *Proc. of 24th International Conference on Thermoelectrics (South Carolina, USA 19-23 June 2005)* 242

[11] Crane D T, Lagradeur J W, Harris F and Bell L E 2009 *Journal of Electronic Materials* **38**: 1375

[12] Kambe M, Jinushi T and Ishijima Z 2011 *Journal of Electronic Materials* **39** 1418

[13] Lin W P and Lee C C 2011 *Proc. of IEEE 61th ECTC (Florida, USA, 31 May-3 June 2011)* 118

[14] Ishii K 2013 *Hitachi Chemicals Technical Report* **55** 50

[15] Fujiwara S, Tohei T, Jinushi T and Ishijima Z 2012 *Proc. of 4th ESTC(Amsterdam, Netherlands, 17-20 September 1993)* 1

[16] Tohei T, Fujiwara S, Jinushi T and Ishijima Z 2013 *J. Jpn. Soc. Powder Powder Metallurgy* **60** 360

[17] Jpn.Weld. Soc. 2003 *YOSETSU SETSUGO handbook* 2nd edition (Maruzen) 434

A 60-GHz-band 2 x 4 planar dipole array antenna module fabricated by 3-D SiP technology

Noriharu Suematsu[1], **Yuya Suzuki**, **Satoshi Yoshida**[2], **Shoichi Tanifuji**[3], **Suguru Kameda**, **Tadashi Takagi**, **Kazuo Tsubouchi**

Research Institute of Electrical Communication (RIEC), Tohoku University,

2-1-1 Katahira, Aoba-ku, Sendai, Miyagi 980-8577, Japan

suematsu@riec.tohoku.ac.jp

Abstract. A 2 x 4 phased array antenna module has been developed for 60-GHz-band short-range high-speed wireless communication terminals. To realize the required vertical distance between the antenna elements, the module is made of five sheets of multi-layered organic substrates vertically stacked with Cu balls, and the 1 x 4 dipole array antenna is placed on both the top and bottom organic substrates. To reduce the mutual coupling between the element antennas, a monolithic microwave integrated circuit (MMIC) is flip-chip mounted on the feed line of each element antenna. The Au-stud bump flip-chip mounting technique helps achieve a lower return loss in the transition section at the MMIC than the Au-wire bonded. The placement accuracy of each antenna element in the vertical direction, 60-GHz signal vertical interconnection between the substrates with Cu balls, and flip-chip mounting of the MMIC are confirmed by 3-D computed tomography (CT) scans.

1. Introduction

60-GHz-band short-range wireless communication supports very high speed (>Gbps) data transmission service for consumers. To extend the communication range with a limited transmission power of 10 dBm, a phase array antenna (PAA) technique was introduced [1-3]. Because the PAA requires a multi-antenna/radio frequency (RF) transceiver configuration, a compact and cost-efficient solution is required. To address this need, we have proposed a 60-GHz-band antenna integrated module using a 3-D system-in-package (SiP) technology [4-7]. The proposed 2 x 4 PAA module consists of five sheets of organic substrates, vertically stacked with Cu balls. This PAA module structure realizes the required element antenna spacing (nearly half wavelength at 60 GHz) in the vertical direction of the PAA module [6-7]. To reduce the re-radiation from each element antenna, the Au-stud bump bonding (Au-SBB) flip-chip mounting technique is employed for monolithic microwave IC (MMIC) mounting on the organic substrate. We investigated the accuracy of the vertical placement of array antenna elements, vertical RF interconnection sections between the stacking organic substrates, and flip-chip mounting of the MMIC chip on the organic substrates by using 3-D computed tomography (CT) scans.

[1] Corresponding author
[2] Now, with ISAS, JAXA, Japan.
[3] Now, with Okinawa National College of Technology, Japan.

Content from this work may be used under the terms of the Creative Commons Attribution 3.0 licence. Any further distribution of this work must maintain attribution to the author(s) and the title of the work, journal citation and DOI.
Published under licence by IOP Publishing Ltd

2. Configuration of the 2 x 4 PAA module

To achieve a 10 dBi antenna gain, a 2 x 4 array antenna was selected [6]. Figure 1 shows the block diagram of the fabricated 2 x 4 PAA module and receiver. In the PAA module, 16 element antennas and down conversion MMIC mixers and a local oscillator (LO) feed circuit with power dividers (T-junctions) are integrated. Because the MMIC mixers are connected to each element antenna, the PAA module has 16 ports of intermediate frequency (IF) output. These ports are connected to the external receiver section and an appropriate phase difference is added to each IF signal in the digital domain.

Figure 2 shows the structure of the fabricated PAA module. The module consists of five sheets of organic substrates (S1-S5), and eight chips of MMIC mixers and connectors for the LO and IF. The 2 x 4 planar dipole antennas are configured on the top (S1) and bottom (S5) substrates. MEGTRON6 (Panasonic) is selected as the organic substrate because of its relatively low dielectric constant ($\varepsilon r = 3.5$) and low loss ($\tan \delta = 0.002$) characteristics. The five sheets of substrates (S1-S5) are vertically stacked by Cu balls ($\phi = 350$ μm) to realize the required vertical element antenna spacing of 0.65 λ [4]. The MMIC mixer chips are flip-chip mounted on the top (S1) and bottom (S5) substrates by Au-SBB.

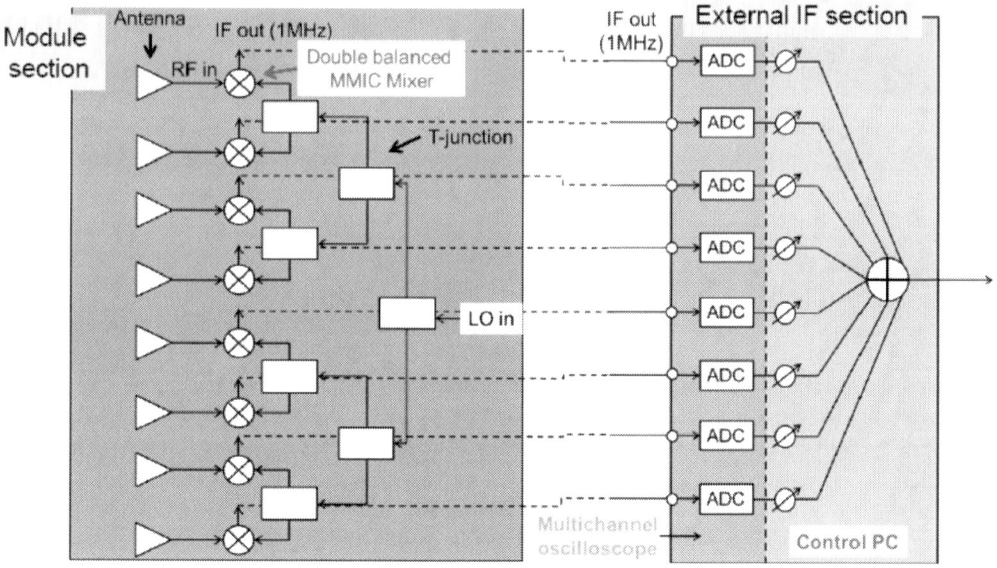

Figure 1. Block diagram of fabricated 2 x 4 PAA module and receiver.

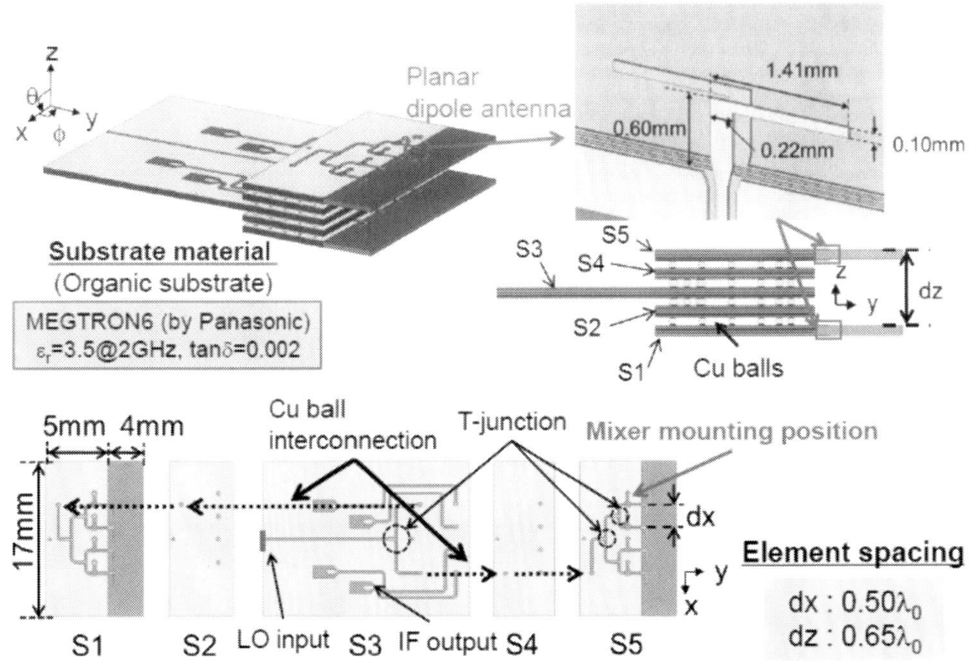

Figure 2. Structure of fabricated 2 x 4 PAA module.

3. Fabrication and measurements

Figure 3 shows the photograph of the fabricated 2 x 4 PAA module. On the top and bottom substrates (S1 and S5), the 1 x 4 planar dipole array antenna and four MMIC mixer chips are integrated. The IF ports and the LO input port are located on the 3rd substrate (S3). Figure 4 shows the side view of the PAA module. The Cu-ball interconnection can realized the and Table 1 is measured antenna distances in the vertical direction of the six module samples. The Cu balls play the role of interconnections between the substrates, as well as that of spacers. As the diameter of each Cu ball is 350 μm and the thickness of each substrate (S1 to S5) is 460 μm, the total thickness becomes 3.24 mm, which is almost equal to 0.65 λ at 60 GHz. As shown in figure 4, a very flat leveling can be obtained through the use of the Cu-ball interconnection for stacking organic substrates up to five sheets. Measured data listed in table 1 show that the deviation of the antenna distance in the vertical direction is less than 20 μm (less than 1.5 degrees at 60 GHz); therefore, this deviation will not affect the PAA's performance. From this result, we can confirm fabrication repeatability.

The placement accuracies of Cu-balls and Au-SBB for MMIC flip-chip mounting on the organic substrates are confirmed by the 3-D CT scan. A micro-focus X-ray inspection system is used. Figure 5 shows the cut view along the surface of the top substrate (S1) of the PAA module. A mixer MMIC chip is mounted on the feed line of the element antenna and Au bumps are well placed between the MMIC chip and lines and the ground metal of the organic substrate. The diameter of the Au bump is set to be around 60 μm and it fits the line width of the substrate. The Cu ball is used in the 60-GHz LO vertical feed and is placed at the center of the island for the vertical feed. Figure 6 shows the cross-sectional view of the PAA module at the MMIC mounting area. From the figure, the Au bump height of 50 μm and the placement accuracies of Cu balls at the 60-GHz vertical feed section are confirmed. The 3-D CT scan is a powerful tool to investigate the internal connections of the PAA module.

To verify the 60-GHz signal transmission performance at the MMIC mixer chip interconnection, three samples are fabricated using conventional wire bonding technology and the Au-ABB flip-chip technology. The measured return loss (S11) is shown in figure 7. The Au-SBB flip-chip bonding exhibits a lower return loss than conventional wiring technology at 60 GHz. The measured samples show good repeatability of fabrication.

Finally, antenna pattern measurement is carried out using the fabricated 2 x 4 PAA module. Figure 8 shows the results of electro-magnetic (EM) simulation and antenna pattern measurement of the 2 x 4 PAA module. The maximum antenna gain obtained through simulation is 10.9 dBi and that obtained through measurement is 11.1 dBi. The measurement result is in good agreement with the EM simulation result. These results confirm the accuracy of antenna placement in both vertical and horizontal directions.

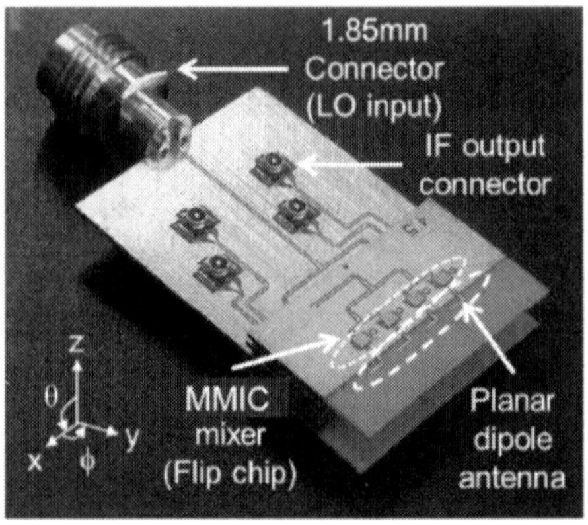

Figure 3. Photograph of fabricated 2 x 4 PAA module.

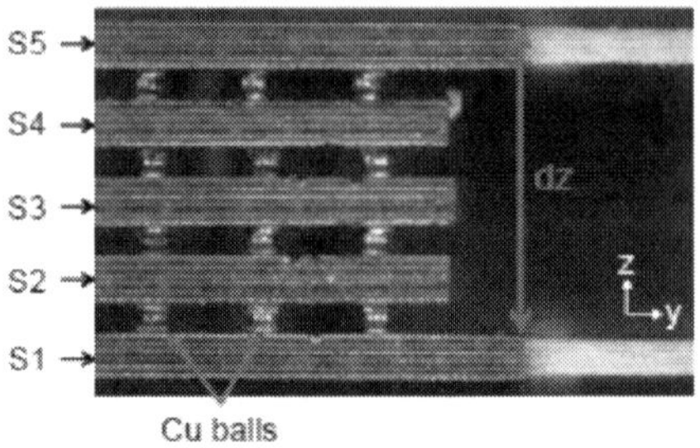

Figure 4. Side view of PAA module.

Table 1. Measured antenna distance in the vertical direction (dz) for six module samples.

	dz (Measurement)	dz (Design)
Sample 1	3.23 mm	3.24 mm
Sample 2	3.24 mm	
Sample 3	3.24 mm	
Sample 4	3.26 mm	
Sample 5	3.25 mm	
Sample 6	3.23 mm	

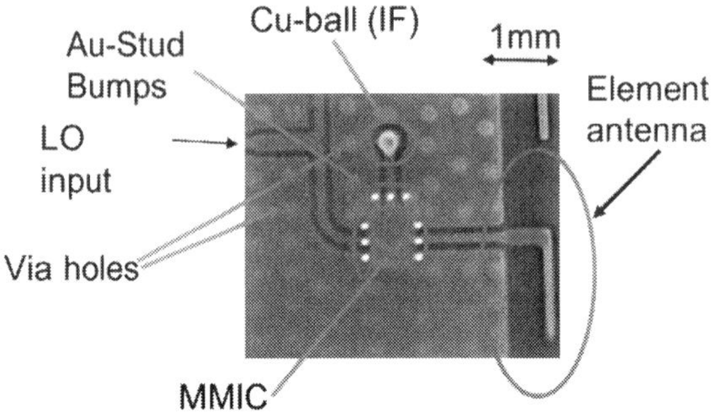

Figure 5. Cut view along the surface of top substrate (S1) of PAA module.

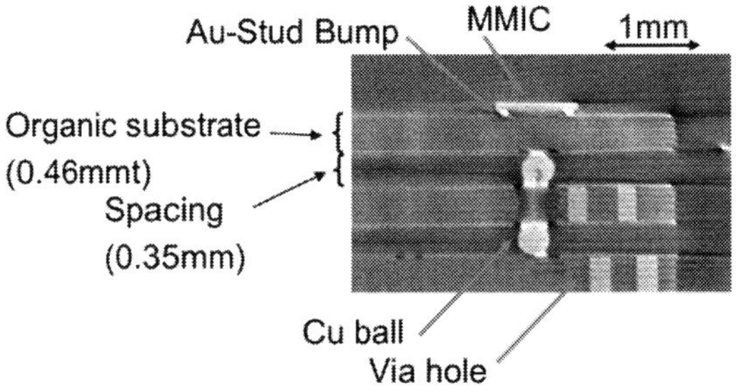

Figure 6. Cross sectional view of PAA module.

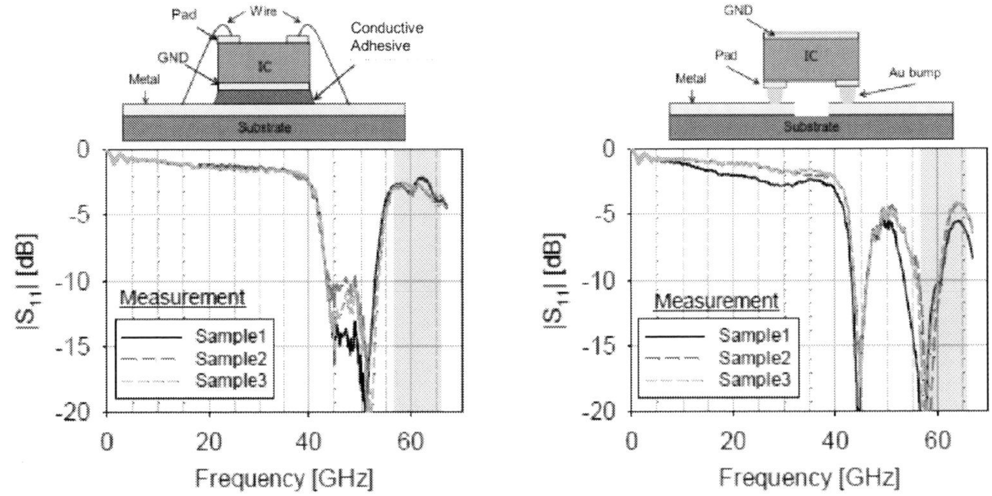

(a) Bonding wire (b) Flip chip

Figure 7. Measured return loss of MMIC mixer mounted on organic substrates. (a) Au bonding wire connection. (b) Flip-chip mounting by Au-ABB.

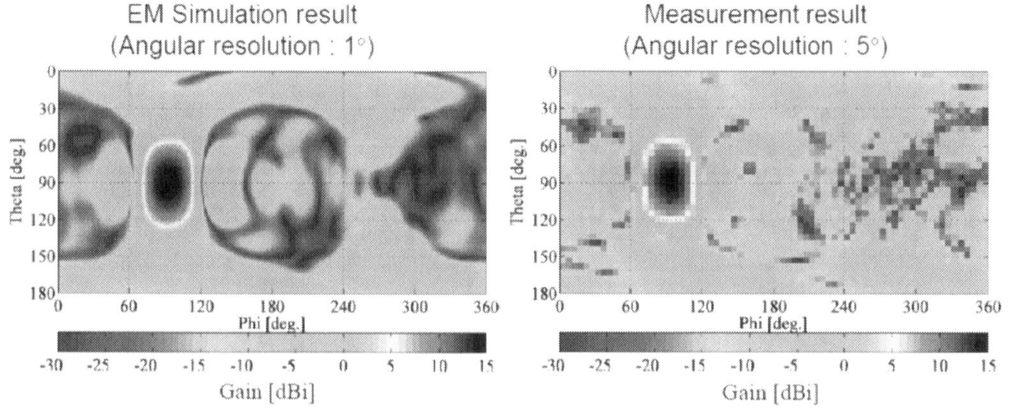

Figure 8. Electro-magnetic (EM) simulation and measurement results of antenna pattern of 2 x 4 PAA module.

4. Conclusions

A 60-GHz-band 2 x 4 PAA module has been fabricated using 3-D SiP technology. To achieve low-loss 60-GHz-signal transmission between the organic substrate and the MMIC chip, the Au-ABB flip-chip bonding technique is employed. To realize the appropriate antenna distance (0.65 λ at 60 GHz) in the vertical direction of the PAA module, a stacked organic substrate configuration using the Cu-ball interconnection is employed. By selecting the number of sheets of the organic substrate and the diameter of the Cu ball, the required distance between the antennas mounted on the top and bottom of the stacked substrates can be achieved. The placement accuracies of the Cu balls and the Au bump for MMIC flip-chip mounting on the organic substrates are confirmed by 3-D CT scans. The element antenna distance in the vertical direction is confirmed by the photograph of the cut view of the PAA module and the measured antenna pattern.

Acknowledgement

The authors would like to thank the Core Research for Evolutional Science and Technology program (CREST) of Japan Science and Technology program (JST) for fund support.

References

[1] Fakharzadeh M, Nezhad-Ahmadi M-R, Biglarbegian B, Ahmadi-Ahokouh J and Safavi-Naeini S 2010 *IEEE Trans. Antennas Propag.*, **58(4)** 1093

[2] Emami S, Wiser R F, Ali E, Forbes M G, Gordon M Q, Xiang Guan, Lo S, McElwee P T, Parker J, Tani J R, Gilbert J M and Doan C H 2011 *IEEE ISSCC* 164

[3] Suga R, Nakano H, Hirachi Y, Hirokawa J and Ando M 2010 *IEEE MTT-S Int. Microw. Symp. Dig.*, 449

[4] Tsubouchi K, Yokoyama M and Nakase H 2001 *IEICE Trans. on Electron.* **E84-C(12)** 1730

[5] Suematsu N, Yoshida S, Tanifuji S, Kameda S, Takagi T and Tsubouchi K 2012 *IEICE Trans. on Electron.* **E95-C(7)** 1141

[6] Yoshida S, Suzuki Y, Ta T T, Kameda S, Suematsu N, Takagi T and Tsubouchi K 2013 *IEEE Trans. Antennas Propag.* **61(7)** 3502

[7] Suzuki Y, Yoshida S, Ta T T, Tanifuji S, Kameda S, Suematsu N, Takagi T and Tsubouchi K 2013 *European Microwave Conference (EuMC)* EUMC/EuRAD03-4

Finite element analysis of influence of passivation layer on Cu/low-k structure during thermosonic Cu wire bonding

Xiaoyi Bai[1, 2], Baohua Chang[1, 2, *], Dong Du[1]

[1] Department of Mechanical Engineering, Tsinghua University, Beijing, 100084, China
[2] State Key Laboratory of Tribology, Tsinghua University, Beijing 100084, China

*E-mail: bhchang@tsinghua.edu.cn

Abstract. An incrementally coupled "mechanical–ultrasonic" finite element model, in which both the ultrasound vibration and the bonding force are considered simultaneously, is used to model the thermosonic Cu wire-bonding process on Cu/low-k structures and to study the influence of the passivation layer on the stress condition in the Cu/low-k layer. Results show that the passivation layer acts as a stress buffer layer during the wire-bonding process. The introduction of a passivation layer alleviates some of the impact effect from the free air ball (FAB). Increasing the passivation thickness can lead to significant stress alleviation. However, as the elastic modulus of passivation increases, the alleviation effect increases notably at first and then decreases.

1. Introduction

In integrated circuit (IC) manufacturing, Cu/low-k structures are being used because they have better electrical and electromigration performance than traditional Al/SiO_2 wafers [1]. However, the poor adhesion of low-k layers and the weak mechanical properties of low-k materials cause severe problems during the wire-bonding process, such as cupping, delamination, and cracks [2]. Meanwhile, Cu wire bonding is more and more widely used in wire bonding to replace traditional Au wire due to its better electrical and mechanical properties and, more importantly, its lower cost. However, because Cu is harder than Au, when it is bonded to the soft Cu/low-k structures, the problems mentioned above will be even more serious [3]. All those problems are closely related to the stress condition in Cu/low-k during the bonding process. Experiments have revealed that passivation can reduce the failure of Cu/low-k structures during wire bonding [4, 5]. However, how the passivation layer design influences the stress condition in Cu/low-k is not yet clear. In the present study, an incrementally coupled mechanical–ultrasonic finite element model is employed, taking both material softening and mechanical oscillation from ultrasound into account, in order to study the influences of passivation layer thickness and elastic modulus on stress in Cu/low-k structures.

2. Finite element model

Figure 1(a) shows the finite element mesh used in elastic–plastic finite element analysis at the initial condition. The Cu/low-k structure consists of three layers of copper metal. Figure 1(b) shows the details of the structure. In this model, the free air ball diameter is 35 μm and the wire

Content from this work may be used under the terms of the Creative Commons Attribution 3.0 licence. Any further distribution of this work must maintain attribution to the author(s) and the title of the work, journal citation and DOI.
Published under licence by IOP Publishing Ltd

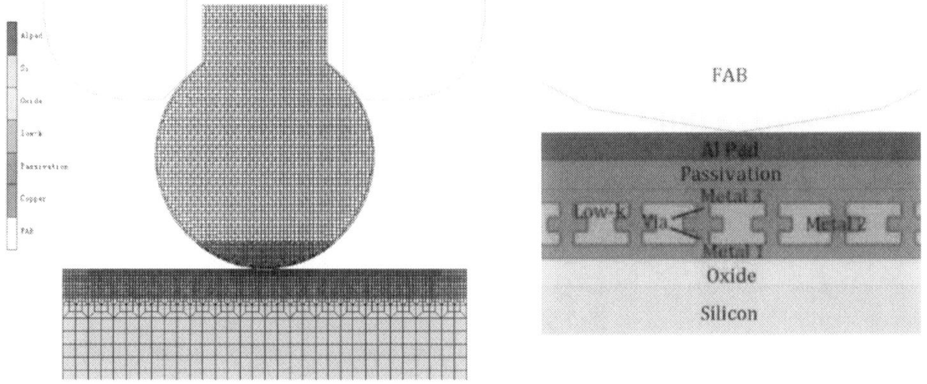

(a) Global finite element mesh. (b) Local structure of Cu/low-k layer.

Figure 1. Finite element mesh.

Table 1. Thicknesses of different parts in finite element analysis model.

	Al pad	Passivation	Cu Structures	Low-k	SiO_2	Si
Thickness (μm)	1	1	0.5	2.5	1	10

diameter is 20 μm. The thicknesses of the different layers in the pad structure are listed in Table 1.

In this model, both material softening and mechanical vibration of ultrasound effects are taken into account by using an "increasingly coupled mechanical–ultrasonic" scheme [6], in which the application of the capillary displacement is divided into many steps (100 steps in the present study). In every step, an axial displacement is applied to simulate the downward movement of the capillary and a transversal displacement parallel to the underpad is applied to simulate the ultrasonic vibration of the capillary. An elastic–plastic mechanical analysis is then carried out to calculate the distortion and stress in the free air ball (FAB) and Cu/low-k structure.

The displacement of ultrasonic vibration follows the following equation:

$$Y = A_0 \sin\left(2\pi ft\right) \tag{1}$$

where A_0 is the ultrasound amplitude and f is the ultrasound frequency (120 kHz in the present study).

The Coulomb friction model is used to describe the friction involved in the wire-bonding process. The friction coefficient between the capillary and the copper ball is 0.22 and the friction coefficient between the copper ball and pad layer is 0.2 [7].

The material properties used here are listed in Table 2. The Cu FAB, Cu via, and Al plating are assumed to be isotropic elastic plastic and power laws are used here to predict the stress. The governing equation is [7]:

$$\sigma = C\varepsilon^n \text{ (MPa)} \tag{2}$$

where σ is the effective stress, ε is the effective strain, C is the strain-hardening coefficient, and n is the strain-hardening exponent. Different values of C and n are used for Cu FAB, Cu via, and Al pad.

Table 2. Material properties.

Contact body	Copper	Aluminum	Passivation	Low-k	Oxide	Silicon
Elastic modulus, E (GPa)	130	71	32	18	71.4	169
Posson's ratio, ν	0.34	0.33	0.24	0.3	0.27	0.3
Density, ρ, (g/cm^3)	8.96	2.7	2.31	2	2.2	2.33

3. Validation of FEA model

To validate the model developed, two different structures are numerically compared first. One structure includes a passivation layer directly under the Al pad and the other does not, as depicted in figure 2.

Figure 3 shows the distributions of the first principal stress and the maximum shear stress in the low-k layer [green part with the square in figure 3(a)]. The highest stresses in the pad low-k layer occur at the last ultrasound vibration cycle when the capillary's parallel movement reaches the maximum ultrasound amplitude. For both structures, the capillary moved down 9.23 μm. The ball diameters for structure 1 and structure 2 are 38.84 μm and 39.2μm. The bonded regions' radii of structure 1 and structure 2 are 10.38 μm and 11.5 μm. The maximum first principal stresses in structure 1 and structure 2 are 60.77 MPa and 267.4 MPa, respectively, and the maximum shear stresses in structure 1 and structure 2 are 80.84 MPa and 104 MPa, respectively. The maximum shear stress concentration is 11 μm from the center for structure 1 and 9 μm for structure 2. For both structures, the first principal stress concentration is 11 μm from the center. It is clear than the underpad of structure 1 suffers less stress than structure 2, and thus the risk of pad damage in structure 1 is smaller than that in structure 2. This result is in agreement with the experimental findings in which the passivation directly under the Al pad strengthened the structure [4]. The predicted stresses from the finite element model developed in the present work are good indications of the tendency of underpad failure.

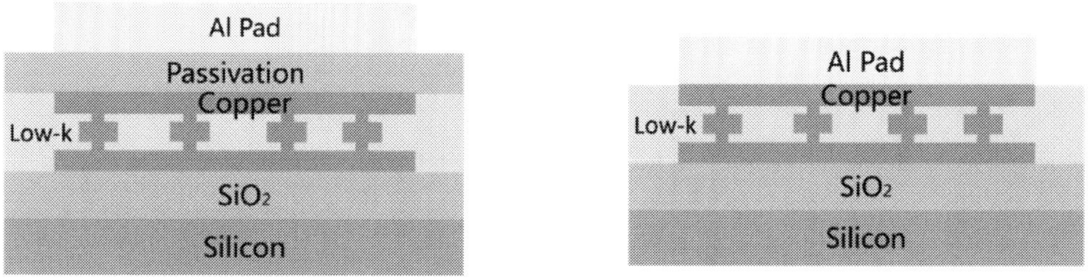

(a) Structure 1: with passivation layer. (b) Structure 2: without passivation layer.

Figure 2. Two different types of pad structures.

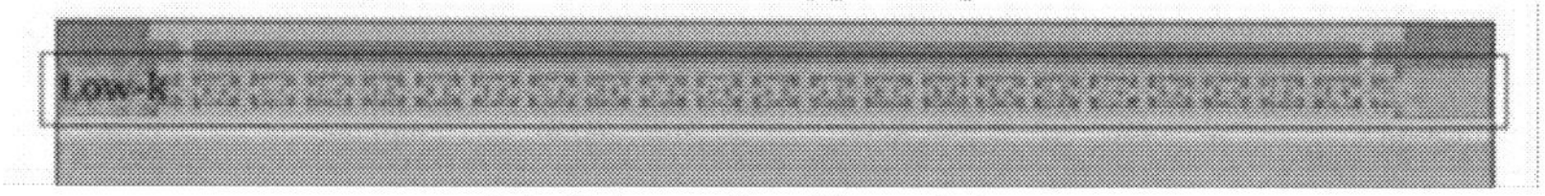

(a) Schematic of Low-k layer

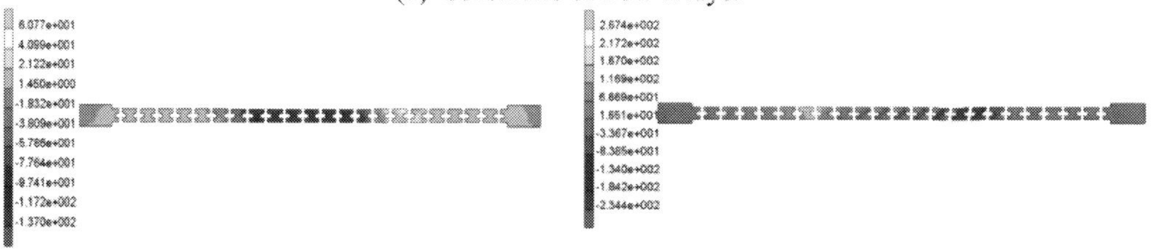

(b) First principal stress of structure 1. (c) First principal stress of structure 2.

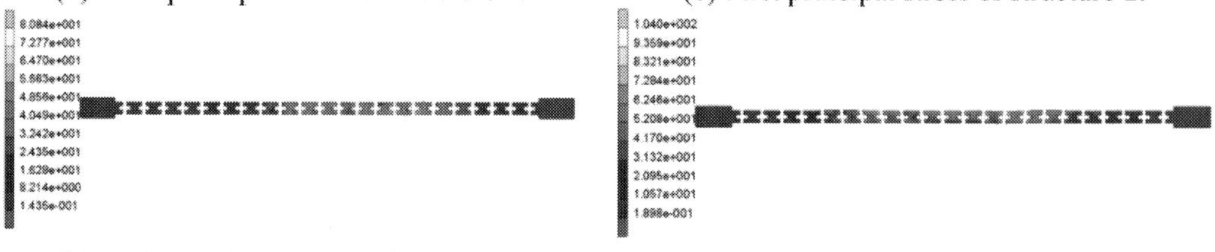

(d) Maximum shear stress of structure 1. (e) Maximum shear stress of structure 2.

Figure 3. Stress conditions in underpad.

4. Results and discussion

Both the experiments and the numerical predictions demonstrated the role of passivation as a stress buffer. To better design and select the passivation layer, the FE model developed was employed to study the influences of thickness and elastic modulus of the passivation layer on stress conditions in Cu/low-k structures.

4.1. Influence of passivation thickness

Stresses and distortions in a Cu/low-k structure are numerically predicted when the thickness of the passivation layer changes from 0 to 2 μm, with a constant elastic modulus of 32 GPa. The FAB is totally suppressed by a distance of 10 μm.

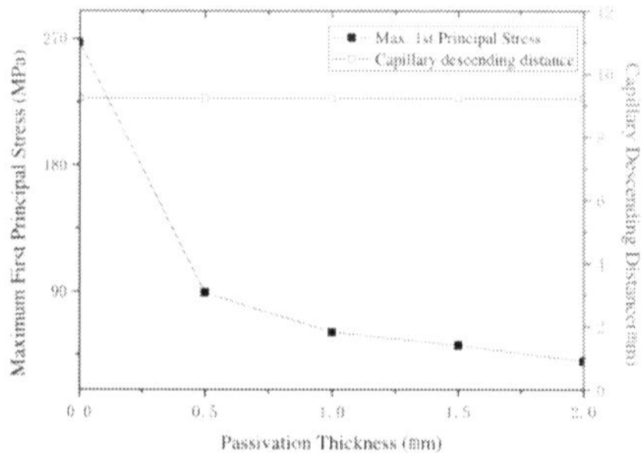

(a) Maximum first principal stress and the capillary descending distance.

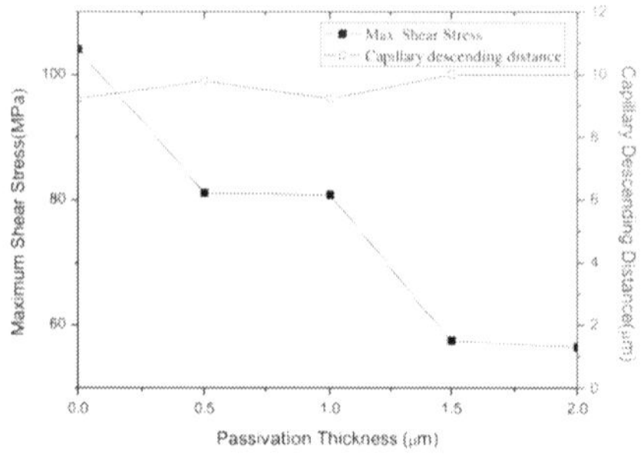

(b) Maximum shear stress and the capillary descending distance.

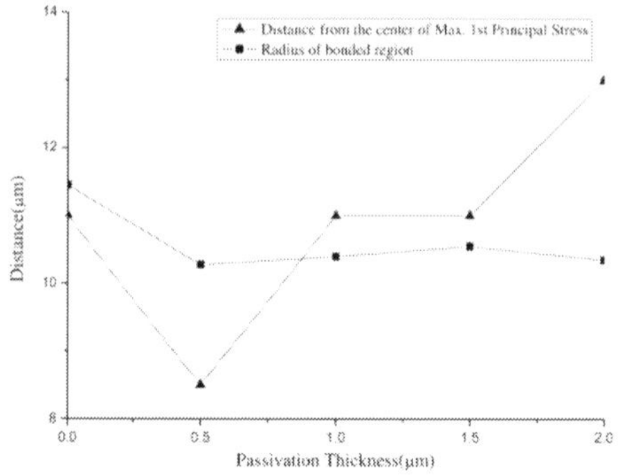

(c) Radii of bonded region and the location of maximum first principal stress.

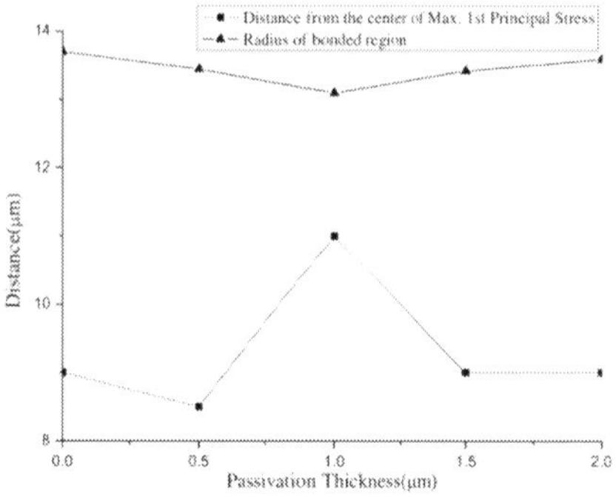

(d) Radii of bonded region and the location of maximum shear stress concentration.

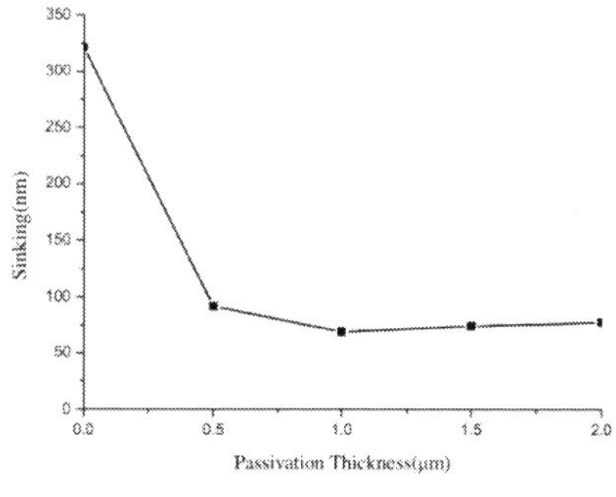

(e) Pad sinking.

Figure 4. Stresses in Cu/low-k layer and pad sinking for different passivation thicknesses.

Figure 4(a) shows the maximum first principal stress and the corresponding capillary descending distance in the low-k layer for different passivation thicknesses. Figure 4(b) shows the maximum shear stress and the corresponding capillary descending distance for different passivation thicknesses. It is clear that when the passivation thickness increases, both the maximum first principal stress and the maximum shear stress decrease. The passivation thickness has little influence on the time it takes for the low-k layer to reach the maximum first principal stress and the maximum shear stress, i.e., when the capillary moves down over 9 μm and the ultrasonic amplitude reaches a maximum. Figures 4(c) and 4(b) show the stress concentration point locations and the corresponding radii of the bonded region of the maximum first principal stress and the maximum shear stress, respectively. For the first principal stress, when the passivation layer thickness is less than 1 μm, the stress concentration points are under the bonded ball; when the thickness reaches 1 μm, the concentration points are beyond the bonded region. For the maximum shear stress, all the stress concentration points are within the bonded region. Figure 4(e) presents the variation of pad sinking with increasing thicknesses of passivation. It can be seen that when a thin layer of passivation with a thickness of 0.5 μm is introduced, the pad sinking decreases significantly. When the passivation thickness is increased beyond 0.5 μm, the pad sinking first decreases and then increases slightly.

The results show that the passivation layer acts as a stress buffer layer. It alleviates some of the impact of the wire-bonding process.

4.2. Influence of elastic modulus of passivation

Stresses and distortions in Cu/low-k structure are given in figure 5 for elastic modulus of passivation ranging from 0.1 GPa to 50 GPa at a constant thickness of 1 μm. The FAB is totally suppressed by a distance of 10 μm.

Figure 5(a) and 5(b) show the variations in maximum first principal stress and the maximum shear stress in the low-k layer with their corresponding capillary descending distance for different elastic moduli of passivation. It can be seen that both the maximum first principal stress and the maximum shear stress decrease notably at first and then increase when the elastic modulus passivation increases from 0.1 GPa to 50 GPa. When the elastic modulus is 20 GPa, the first principal stress is the smallest, whereas the maximum shear stress is the smallest when the elastic modulus is 10 GPa. Also, the passivation thickness has little influence on the time it takes for the low-k layer to reach the maximum first principal stress and the maximum shear stress, i.e., when the capillary moves down over 9 μm and the ultrasonic amplitude reaches a maximum.

Figure 5(c) and 5(b) show the stress concentration point locations and the corresponding radii of the bonded region of the maximum first principal stress and the maximum shear stress, respectively. For the first principal stress, when the modulus is 0.1 GPa, the concentration point is far away from the bonded region; when the modulus is greater than 0.1 GPa, the concentration points are very close to the bonded region. For the maximum shear stress, there are no obvious trends in the locations of the concentration points. Figure 5(e) shows how the pad sinking changes for different passivation moduli.

It is predicted that the pad sinking decreases dramatically at first and then continues to decrease slightly for increased elastic moduli.

Increasing the elastic modulus of the passivation layer can lead to stress alleviation at first. However, by increasing the elastic modulus too much, beyond 20 GPa in the present work, the stress improvement will decrease.

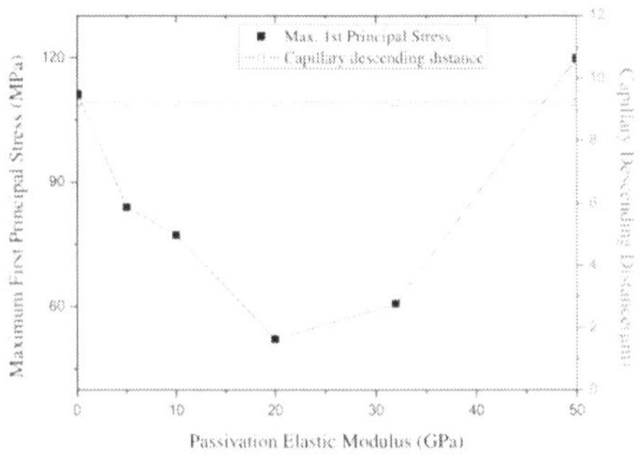

(a) Maximum first principal stress.

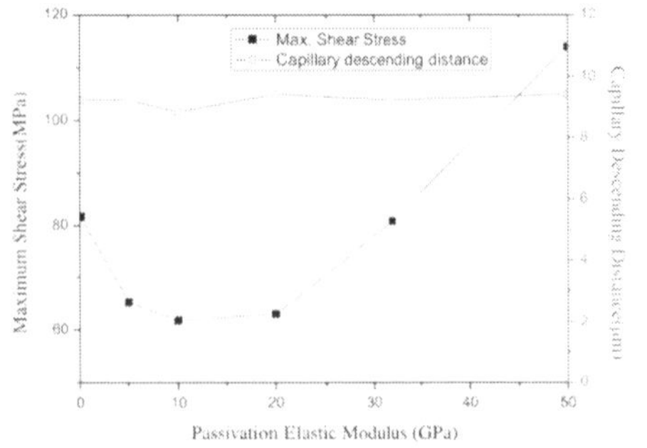

(b) Maximum shear stress.

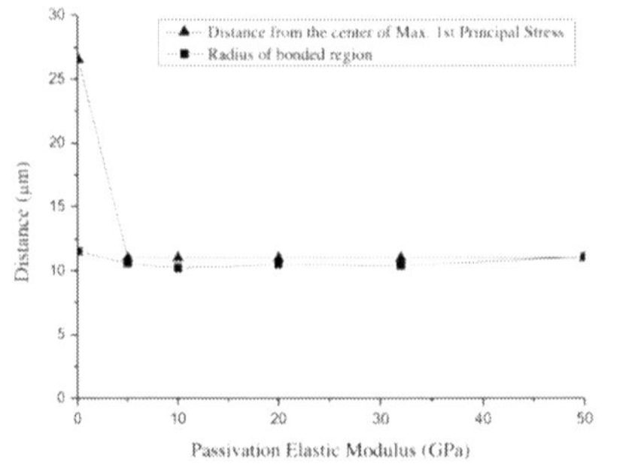

(c) Radii of bonded region and the location of maximum first principal stress.

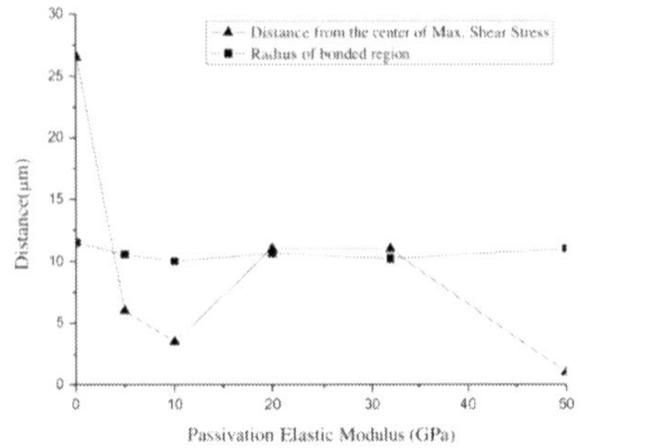

(d) Radii of bonded region and the location of maximum shear stress concentration.

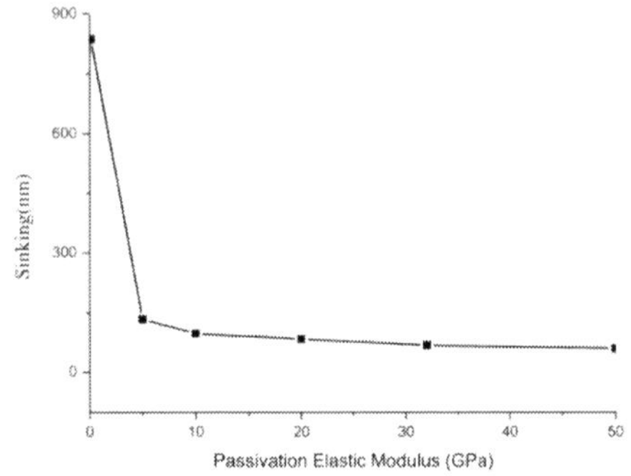

(e) Pad sinking.

Figure 5. Stresses in Cu/low-k layer and pad sinking for different passivation moduli.

5. Conclusions

(a) For a passivation layer with an elastic modulus of 32 GPa, the maximum first principal stress and the maximum shear stress decrease when the passivation thickness increases from 0 to 2 μm.

(b) The pad sinking decreases significantly when the passivation thickness increases from 0 to 0.5 μm; when the thickness is further increased beyond 0.5 μm, the pad sinking first decreases and then increases slightly.

(c) For a passivation layer with thickness of 1 μm, the first principal stress decreases notably and then increases when the elastic moduli of passivation increase from 0.1 GPa to 50 GPa. The proper value would be between 10 GPa and 20 GPa in the present study.
(d) The pad sinking decreases dramatically at first and then continues to decrease slightly for increased elastic moduli of passivation.

Acknowledgement
The authors gratefully appreciated the financial support from Tsinghua University Initiative Scientific Research Program (No. 20111081006).

References
[1] Wang ZJ, Huang V, Yao SY, He ZL, Jiang YW, Zhang CL, et al. 2007 Cu low k device wire bonding process modelling *Electronic Packaging Technology, 2007. ICEPT 2007. 8th Int. Conf. on (Shanghai, China, 14-17 August 2007)* 1-6
[2] Van Driel, W D 2007 Facing the challenge of designing for Cu/low-k reliability *Microelectron Reliab.* **47**(12) 1969-74
[3] Degryse D, Vandevelde B and Beyne E 2004 Mechanical FEM simulation of bonding process on Cu LowK wafers *Components and Packaging Technologies, IEEE Transactions on* **27**(4) 643-50.
[4] Brunner J W, Keller F and Pan T 2003 Optimization of wire bonding over Cu-low K pad stack *Int. Symp. on microelectronics (Boston MA, 18-20 November 2003) (Society of Photo-Optical Instrumentation Engineers, SPIE proceedings series vol 5288)* 136-140
[5] Yeh C and Lai Y2006 Comprehensive dynamic analysis of wirebonding on Cu/low-k wafers *Advanced Packaging, IEEE Transactions on* **29**(2) 264-70
[6] Jiang W, Chang BH, Du D, Huang H and Zhou YH. 2012 The Influence of materials on the substrate stress in wirebonding *Transactions of the China welding institution* **33**(003) 13-6
[7] Huang H, Chang BH and Du D 2011 Effect of superimposed ultrasound on mechanical properties of copper. *Mater. Sci. Tech.* **27**(7) 1117-22

IOP Conf. Series: Materials Science and Engineering **61** (2014) 012038 doi:10.1088/1757-899X/61/1/012038

Microstructure of lead-free solder bumps using laser reflow soldering

Hiroshi Nishikawa[1], **Noriya Iwata**[2] **and Shinya Kubota**[2]

[1] Joining and Welding Research Institute, Osaka University, 11-1 Mihogaoka, Ibaraki, 567-0047, Japan

[2] Graduate school of Engineering, Osaka University, 2-1 Yamadaoka, Suita, 565-0871, Japan

E-mail: nisikawa@jwri.osaka-u.ac.jp

Abstract. Compared with conventional reflow soldering using a furnace, laser reflow soldering brings advantages such as localized heating, rapid rise and fall in temperature, non-contact soldering and the fact that it is an easily automated process. In this study, to elucidate the characteristics of laser reflow soldering, we investigated the microstructures of a Sn–Ag–Cu solder bump and a Sn–Bi solder bump on a Cu pad after reflow and aging. In the as-soldered condition, we found obvious microstructural refinement and a thin intermetallic compound (IMC) layer at the interface for both the Sn–Ag–Cu solder bump and the Sn-Bi solder bump using laser reflow soldering. Also, during isothermal aging, the total thickness of the IMC layer increased, and a distinct second layer was observed at the interface between the Cu pad and the first layer, regardless of the soldering method. In particular, the growth of the IMC layer was faster in the case of the laser reflow soldering than in the case of the conventional reflow soldering.

1. Introduction

Lead-free soldering technology is commonly being used in the electronics industry all over the world. Among the various lead-free solders available, Sn–Ag–Cu solders are considered the most promising for both conventional wave and reflow soldering processes. Reflow soldering using a furnace has been widely applied in electronics packaging and assembly processes. However, with the miniaturization of electronic devices and the use of heat-sensitive electronic components, the use of the conventional reflow soldering process often gives rise to difficulties. Recently, a laser reflow soldering process was introduced in industry for practical use because of its low cost and its unique properties such as localized and non-contact heating, rapid rise and fall in temperature, and ease of automation as compared to the reflow soldering process. However, there has been limited discussion on the basic phenomena of the laser reflow soldering process and the performance of the joints soldered by this process [1-4].

Lowering the reflow temperature reduces thermal stresses on components in electronic products, and some heat-sensitive materials require low-temperature assembly. Thus, it is very important to study low-melting-temperature solders. Sn–Bi eutectic solder has been considered a promising lead-free solder for low-temperature applications due to its low melting point (138 °C). Several studies on Sn-Bi eutectic solder have been conducted to investigate the basic characteristics of the solder joint [5-10]. However, there is a lack information about laser reflow soldering for Sn–Bi eutectic solder.

Content from this work may be used under the terms of the Creative Commons Attribution 3.0 licence. Any further distribution of this work must maintain attribution to the author(s) and the title of the work, journal citation and DOI.
Published under licence by IOP Publishing Ltd

The solder microstructure and the growth of an intermetallic compound (IMC) at the solder/metallization interface significantly influences solder joint reliability. So, in this study, to elucidate the characteristics of the laser reflow soldering, we investigated the effect of heating method on the solder microstructure and the growth of IMC at the interface of a Sn–Ag–Cu solder bump and a Sn–Bi solder bump on a Cu pad.

2. Experimental procedure

A Sn–3.0 mass%Ag–0.5 mass%Cu solder ball with a diameter of 1 mm and a Sn–58 mass% solder ball with a diameter of 0.76 mm were used in this study. Cu pads fabricated on an FR-4 printed circuit board were used as the substrate. Commercially available fluxes were also used. Prior to the soldering, the Cu pads were immersed in a 4 % HCl solution for 120 s. Then, the Cu pads and the solder balls were ultrasonically rinsed in an ethanol solution. After the solder balls had been dipped in the flux, they were placed on the Cu pads. Figure 1 shows a schematic illustration of a sample soldered joint being formed on a Cu pad. After being placed on the Cu pads, the solder balls were heated by laser in air using a laser soldering system (UNIX-413L2, JAPAN UNIX Co., Ltd.). This laser soldering system consisted of a high-power diode laser (λ = 940 nm) with a 1.2 mm beam and a 120 mm focal-length objective, and it had a maximum laser power output of 50 W. As a reference, the solder balls were also heated by using reflow soldering at 523 K for the Sn–Ag–Cu solder and 443 K for the Sn-Bi solder for 60 s in a furnace under nitrogen atmosphere. Some of the soldered joints were then subjected to isothermal aging at 423 K for the Sn–Ag–Cu solder and at 353 K for the Sn–Bi solder. Then, some of these joints were cut and their cross sections were polished so that the solder/Cu pad interface and the microstructure of the solder matrix could be observed by scanning electron microscopy (SEM) and the thickness of the IMC layer at the interface measured.

3. Results and discussion

Figure 2 shows SEM micrographs of the solder matrix just after reflow soldering and after isothermal aging for 504 h. Figure 2(a) shows the Sn–Ag–Cu solder matrices and Figure 2(b) shows the Sn–Bi solder matrices. In the as-soldered condition in the case of Sn–Ag–Cu solder, the microstructures after reflow were composed of dendrites of a β-Sn phase, which formed as the primary phase, and a network band, which was the β-Sn/intermetallic compound particles eutectic phase. The size of the primary β-Sn phase for the laser reflow soldering was smaller than that of the conventional reflow soldering. This is because the cooling rate of the laser soldering is much faster as compared with the conventional reflow soldering. After isothermal aging of both solders for 504 h, microstructural coarsening occurred during the aging. The microstructure of the solder matrix for the laser reflow soldering was similar to that for the conventional reflow soldering. In the case of Sn-Bi solder,

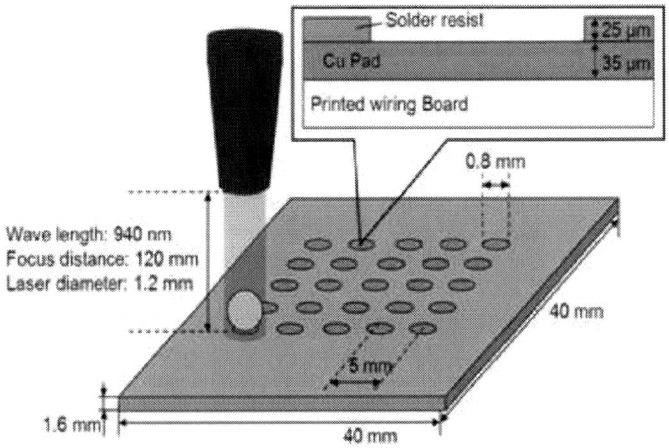

Figure 1. Schematic experimental setup of laser soldering with a Cu pad.

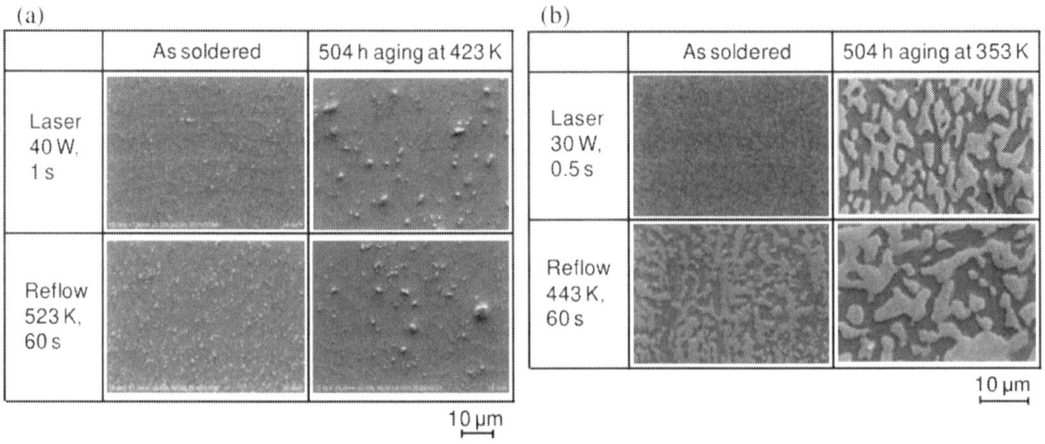

Figure 2. Typical SEM micrographs of the solder matrix after reflow soldering and after isothermal aging for 504 h. (a) Sn-3.0Ag-0.5 Cu solder, (b) Sn-58Bi solder

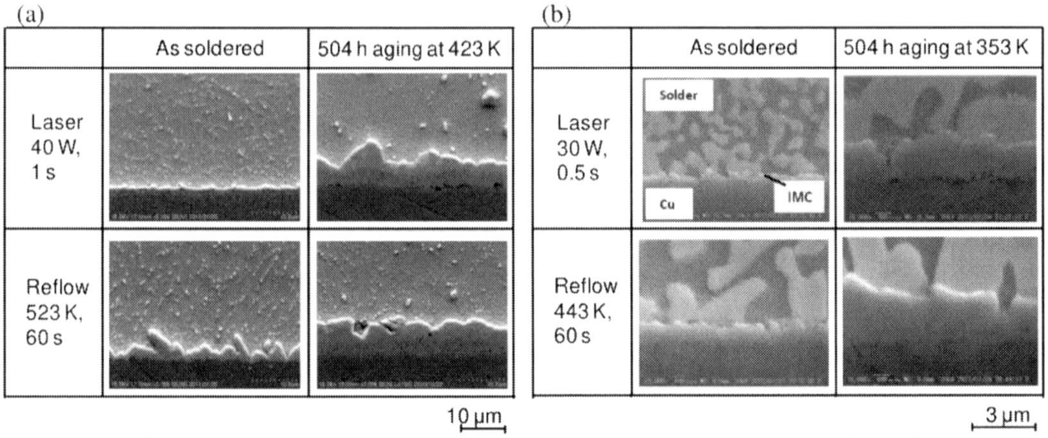

Figure 3. SEM micrographs of the interface between the solder and Cu pad after reflow soldering and isothermal aging for 504 h. (a) Sn-3.0Ag-0.5Cu solder, (b) Sn-58Bi solder

the original Sn and Bi phases were finely dispersed, and the sizes of the Sn and Bi phases for the laser reflow soldering were much smaller than those of the conventional reflow soldering. These fine matrices were eventually superseded by growing islands of Sn and Bi after aging for 504 h. The Sn and Bi phases of the solder for the laser reflow soldering were somewhat smaller than those for the conventional reflow soldering. Thus, the heating method in particular affected the microstructure of the solder matrix only in the as-soldered condition

Figure 3 shows SEM images of the interface between the solder and the Cu pad just after soldering and after aging for 504 h. Figure 3(a) shows the Sn–Ag–Cu solder interface and Figure 3(b) shows the Sn-Bi solder interface. For both solders and in the as-soldered condition, a relatively thin IMC layer was formed at the interface in the case of the laser reflow soldering. The thickness of this IMC layer was less 1 μm. This is because the duration for which the solder was heated to a temperature greater than its melting temperature was much shorter in the case of laser reflow soldering than it was for conventional reflow soldering. After isothermal aging for 504 h, a distinct second layer was noticed at the interface between the Cu pad and the first layer, and the total thickness of the IMC layer increased, regardless of the heating method. As shown in Figure 3, it is clear that there was some effect of the heating method on the IMC layer growth at the interface, regardless of solder composition.

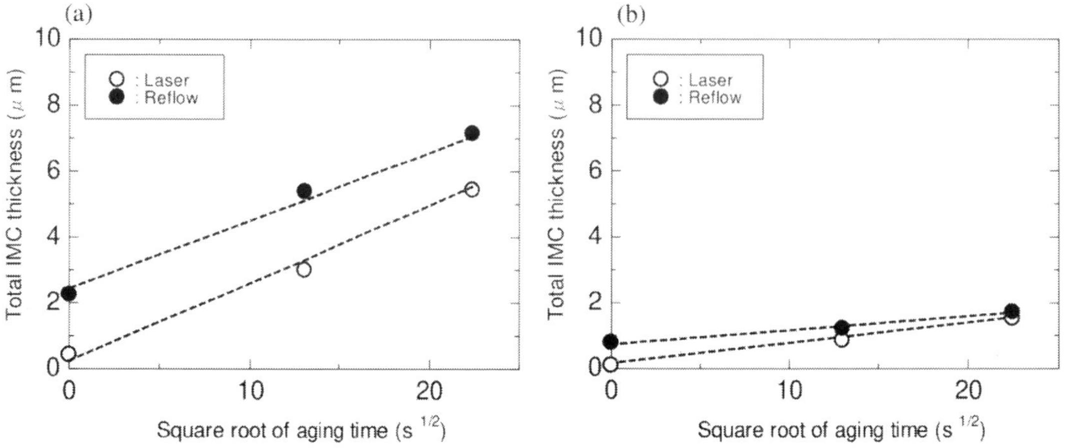

Figure 4. Effect of isothermal aging on the total IMC thickness at the interface.
(a) Sn-3.0Ag-0.5 Cu solder, (b) Sn-58Bi solder

Figure 4 shows the effect of aging time on the growth of the IMC layer during the isothermal aging. The total IMC thickness for the Sn–Ag–Cu solder is shown in Figure 4(a) and that of the Sn–Bi solder is shown in Figure 4(b). Regardless of solder composition and for both the laser reflow soldering and the conventional reflow soldering, the total IMC layer thickness increased with increasing aging time. The thickness is proportional to the square root of aging time. Accordingly, the growth of the IMC layer during the isothermal aging process can be considered a diffusion control process for both solders. The growth rate of the IMC layer for the laser reflow soldering was faster than that for the conventional reflow soldering.

4. Conclusions

In this study, the effect of heating method on the solder microstructure and the growth of the IMC layer at the interface of the Sn–Ag–Cu solder bump and the Sn–Bi solder bump on a Cu pad was investigated in order to elucidate the characteristics of laser reflow soldering. The main results obtained in this study are summarized as follows:

(1) The heating method affected the microstructure of the solder matrix only in the as-soldered condition, regardless of solder composition. For example, in the case of the Sn–Bi solder, Sn and Bi phases were finely dispersed, and the sizes of the Sn and Bi phases for the laser reflow soldering were smaller than those of the conventional reflow soldering.

(2) For both solders and in the as-soldered condition, a relatively thin IMC layer was formed at the interface in the case of the laser reflow soldering. Thus, the heating method had an effect on the formation and growth of the IMC layer at the interface during the isothermal aging.

References

[1] Lee J H, Lee Y H and Kim Y S 2000 *Script Mater.* **42** 789
[2] Li M Y, Wang C Q, Bang H S and Kim Y P 2005 *J. Mater. Processing Technol.* **168** 303
[3] Tan C W, Chan Y C, Leung B and Liu H D 2008 *Opt. Laser. Eng.* **46** 75
[4] Britten S, Olowinsky A and Gillner A 2013 *Physics Procedia* **41** 153
[5] Miao H W and Duh J G 2001 *Mater. Chem. Phys.* **71** 255
[6] Li J F, Mannan S H, Clode M P, Whally D C and Hutt D A 2006 *Acta Mater.* **54** 2907
[7] Zang L, Yuan Z, Zhao H. and Zhang X *Mater. Lett.* **63** 2067
[8] Zou H F, Zhang Q K and Zhang Z F 2009 *Scripta Mater.* **61** 308
[9] Zhang Q K and Zhang Z F 2011 *Mater. Sci. Eng.* **A 528** 2686
[10] Yen Y W, Liou W K, Chen C-M, Lin C K, Hung M K. 2011 *Mater. Chem. Phys.* **128** 233

Plasma-spraying synthesis of high-performance photocatalytic TiO_2 coatings

Yasuo Takahashi[1], Yoshitaka Shibata[2], Masakatsu Maeda[1], Yasuyuki Miyano[3], Kensuke Murai[4] and Akira Ohmori[1]

[1]Joining and Welding Research Institute, Osaka University, Japan
E-mail: takasy@jwri.osaka-u.ac.jp

[2] Production Division, Komatsu Ltd. Hirakata, Osaka, Japan
[3] Akita University, Akita, Japan
[4] AIST Kansai, Ikeda, Hyogo, Japan

Abstract. Anatase (A-) TiO_2 is a photocatalytic material that can decompose air-pollutants, acetaldehyde, bacteria, and so on. In this study, three kinds of powder (A-TiO_2 without HAp, TiO_2 + 10mass%HAp, and TiO_2+30mass%HAp, where HAp is hydroxyapatite and PBS is polybutylene succinate) were plasma sprayed on biodegradable PBS substrates. HAp powder was mixed with A-TiO_2 powder by spray granulation in order to facilitate adsorption of acetaldehyde and bacteria. The crystal structure was almost completely maintained during the plasma spray process. HAp enhanced the decomposition of acetaldehyde and bacteria by promoting adsorption. A 10mass% HAp content was the most effective for decomposing acetaldehyde when plasma preheating of the PBS was not carried out before the plasma spraying. The plasma preheating of PBS increased the yield rate of the spray process and facilitated the decomposition of acetaldehyde by A-TiO_2 coatings without HAp. HAp addition improved photocatalytic sterilization when plasma preheating of the PBS was performed.

1. Introduction

Anatase titanium dioxide (A-TiO_2) is widely known as a photocatalytic material [1]. Photocatalytic A-TiO_2 coatings help ultraviolet rays decompose toxic substances such as acetaldehyde and sterilize bacteria. The plasma spraying of A-TiO_2 powder is often used to fabricate these materials, but A-TiO_2 transforms to rutile (R-) TiO_2 at temperatures higher than 1188 K [2]. Therefore, the spraying conditions should be controlled so that the A-TiO_2 phase cannot transform into the R-TiO_2 phase during plasma spraying. PBS (polybutylene succinate) plastic is a better substrate for this procedure than metals with high melting points, because the melting point of PBS is 387 K, and so PBS can adhere to A-TiO_2 more easily than metals at low temperatures (with low heat-input). In other words, it is very easy to obtain A-TiO_2 coatings on PBS by plasma spraying, and it is possible to retain the A-TiO_2 structure after spraying [2, 3]. PBS is a biodegradable plastic, and TiO_2 is harmless. Therefore, it is environmentally friendly to make photocatalytic A-TiO_2 coatings on PBS. HAp (hydroxyapatite) adsorbs acetaldehyde, mold, and bacteria. Thus, A-TiO_2 powder mixed with HAp powder can enhance the decomposition rate. It is therefore important to investigate the plasma-spraying synthesis of A-TiO_2 coatings mixed with HAp powder and to examine the photocatalytic properties of the resulting materials [4]. The purpose of the present study is to synthesize high-performance photocatalytic A-TiO_2 coatings by plasma spraying and to verify the effect of HAp on the photocatalysis.

Content from this work may be used under the terms of the Creative Commons Attribution 3.0 licence. Any further distribution of this work must maintain attribution to the author(s) and the title of the work, journal citation and DOI.
Published under licence by IOP Publishing Ltd

2. Experimental procedure

The size (coating area) of the PBS substrate for plasma spraying was 50 mm × 50 mm, and the thickness was 9 mm. The three kinds of spraying materials were A-TiO_2 powder, A-TiO_2 + 10mass%HAp powder (TiO_2+10%HAp), and A-TiO_2 + 30mass%HAp (TiO_2+30%HAp) powder. The mixed powder was made by spray granulation. The average diameters of the A-TiO_2 and HAp powders before spray-granulating were 200 nm and 5 μm, respectively. The plasma spraying conditions are shown in Table 1. Figure 1 shows a cross sectional illustration of the plasma torch used in this study. The plasma spraying was carried out either without preheating the PBS or immediately after preheating. The preheating of PBS was performed in a one-pass traverse to increase the yield rate of the sprayed coatings and to increase the anatase ratio (abundance fraction of the A-TiO_2 phase). The preheating was carried out under an arc current of 300 A to increase the coating yield rate. The traverse speed was 50 mm/s. The coatings were observed with a scanning electron microscope (SEM). The coating structures were measured by X ray diffraction (XRD). The anatase ratio f (parentage fraction) in the coatings is determined by

$$f = \frac{1}{1 + 1.265\,\dfrac{I_R}{I_A}} \times 100, \tag{1}$$

where I_A and I_R are, respectively, the maximum peak values of A-TiO_2 and R-TiO_2 in the XRD graphs [5]. The photocatalytic properties of the materials were investigated by examining the decomposition (or sterilization) of acetaldehyde or bacteria (*Staphylococcus aureus*: IFO 12732). For investigations of the decomposition of acetaldehyde, air (temperature: 25°C; humidity: 50%; volume: 2×10^{-3} m^3) containing 100 ppm acetaldehyde was prepared in a glass vessel (flask) in which a specimen with a coated area of 2500 mm^2 was hung, as shown in Figure 2. Ultraviolet light with a peak wavelength of 352 nm was applied at 20 W to the specimen from the bottom of the flask. The distance between the coatings and the light source was 10 mm. The amount of acetaldehyde remaining was measured by a gas detector after the designated irradiation time t. On the other hand, for the estimation of the sterilizing power, a loopful of inoculum (*S. aureus*) was put into 150 cm^3 of the 100 % NB (nutrient broth) medium and the fungus liquid (solution) was cultivated for 20 hours at 37°C. Then, 1 cm^3 of the fungus liquid was diluted to 0.5% in 200 cm^3 of the NB medium to obtain the testing solution (10^5 cfu[colony forming units]/cm^3), as shown in Figure 3. For each experiment, 200 cm^2 of the testing solution was poured into a glass flask and the coated specimen was hung (dipped) into the solution, as shown in Figure 4. The sample was irradiated with ultraviolet light (black light) with a peak

Table 1. Preheating and plasma spraying conditions.

Parameters (unit)	Values	
	Non-preheating	With preheating
Ar pressure (MPa)	0.48	0.41
Distance between gun and substrate (mm)	100	100
Arc current (A)	500	300
Traverse speed (mm/s)	165	50
Stepping width (mm)	4.0	5.0
Spraying pass	2	2

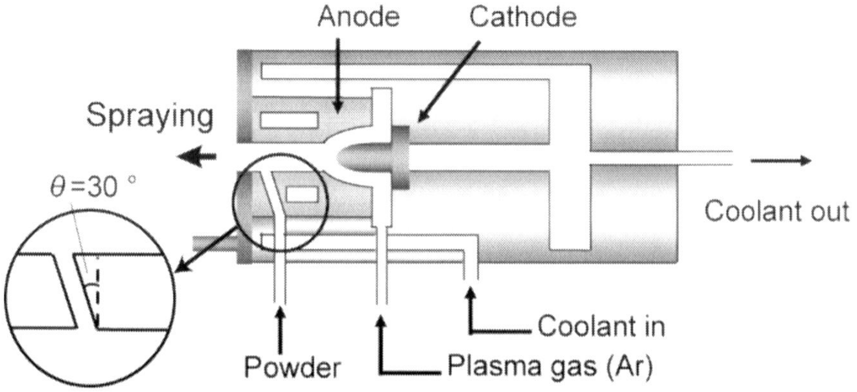

Figure 1. Schematic diagram (cross section) of plasma spray torch.

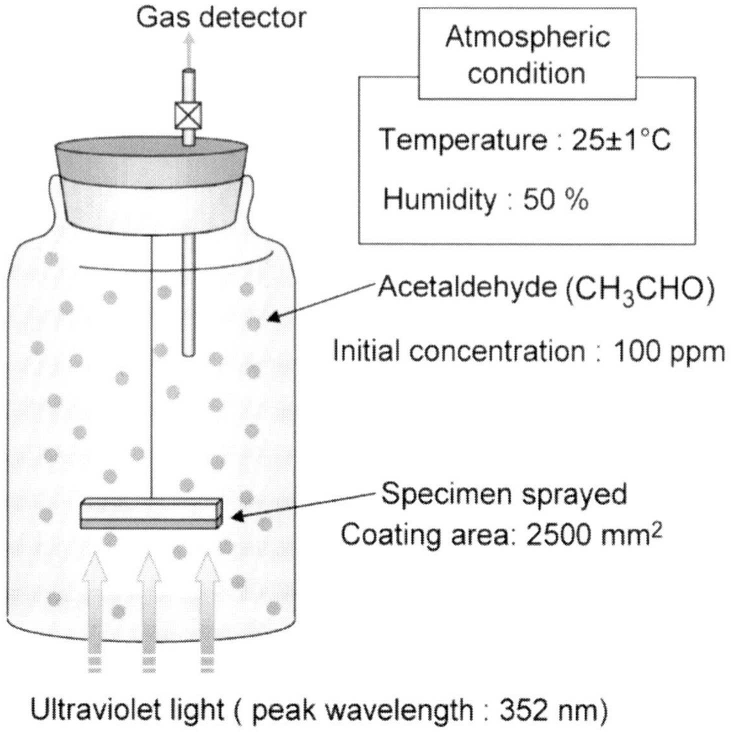

Figure 2. Experimental method for photocatalytic acetaldehyde decomposition.

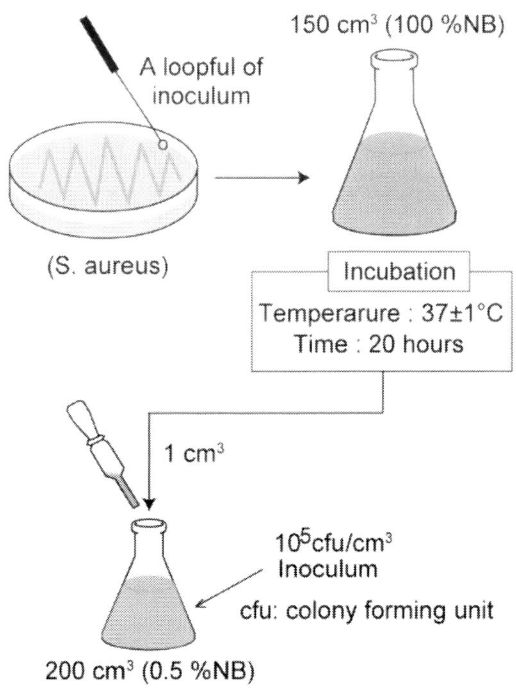

Figure 3. Preparation of fungus liquid from inoculum (*S. aureus*).

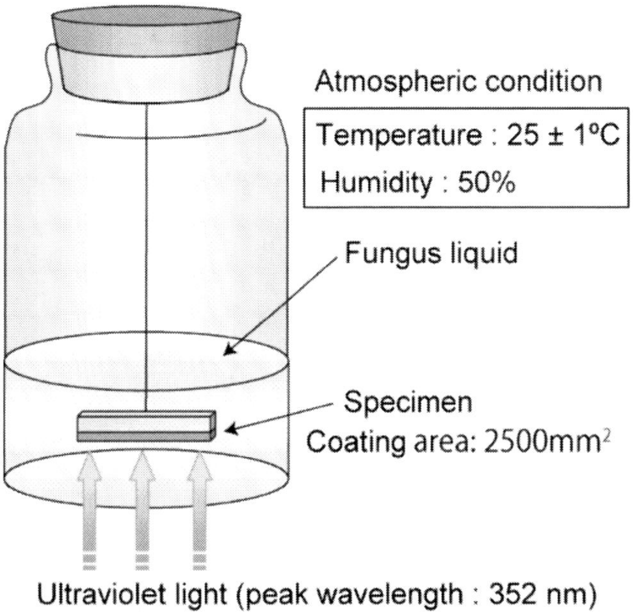

Figure 4. Experimental method for photocatalytic sterilization of *S. aureus*.

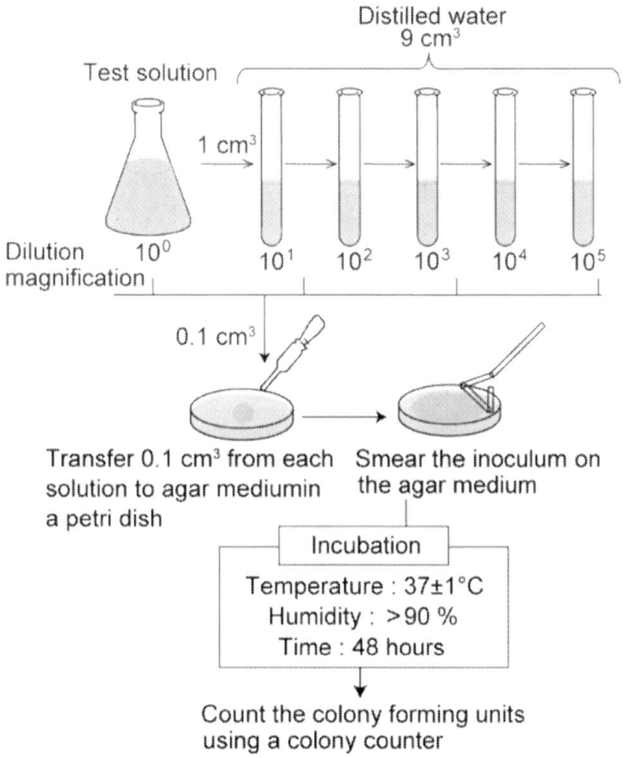

Figure 5. Outline of the spread plate method.

wavelength of 352 nm at 20 W from a distance of 10 mm for the designated irradiation time t. The total viable bacterial count C per unit cm^3 (C_0 at $t = 0$ s) was measured using the spread plate method (incubation conditions: 37°C, humidity > 90%, 48 hours) to estimate the sterilizing ability of the material. The spread plate method is illustrated in Figure 5. The bacteria-adsorption characteristics of HAp were also estimated by using fungus liquid with a concentration of 10^6 cfu/cm^3. In these experiments, 3 cm^3 of the fungus liquid was poured into a test tube, and a specimen with a coated area of 400 mm^2 was dipped in the fungus liquid and stirred at 70 rpm for 24 hours at 4°C in an atmosphere with 50% humidity without light. After that, the total viable bacterial count was measured in the same way using the spread plate method.

3. Results and discussion

Figure 6 shows the surface morphology and XRD results for the spraying powder after spray granulation. The surface morphologies of the different powders are not so different just after granulation, as seen in Figure 6. The average diameters were about 34 µm. R-TiO_2 peaks are not observed. The TiO_2 powder has no HAp peaks.

Figure 7 shows cross sectional views of the sprayed coatings. The photos in the left column were obtained after plasma spraying without preheating of the PBS, and those in the right column were obtained with preheating of the PBS. As seen in Figure 7, the preheating makes it easier for the sprayed particles to implant into the PBS substrate, since the substrate surface is premelted. This increases the coating yield rate.

Figure 8 shows the surface morphology obtained after spray coating. The surface morphology is not strongly influenced by the preheating or the addition of HAp.

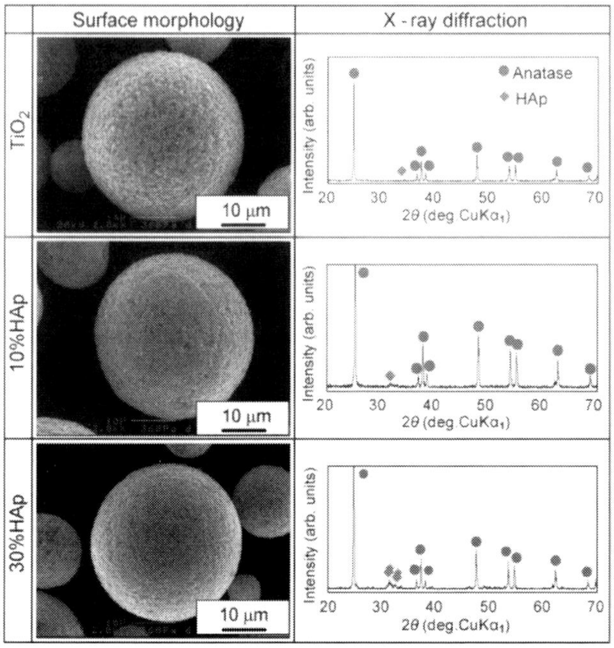

Figure 6. SEM images showing the surface morphology and XRD patterns of the three kinds of spraying powders.

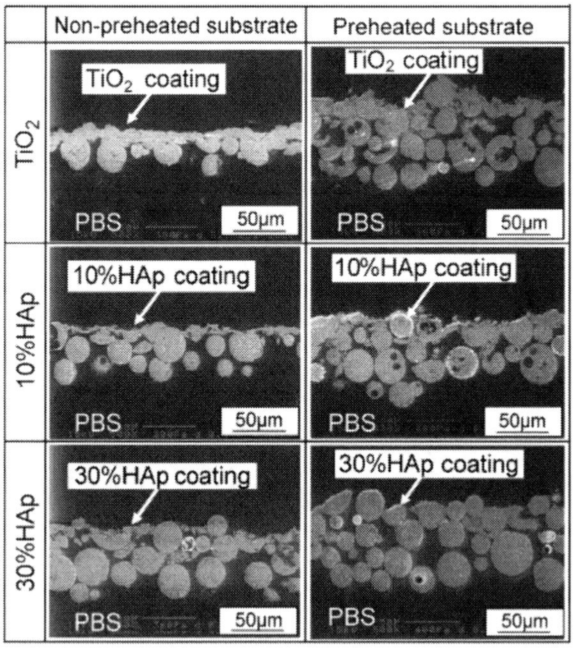

Figure 7. Cross-sectional SEM images of sprayed coatings.

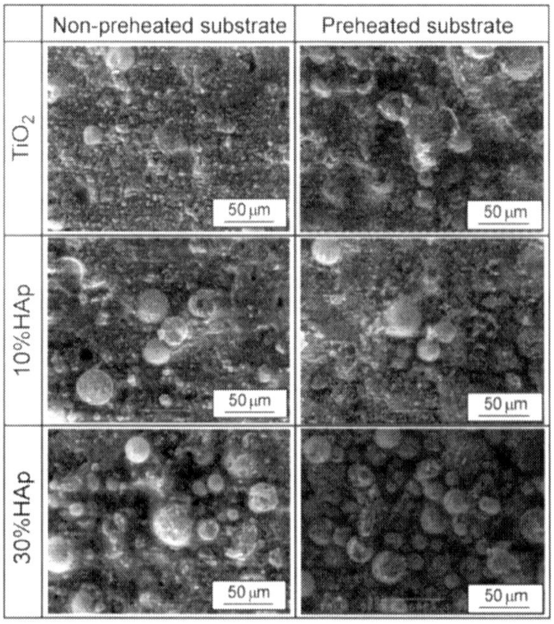

Figure 8. Surface morphologies (SEM images) of sprayed coatings.

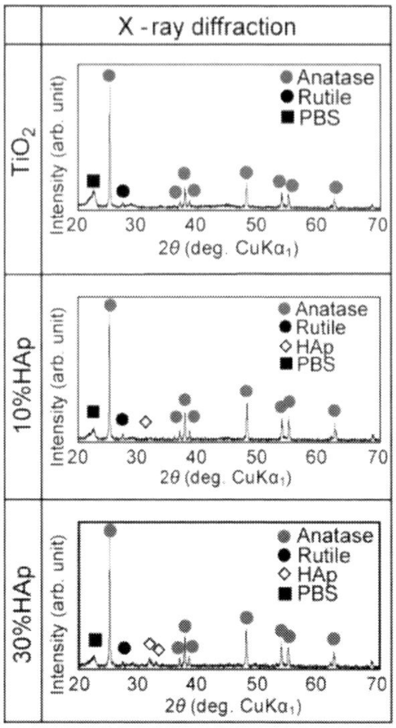

Figure 9. XRD patterns of sprayed coatings with different HAp compositions.

Table 2. Anatase ratio obtained for different HAp contents with and without preheating.

Plasma preheating	TiO_2	10%HAp	30%HAp
Without preheating	94.4	95.1	95.5
With Preheating	96.3	95.8	100

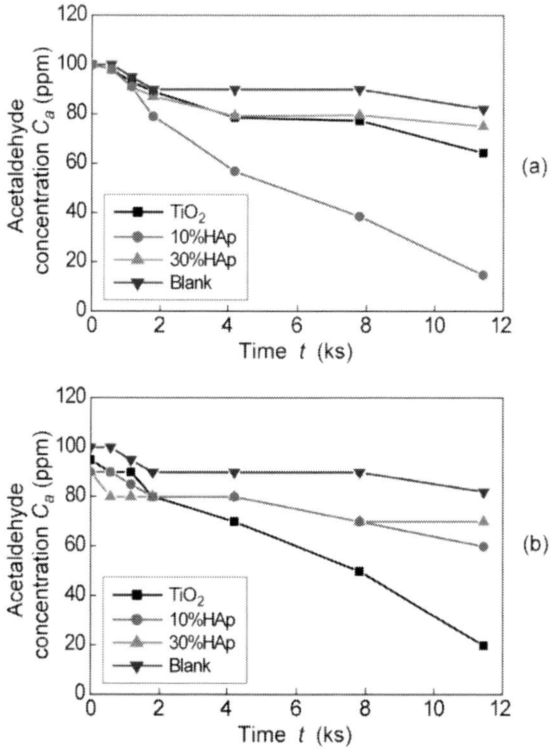

Figure 10. Photocatalytic decomposition of acetaldehyde by sprayed coatings with different HAp contents fabricated (a) without preheating and (b) with preheating.

Figure 9 shows the XRD results for the coating plasma sprayed without preheating. The R-TiO_2 peaks are very small, and a small HAp peak was also observed when HAp was mixed with A-TiO_2. The anatase ratio f of each coating is greater than 94%, as shown in Table 2. Preheating can increase the anatase ratio.

Figure 10 shows the changes in acetaldehyde concentration with time for the sprayed coatings fabricated (a) without preheating and (b) with preheating. It is found that the A-TiO_2+10mass%HAp coatings are the most effective for decomposing acetaldehyde without preheating. If the substrate is

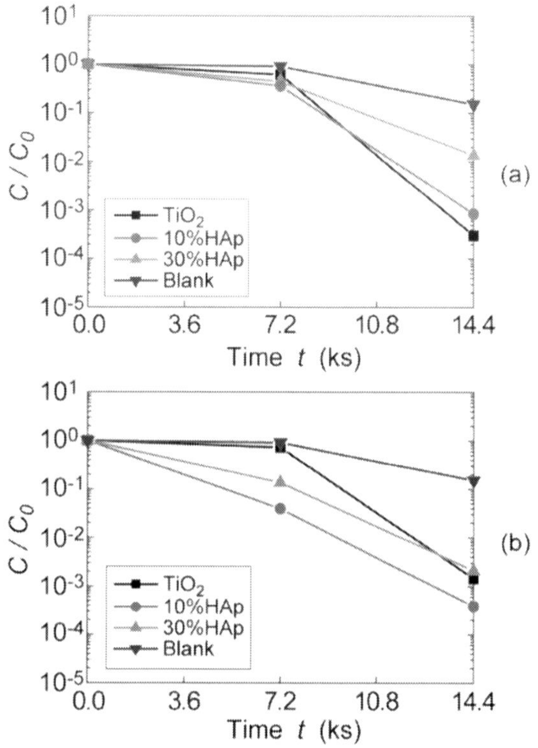

Figure 11. Photocatalytic decomposition of acetaldehyde by sprayed coatings with different HAp contents fabricated (a) without preheating and (b) with preheating.

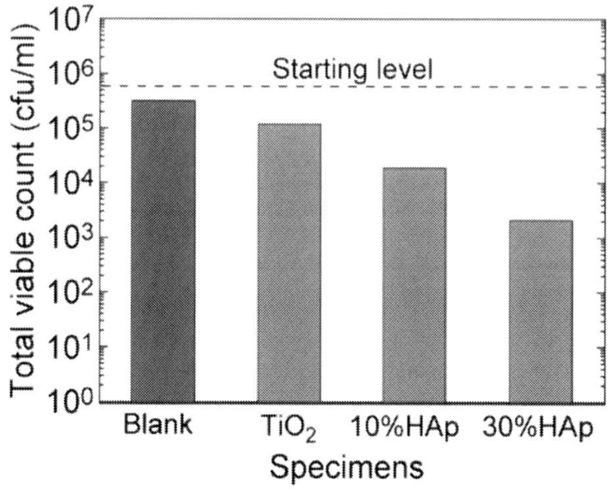

Figure 12. Adsorption ability of HAp.

preheated, A-TiO_2 without HAp shows the highest decomposition rate. This is a very interesting result, but the reason for this phenomenon is not known.

Figure 11 shows the changes in C/C_0 with time for sprayed coatings fabricated (a) without preheating and (b) with preheating. HAp has a large effect on the sterilization ability of the coating. In particular, HAp can improve the sterilization performance of the coatings fabricated with preheating, as seen in Figure 11 (b). Figure 12 shows experimental results on the bacteria adsorption function of HAp. The total viable count decreases with increasing HAp content. It is found that HAp improves the photocatalytic properties of A-TiO_2 through its bacteria-adsorption function.

4. Conclusion

Photocatalytic A-TiO_2–HAp coatings were fabricated by plasma spraying. The acetaldehyde decomposition and sterilization performance of the coatings were then investigated. The main results are as follows:

1) HAp addition is useful for obtaining high-performance photocatalytic TiO_2 coatings fabricated by plasma spraying.
2) Preheating the PBS increases the coating yield rate and the anatase ratio.
3) The sprayed particles are implanted into the PBS by plasma spraying.
4) TiO_2-10%HAp coatings are the most effective catalysts for acetaldehyde decomposition.
5) If preheating is carried out, TiO_2 coatings without HAp decompose acetaldehyde very well.
6) HAp improves the photocatalytic sterilization performance of the coating through its bacteria-adsorption function.

References

[1] Bonsen E M, Schroeter S, Jacobs H and Broekaert J A C 1997 *Chemosphere* **35** 1431
[2] Kanazawa T, Takahashi Y and Omori A 2004 *Proc. Inter. Thermal Spray Conf. & Expo.*
[3] Omori A, Matsusaka S and Toki Y 2001 *High Temp. Soc. Jpn* **27** 274
[4] Omori A, Nakade K, Mizutani S and Yasuoka J 2003 *Trans. JWRI* **32** 319
[5] Spurr R A and Myers H 1957 *Analytical Chemistry* **29** 760

Simple evaluation method for temperature drop at contact interface between rough surfaces under low contact pressure conditions

Toshio Tomimura[1, 3], Yasuo Takahashi[2], TaeWan Do[1], Kensei Shigyo[1] and Yasushi Koito[1]

[1] Department of Advanced Mechanical Systems, Kumamoto University, 2-39-1 Kurokami, Chuo-ku, Kumamoto, 860-8555, JAPAN

[2] Joining and Welding Research Institute, Osaka University, 2-1Yamada-Oka, Suita, Osaka, 565-0871, JAPAN

[3] To whom any correspondence should be addressed.

E-mail: tomi@mech.kumamoto-u.ac.jp

Abstract. For heat removal from systems such as electronic equipment, satellite thermal control systems, and nuclear reactors, reduction of thermal contact resistance (TCR) is the most crucial issue to be addressed. Several studies have attempted to propose evaluation equations for predicting TCR for flat rough surfaces. However, as is well known, there are still wide discrepancies among measured results, even for similar materials. In this study, based on the conventional unit cell model for flat surfaces with roughness and the newly proposed contact surface model for wavy surfaces with roughness, thermal contact resistance under a low contact pressure of 0.1–1.0 MPa is investigated theoretically and experimentally. Comparison of the measured and calculated results shows that the measured temperature drop at the interface (that is, the thermal contact resistance) between flat surfaces with roughness lies between the values evaluated by the unit cell model for the cases with and without the heat flow constriction. Furthermore, when the rough surface has waviness, the introduction of macroscopic constriction resistance is shown to be important for evaluating the temperature drop at the interface.

1. Introduction

In electronic or electric equipment with high heat fluxes, heat transfer modes such as heat conduction, convection, and thermal radiation play an important role in heat removal from the high-temperature regions of the rquipment. In addition to these heat transfer modes, thermal contact resistance (TCR) inevitably occurs at the solid–solid contact interface, and these four modes are apt to be coupled intricately.

Several studies have been conducted to propose evaluation equations for predicting TCR for flat rough surfaces. However, as pointed out by Fletcher [1], Torii and Yanagihara [2], Torii [3], Okada and Matsumoto [4], and Madhusudana [5] et al., there are still wide discrepancies among measured TCRs, even for similar materials. In our previous study [6], to investigate the key factors for the

Content from this work may be used under the terms of the Creative Commons Attribution 3.0 licence. Any further distribution of this work must maintain attribution to the author(s) and the title of the work, journal citation and DOI.
Published under licence by IOP Publishing Ltd

discrepancies, a fundamental analysis was undertaken using a simple contact surface model, which was composed of the unit cell model proposed by Tachibana [7] and Sanokawa [8] and the Hertzian contact model [9]. Based on a series of numerical calculations and measurements, the effects of surface waviness and measured positions of specimen temperatures on TCR measurement were clarified. Furthermore, based on our study using a two-dimensional microscopic surface model, which consists of random numbers and Abbott's bearing area curve [10], the effects of surface waviness and roughness on the temperature fields near the contact interface have been clarified microscopically.

In this study, to obtain a simple evaluation equation for TCR for flat or wavy rough surfaces, theoretical and experimental investigations were conducted under a lower mean nominal contact pressure of 0.1–1.0 MPa.

2. Physical models for flat and wavy surfaces with roughness

In this section, two types of surface models are investigated theoretically. The first model is the conventional unit cell model for macroscopically flat surfaces with roughness; the resultant equations for evaluating TCR and thermal contact conductance (TCC), which is the reciprocal of TCR, are shown in an organized form. The second model is a newly proposed simple model for macroscopically wavy surfaces with roughness. In this model, by assuming a parallel heat flow through the central solid–solid contact zone and the outer torus-shaped air zone surrounding the central zone, simple evaluation equations for TCC and TCR are derived.

2.1. Unit cell model for flat surfaces with roughness

The unit cell model proposed by Tachibana [7] was introduced according to the process explained in Figure 1. Figure 1(a) shows a magnified image of the solid–solid contact interface. Macroscopically, a discontinuous temperature drop ΔT (= $T_{ih} - T_{ic}$) is observed at the contact interface; this phenomenon is called the thermal contact resistance. Here, T_{ih} and T_{ic} are the extrapolated interface temperatures obtained based on the measured temperatures for each solid. Microscopically, however, the temperature is supposed to change smoothly in the vicinity of the interface, as indicated at the broken line. On the other hand, Figure 1(b) shows the contact surface model for the interface of Figure 1(a). In this model, the solid–solid contact through the surface roughness is physically expressed by cylinders with the same height of the surface roughness as that shown in Figure 1(a). Furthermore, these cylinders are assumed to be uniformly distributed over the solid surface and have the same whole contact area as the true contact portions. Since TCR based on the heat flow rate per unit area (that is, the heat flux) is equivalent to that of the unit cell shown in Figure 1(c), it is sufficient to consider the heat flow through the unit cell.

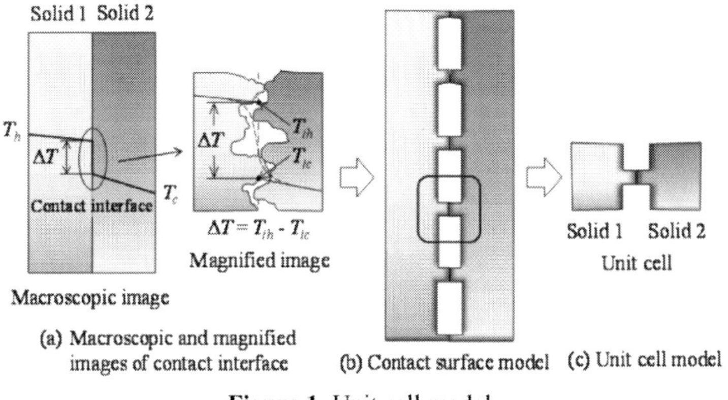

Figure 1. Unit cell model.

As shown in Figure 2, the TCR of the unit cell model R is obtained by taking into account the parallel connection of the thermal resistances caused by the cylinder and the intervening substance (in this study, the air), denoted by R_s and R_f, respectively. In the figure, Q, Q_s, and Q_f are the heat flow rates through the unit cell, cylinder and intervening substance, respectively; A and a are the cross-sectional areas of the unit cell and cylinder, respectively; λ is the effective thermal conductivity of the contacting portion surrounded by the broken line; and λ_f is the thermal conductivity of the intervening substance. In addition, λ, δ, T and H are the thermal conductivity, maximum surface roughness of the solid surface, surface temperature, and hardness (Brinell or Vickers [11]) of the solid, respectively, and the subscripts 1, 2, and i denote Solid 1, Solid 2 and the contacting portion of the cylinders, respectively. Here, the relation between the maximum surface roughness of the solid surface δ and the center line average roughness Ra is given by $\delta \cong 4Ra$ [11].

Based on Fourier's law of heat conduction and the energy conservation law, which should be satisfied for Q, Q_s, and Q_f, and under the assumption that all the isothermal lines are parallel to the solid–solid contact interface, Tachibana [7] obtained the following evaluation equation for the TCR r [m^2·K/W] for flat surfaces with roughness.

$$\frac{1}{r} = \frac{1}{\dfrac{\delta_1}{\lambda_1} + \dfrac{\delta_2}{\lambda_2}} \frac{a}{A} + \frac{\lambda_f}{\delta_1 + \delta_2} \frac{A-a}{A} \quad (1)$$

Here, the relation between the TCR R [K/W] and the resistance r, which is derived based on the heat flux (that is, Q/A, Q_s/a and $Q_f/(A-a)$) is given by

$$r = AR\,. \quad (2)$$

On the other hand, to take into account the increase in the thermal resistance due to the contraction and expansion of heat flow through the cylindrical portion of the unit cell, Sanokawa [8] proposed the following equation. In this equation, the additional thermal resistance $1/h_0$ caused by the constriction of the heat flow is introduced in the denominator of the first term on the right-hand side of equation (1).

$$\frac{1}{r} = \frac{1}{\dfrac{\delta_1}{\lambda_1} + \dfrac{1}{h_0} + \dfrac{\delta_2}{\lambda_2}} \frac{a}{A} + \frac{\lambda_f}{\delta_1 + \delta_2} \frac{A-a}{A} \quad (3)$$

Here, based on a series of theoretical and experimental studies, the additional term $1/h_0$ is given by

$$\frac{1}{h_0} = 2.3 \times 10^{-5} \left(\frac{1}{\lambda_1} + \frac{1}{\lambda_2} \right). \quad (4)$$

The ratio of the true contact area to the apparent contact area a/A included in the right-hand side of equations (1) and (3) can be replaced with the following equation [7, 8].

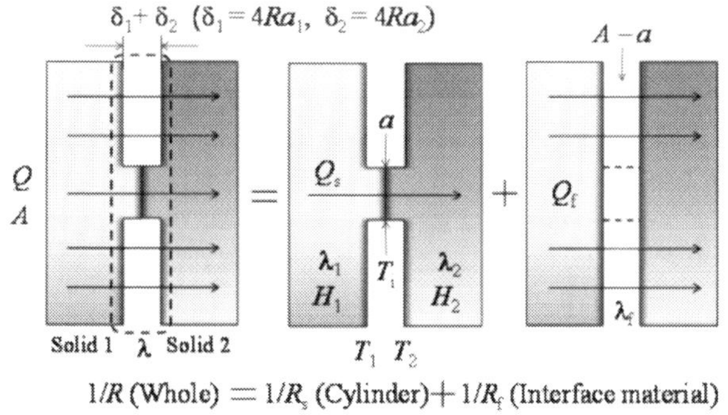

Figure 2. Thermal contact resistance of unit cell model.

$$\frac{a}{A} \cong \frac{p_{\mathrm{m}}}{H_{\min}} \tag{5}$$

Here, p_{m} is the mean nominal contact pressure defined by dividing the axial load f by the apparent contact area A, and $H_{\min}$ denotes the smaller of the hardness values of Solid 1 and Solid 2. Based on the above discussion, the following equations can be obtained for the TCC k, which is the reciprocal of the resistance r.

$$k_{\mathrm{TE}} = \frac{1}{\dfrac{\delta_1}{\lambda_1} + \dfrac{\delta_2}{\lambda_2}} \frac{p_{\mathrm{m}}}{H_{\min}} + \frac{\lambda_{\mathrm{f}}}{\delta_1 + \delta_2}\left(1 - \frac{p_{\mathrm{m}}}{H_{\min}}\right) \quad \text{(Tachibana's equation)} \tag{6}$$

$$k_{\mathrm{SE}} = \frac{1}{\dfrac{\delta_1}{\lambda_1} + 2.3\times 10^{-5}\left(\dfrac{1}{\lambda_1} + \dfrac{1}{\lambda_2}\right) + \dfrac{\delta_2}{\lambda_2}} \frac{p_{\mathrm{m}}}{H_{\min}} + \frac{\lambda_{\mathrm{f}}}{\delta_1 + \delta_2}\left(1 - \frac{p_{\mathrm{m}}}{H_{\min}}\right) \quad \text{(Sanokawa's equation)} \tag{7}$$

Furthermore, with the introduction of the three types of thermal conductance with regard to the cylindrical portion without and with heat flow constriction, denoted by $k_{\mathrm{TE,s}}$ and $k_{\mathrm{SE,s}}$ respectively, and the intervening substance, denoted by k_{f}, equations (6) and (7) can be expressed by the following equations, respectively.

$$k_{\mathrm{TE}} = k_{\mathrm{TE,s}} + k_{\mathrm{f}} \quad \text{(Tachibana's equation)} \tag{8}$$

$$k_{\mathrm{SE}} = k_{\mathrm{SE,s}} + k_{\mathrm{f}} \quad \text{(Sanokawa's equation)} \tag{9}$$

Here, $k_{\mathrm{TE,s}}$, $k_{\mathrm{SE,s}}$ and k_{f} are given as follows:

$$k_{\mathrm{TE,s}} = \frac{1}{\dfrac{\delta_1}{\lambda_1} + \dfrac{\delta_2}{\lambda_2}} \frac{p_{\mathrm{m}}}{H_{\min}} \quad \text{(Cylindrical portion without heat flow constriction),} \tag{10}$$

$$k_{\mathrm{SE,s}} = \frac{1}{\dfrac{\delta_1}{\lambda_1} + 2.3\times 10^{-5}\left(\dfrac{1}{\lambda_1} + \dfrac{1}{\lambda_2}\right) + \dfrac{\delta_2}{\lambda_2}} \frac{p_{\mathrm{m}}}{H_{\min}} \quad \text{(Cylindrical portion with heat flow constriction)} \tag{11}$$

and

$$k_{\mathrm{f}} = \frac{\lambda_{\mathrm{f}}}{\delta_1 + \delta_2}\left(1 - \frac{p_{\mathrm{m}}}{H_{\min}}\right) \quad \text{(Intervening substance).} \tag{12}$$

Equation (4), which gives the additional term $1/h_0$ is obtained under a mean nominal contact pressure of approximately 1.0–10 MPa. This pressure range is almost one order of magnitude higher than that targeted here. Therefore, in this study, the extension and applicability of Tachibana's equations (6) or (8) and Sanokawa's equations (7) or (9) to a pressure range of 0.1–1.0 MPa have been investigated by comparing them with the measured results obtained from the experiments with flat rough surfaces.

2.2. Solid–solid contact model for wavy surfaces with roughness

A pair of solids, Solid 1 and Solid 2, corresponding to the present cylindrical test specimens with radius r_{o}; radius of curvature r_{s1}, r_{s2}; maximum surface roughness δ_1, δ_2; and the same spherical waviness Δ, are placed face to face, as shown in Figure 3(a). The solids have thermal conductivity λ_1, λ_2; harness H_1, H_2; Young's modulus E_1, E_2; and Poisson's ratio ν_1, ν_2. After the two solids are pressed together with axial load f or mean nominal contact pressure p_{m} ($= f / (\pi r_{\mathrm{o}}^2)$), the system is deformed elastically, as shown in Figure 3(b). From a geometrical consideration, the thickness of the intervening substance z_{r} with thermal conductivity λ_{f} at an arbitrary radial position r is expressed using the following equation.

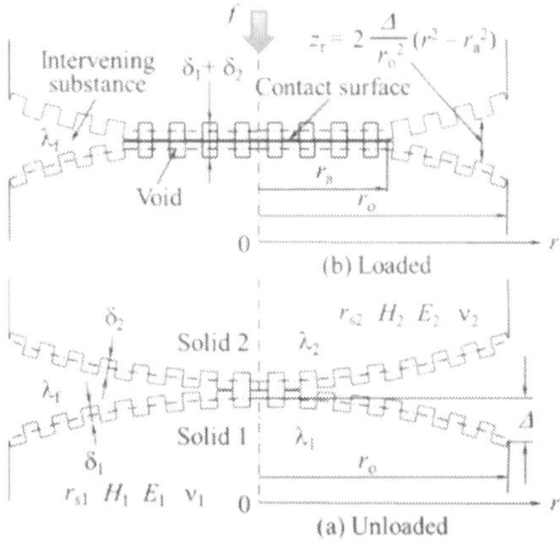

Figure 3. Contact of wavy rough surfaces.

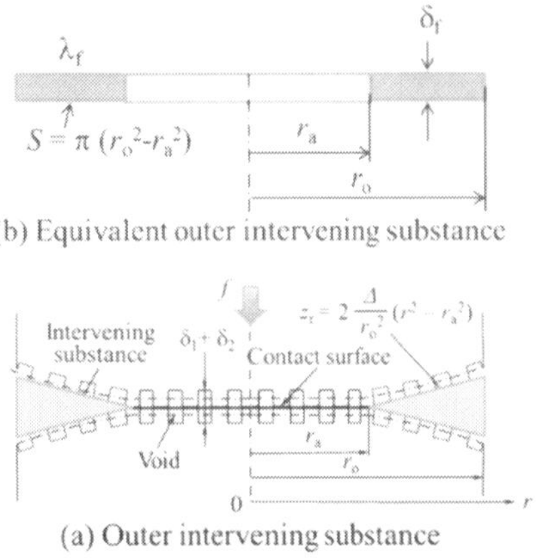

Figure 4. Model of outer intervening substance.

$$z_r = 2\frac{\Delta}{r_0^2}\left(r^2 - r_a^2\right) \quad \left(r_a \le r \le r_0\right) \tag{13}$$

Here, r_a is the radius of the contact surface and is expressed by

$$r_a = \left[\frac{3}{4} f \frac{\left(1-\nu_1^2\right)/E_1 + \left(1-\nu_2^2\right)/E_2}{r_{s1} + r_{s2}} r_{s1} r_{s2}\right]^{1/3} \tag{14}$$

from the Hertzian analysis [9].

In the present study, as the first step for developing a simple evaluation equation for the macroscopically wavy contact surfaces with roughness under lower pressure conditions, Tachibana's equations (6) or (8) and Sanokawa's equations (7) or (9) have been applied at the central solid–solid contact zone ($0 \le r \le r_a$) by introducing the following modified mean nominal contact pressure p_{ma}, evaluated using the central contact area πr_a^2.

$$p_{ma} = \frac{r_0^2}{r_a^2} p_m \tag{15}$$

On the other hand, in the case of the outer intervening substance surrounding the central solid–solid contact zone similar to that shown in Figure 4(a), an equivalent thickness δ_f of the intervening substance has been introduced as the first step for clarifying the effect of the macroscopic waviness on the overall thermal resistance. The value of δ_f shown in Figure 4(b) is obtained from

$$\delta_f = \frac{1}{S}\int_{r_a}^{r_0} z_r \, 2\pi r dr \, . \tag{16}$$

Here, S is the annular projected area of the outer intervening substance and is expressed as

$$S = \pi\left(r_0^2 - r_a^2\right). \tag{17}$$

Accordingly, by substituting equations (13) and (17) into equation (16), and then performing the integration, the following equation for the equivalent thickness δ_f is obtained.

$$\delta_f = \Delta\left[1 - \left(\frac{r_a}{r_0}\right)^2\right] \tag{18}$$

From the equations derived so far, and by assuming a parallel heat flow through the central solid-solid contact zone and its outer intervening substance, the thermal contact conduction of the present wavy surfaces with roughness is obtained as follows, based on the equation applied at the central contact zone, that is, Tachibana's equation or Sanokawa's equation.

$$k_{\mathrm{WTE}} = \left(\frac{r_{\mathrm{a}}}{r_0}\right)^2 k_{\mathrm{CTE}} + \left[1 - \left(\frac{r_{\mathrm{a}}}{r_0}\right)^2\right] k_{\mathrm{Of}} \tag{19}$$

$$k_{\mathrm{WSE}} = \left(\frac{r_{\mathrm{a}}}{r_0}\right)^2 k_{\mathrm{CSE}} + \left[1 - \left(\frac{r_{\mathrm{a}}}{r_0}\right)^2\right] k_{\mathrm{Of}} \tag{20}$$

Here, k_{CTE}, k_{CSE} and k_{Of} are given as follows:

$$k_{\mathrm{CTE}} = \frac{1}{\dfrac{\delta_1}{\lambda_1} + \dfrac{\delta_2}{\lambda_2}} \frac{p_{\mathrm{ma}}}{H_{\mathrm{min}}} + \frac{\lambda_f}{\delta_1 + \delta_2}\left(1 - \frac{p_{\mathrm{ma}}}{H_{\mathrm{min}}}\right), \tag{21}$$

$$k_{\mathrm{CSE}} = \frac{1}{\dfrac{\delta_1}{\lambda_1} + 2.3 \times 10^{-5}\left(\dfrac{1}{\lambda_1} + \dfrac{1}{\lambda_2}\right) + \dfrac{\delta_2}{\lambda_2}} \frac{p_{\mathrm{ma}}}{H_{\mathrm{min}}} + \frac{\lambda_f}{\delta_1 + \delta_2}\left(1 - \frac{p_{\mathrm{ma}}}{H_{\mathrm{min}}}\right) \tag{22}$$

and

$$k_{\mathrm{Of}} = \frac{\lambda_{\mathrm{f}}}{\delta_{\mathrm{f}}}. \tag{23}$$

Furthermore, by substituting equation (18) into equations (19) and (20), the following final equations are obtained.

$$k_{\mathrm{WTE}} = k_{\mathrm{CTE}}(p_{\mathrm{ma}}) + k_{\mathrm{Of}}(\Delta) \quad \text{(Without heat flow constriction at the contact surface)} \tag{24}$$

$$k_{\mathrm{WSE}} = k_{\mathrm{CSE}}(p_{\mathrm{ma}}) + k_{\mathrm{Of}}(\Delta) \quad \text{(With heat flow constriction at the contact surface)} \tag{25}$$

Here, $k_{\mathrm{CTE}}(p_{\mathrm{ma}})$, $k_{\mathrm{CSE}}(p_{\mathrm{ma}})$ and $k_{\mathrm{Of}}(\Delta)$ are given as follows:

$$k_{\mathrm{CTE}}(p_{\mathrm{ma}}) = \left(\frac{r_{\mathrm{a}}}{r_0}\right)^2 k_{\mathrm{CTE}} \tag{26}$$

$$k_{\mathrm{CSE}}(p_{\mathrm{ma}}) = \left(\frac{r_{\mathrm{a}}}{r_0}\right)^2 k_{\mathrm{CSE}} \tag{27}$$

$$k_{\mathrm{Of}}(\Delta) = \frac{\lambda_{\mathrm{f}}}{\Delta} \tag{28}$$

3. Experiment

The validity of the evaluation equations (8) and (9) for flat rough surfaces and equations (24) and (25) for wavy rough surfaces, derived in the previous section, was experimentally confirmed using two pairs of cylindrical brass specimens.

3.1. Experimental apparatus

Figure 5 shows a schematic of the experimental apparatus. A film heater of 40 mm (= $2r_0$) diameter, 0.5 mm thickness and 20 W (= Q) heat output is attached on the top surface of the upper test specimen. A copper cooling block, which is maintained at 20 °C by circulating water from a constant temperature water bath, is placed under the bottom surface of the lower test specimen. The test specimens are placed in a hollow cylinder with a 50–mm–thick styrene foam wall. Under the bottom of the film heater, an acrylic resin block of 20 mm thick is placed for thermal insulation. The axial load f is applied to the test specimens via a lever and hanging weight arrangement. Here, a balancing weight is set at the opposite end of the lever to obtain the no–load condition.

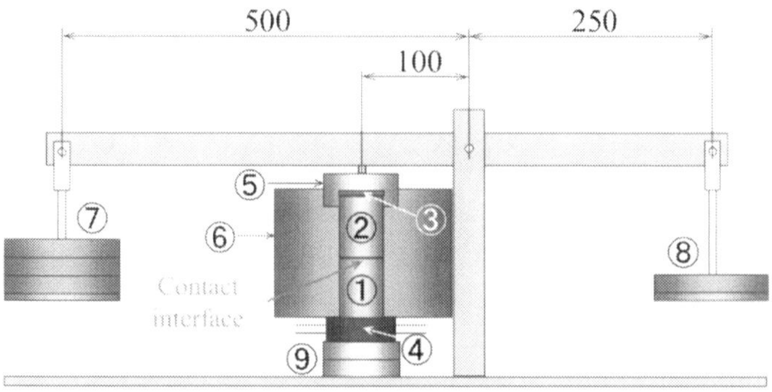

① Lower test specimen ② Upper test specimen ③ Film heater
④ Cooling block ⑤ Acrylic resin block ⑥ Thermal insulation
⑦ Weight ⑧ Balancing weight ⑨ Spacer

Figure 5. Schematic of experimental apparatus.

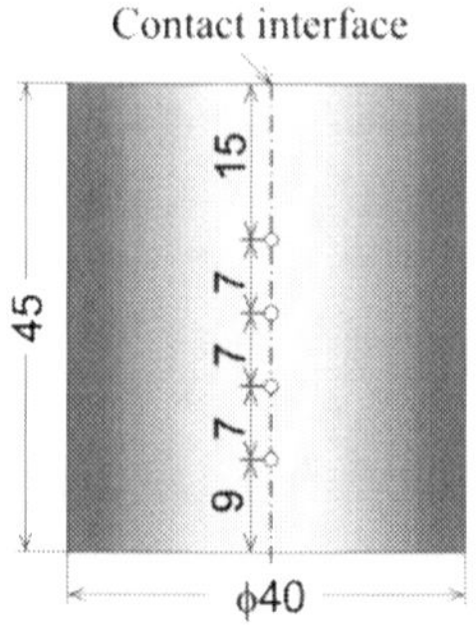

Figure 6. Test specimen.

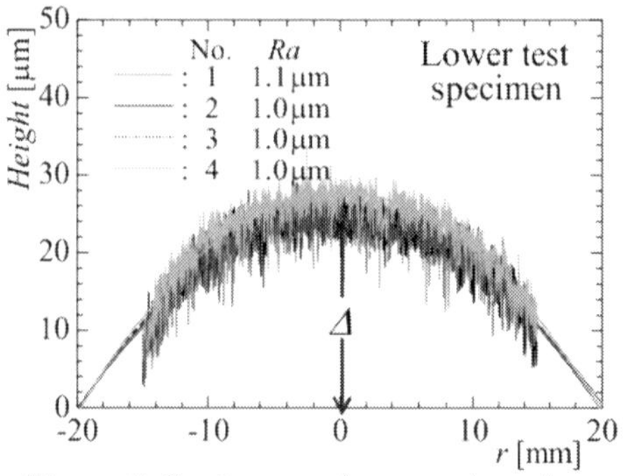

Figure 7. Surface roughness and waviness

3.2. Test specimen

In the present study, two pairs of brass specimens are tested in still air and at room temperature. One specimen has a flat rough surface, and the other has a wavy rough surface. As shown in Figure 6, the diameter d (=$2r_o$) and length l of each specimen are 40 mm and 45 mm, respectively. Each specimen has four small holes of 0.6 mm diameter and 10 mm depth, located at 15, 22, 29, and 36 mm from the contact surface, for temperature measurements. Copper–constantan–sheathed thermocouples of 0.5 mm diameter are inserted in each hole by applying a thermally conductive silver paste.

The contact surface of the specimen is finished by polishing after being turned by a lathe turning machine. Figure 7 shows examples of the center line average surface roughness Ra and the macroscopic waviness Δ of such a finished surface, measured using a stylus profilometer. The mean values of Ra and the maximum values of Δ, measured at four sections by rotating the specimens in steps of 45°, are listed in Table 1.

4. Results and discussion

In this section, the thermal contact resistance at the solid–solid contact interface has been discussed by comparing the theoretical and experimental results. First, in the case of the solid–solid contact of flat rough surfaces, the extension and applicability of Tachibana's equation (6) or (8) and Sanokawa's equation (7) or (9) to the low pressure range has been investigated. Next, for wavy surfaces with

Table 1. Measured *Ra* and Δ of test specimen.

Test specimen		*Ra* [μm]	Δ [μm]
Flat rough surface	Upper	0.3	0
	Lower	0.4	0
Wavy rough surface	Upper	1.0	25
	Lower	1.0	25

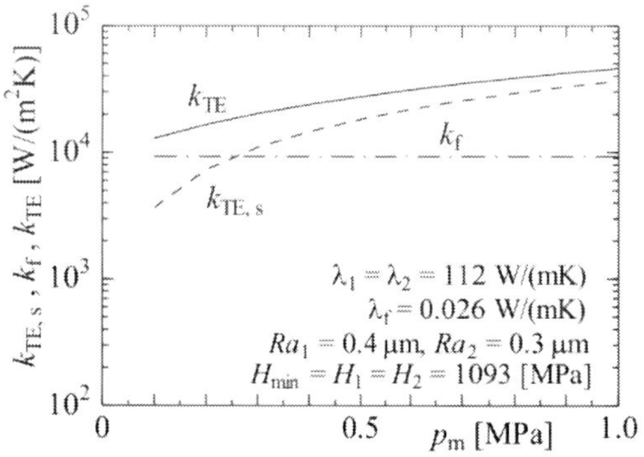

Figure 8. Thermal contact conductance by Tachibana's equation (8).

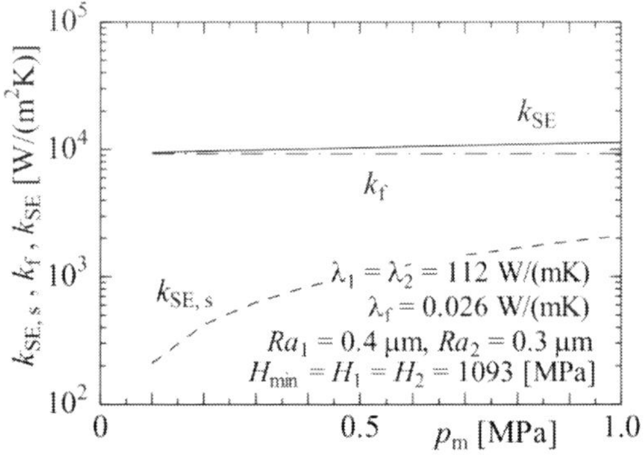

Figure 9. Thermal contact conductance by Sanokawa's equation (9).

roughness, the validity of the simple evaluation equations (24) and (25) for thermal contact conductance has been examined.

In general, the relationship among ΔT at the solid–solid contact interface, the heat flux flowing through the interface q (= Q/A, Q: heat rate flowing through the interface, A: cross-sectional area of the solid), the thermal contact resistance r, and the thermal contact conductance k is given by the following equation.

$$q = \frac{\Delta T}{r} = k\,\Delta T \tag{29}$$

In the present study, the heat output of the film heater attached on the top surface of the upper test specimen Q, which can be regarded as the heat rate flowing through the solid–solid contact interface, was kept constant at 20 W throughout the experiments. Hence, for facilitating intuitive understanding of the behavior of the thermal contact resistance r or the thermal contact conductance k at the interface, the temperature drop ΔT, which is proportional to r or inversely proportional to k as can be seen from equation (29), has been used for the comparison between the experimental and theoretical results.

4.1. Solid–solid contact of flat surfaces with roughness

Figures 8 and 9 illustrate the effect of p_m on the thermal contact conductance k_{TE} and k_{SE}, respectively, calculated under the conditions corresponding to the present experiment. In the figures, $k_{TE,s}$ and $k_{SE,s}$ are indicated by the broken line and the component k_f by the dash-dotted line. The thermal conductivity λ of the test specimen was measured by using a 90–mm–long test cylinder, cut from the same cylinder. In addition, the Vickers hardness H_V of the test specimen was also measured with a portable hardness testing machine. In the case of Tachibana's equation (8), $k_{TE,s}$ by the cylindrical portion is dominant and a higher thermal contact conductance is obtained. On the other hand, in the case of Sanokawa's equation (9), the component $k_{SE,s}$ of the cylindrical portion is quite low because of the contraction and expansion of the heat flow through the cylindrical portion of the unit cell. As a result, k_f owing to the intervening substance (in this case, the air layer) accounts for most of the conductance, and k_{SE} is several times lower than k_{TE}.

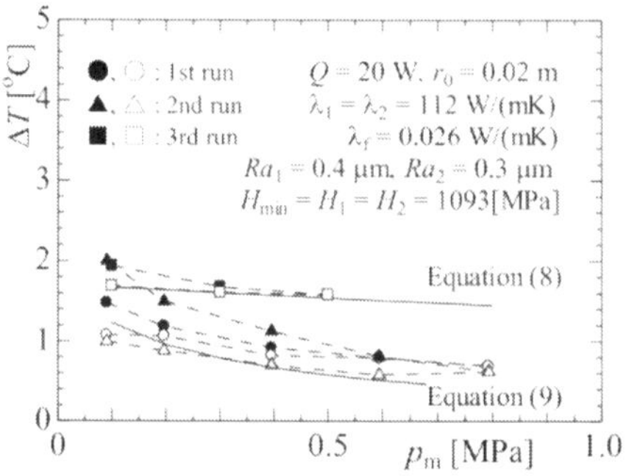

Figure 10. Comparison between measured and calculated temperature drop at contact interface for flat rough surfaces.

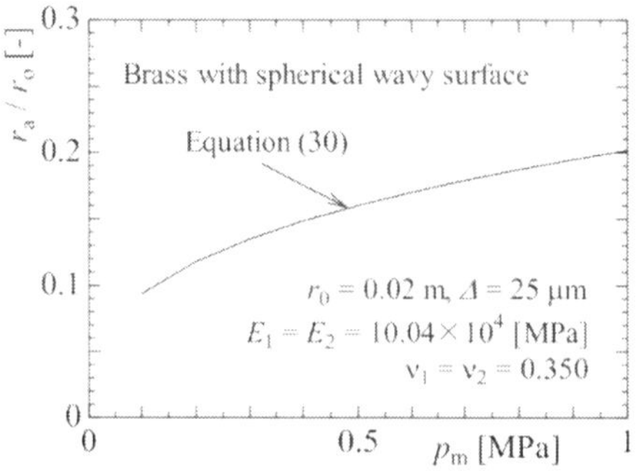

Figure 11. Effect of mean nominal contact pressure on radius of contact surface (for brass specimen).

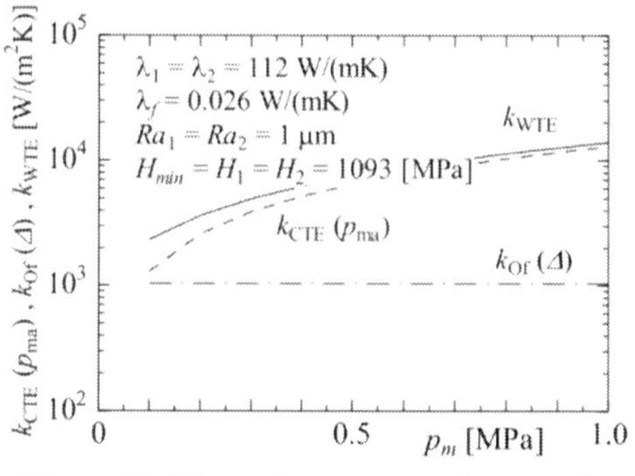

Figure12. Thermal contact conductance by evaluation equation (24) based on Tachibana's equation (6).

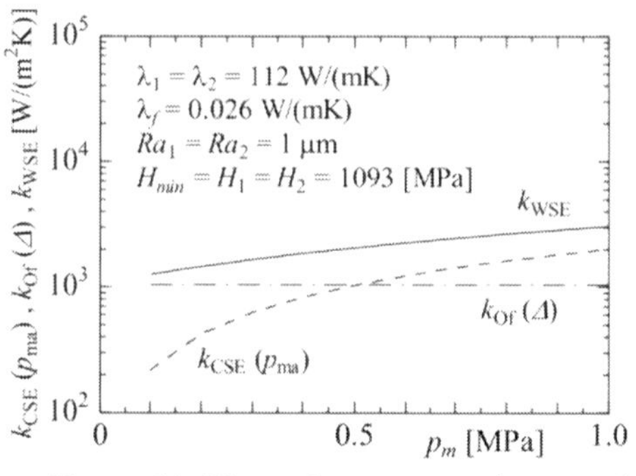

Figure 13. Thermal contact conductance by evaluation equation (24) based on Sanokawa's equation (7).

In Figure 10, the experimental and theoretical results are shown in terms of ΔT at the solid–solid contact interface. Here, the symbols (circles, triangles, and squares) denote the measurement results; the closed symbols correspond to the loading process and the open symbols to the unloading process. In each measurement, reassembling and assembling of the lower and upper specimens was repeated. In the figure, the hysteresis phenomenon is observed during the loading and unloading processes, and almost all of the experimental results exist between the results obtained using Sanokawa's equation extrapolated to the lower p_m and those obtained using Tachibana's equation.

4.2. Solid–solid contact of wavy surfaces with roughness

Figure 11 shows the effect of p_m on the radius of the contact surface r_a for the present brass test specimen. Here, equation (14) has been rearranged as the following equation using the relations $r_{s1} = r_{s2} \cong r_0^2/(2\Delta)$, $E = E_1 = E_2$, $\nu = \nu_1 = \nu_2$ and $p_m = f/(\pi r_o^2)$.

$$\frac{r_a}{r_0} = \left[\frac{3}{8}\pi\frac{\left(1-\nu^2\right)}{E}\cdot\frac{r_0}{\Delta}p_m\right]^{1/3} \tag{30}$$

In the figure, the Young's modulus E and the Poisson's ratio ν are considered as 10.06×10^4 MPa and 0.350 [12], respectively. As seen from the figure, only a 1–4% macroscopic contact area against the cross-sectional area of the cylinder has been realized under the present lower p_m conditions.

Figures 12 and 13 illustrate the effect of p_m on the thermal contact conductance k_{WTE} and k_{WSE}, respectively, calculated under the conditions corresponding to the present expe–iment. In the figures,

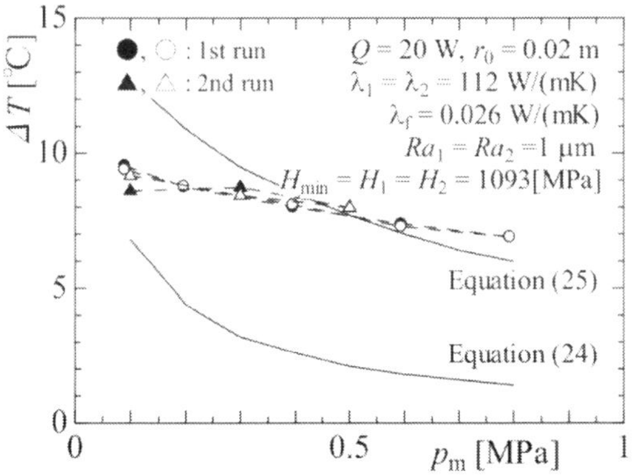

Figure 14. Comparison between measured and calculated temperature drop at contact interface for wavy rough surfaces.

k_{CTE} (p_{ma}) and k_{CSE} (p_{ma}) are indicated by the broken line, and the component k_{Of} (Δ) by the dash-dotted line. In the case of evaluation equation (24) based on Tachibana's equation (6), k_{CTE} (p_{ma}) at the central solid–solid contact zone is dominant and a high thermal contact conductance is obtained. On the other hand, in the case of evaluation equation (25) based on Sanokawa's equation (7), k_{CSE} (p_{ma}) at the central solid–solid contact zone is low because of the contraction and expansion of the heat flow through the cylindrical portion of the unit cell. However, in the present model, since the macroscopic constriction of the heat flow at the solid–solid contact interface has not been introduced, k_{CSE} (p_{ma}) shows roughly the same effect as k_{Of} (Δ) on the whole thermal contact conductance k_{WSE}. On the other hand, as previously shown in Fig. 9, $k_{SE,\,s}$ for the flat rough surfaces, in which the microscopic constriction of the heat flow is introduced at the contact interface formed by the surface roughness, has a much smaller effect than k_f on the whole thermal contact conductance k_{SE}.

The experimental and theoretical ΔT values t the solid–solid contact interface are shown in Figure 14. Here, the symbols (circles, triangles) denote the measured results; the closed symbols correspond to the loading process and the open symbols to the unloading process. In contrast to case of the flat rough surfaces, almost no hysteresis is observed during the loading and unloading processes. Equation (24), which does not consider the additional thermal resistance at the cylindrical portion of the unit cell, considerably underestimates ΔT under the present low pressure range. Unfortunately, however, even if the additional resistance is taken into account as in equation (25), the agreement between the experimental and theoretical results is not so good. This may be due to the absence of macroscopic constriction of the heat flow at the contact interface; this is the major problem for consideration in future research.

5. Conclusions

From the present theoretical and experimental studies, the following findings have been obtained.

(1) It has been shown that, in the case of solid–solid contact of flat surfaces with roughness, almost all of the measured temperature drops caused by the thermal contact resistance at the contact interface exist between the values evaluated by the unit cell model with and without the heat flow constriction. This implies that Sanokawa's equation can be successfully extrapolated to the lower mean nominal contact pressure of 0.1–1.0 MPa.

(2) When the rough surface has waviness, the newly proposed evaluation equation based on Tachibana's equation has been proved to underestimate the temperature drop at the contact interface. On the other hand, although another proposed equation based on Sanokawa's equation is consistent

with the measured results to some extent, the introduction of constriction resistance due to the macroscopic waviness seems to be necessary for a more accurate evaluation of the temperature drop.

References

[1] Fletcher L S 1988 *Trans. ASME, J. Heat Trans*fer **110** 1059
[2] Torii K and Yanagihara J I 1989 *J. Heat Transfer Soc. Jpn.* **28** 79
[3] Torii K 1993 *J. Jpn. Soc. Mech. Eng.* **96** 198
[4] Okada M and Matsumoto K 1993 *J. Heat Transfer Soc. Jpn.* **32** 25
[5] Madhusudana C V 1996 *Thermal Contact Conductance* (New York: Springer) pp 1-8
[6] Tomimura T and Fujii M 1999 *Proc. Symp. on Advances in Packaging, APACK '99* (Singapore: Gintic Institute of Manufacturing Technology) pp 216-219
[7] Tachibana F 1952 *J. Jpn. Soc. Mech. Eng.* **55** 102
[8] Sanokawa Y1967 *Trans. Jpn. Soc. Mech. Eng.* **33** 1131
[9] Yamamoto Y and Kaneta M 1999 *Tribology* (Tokyo: Rikougaku-sha) pp 20-37
[10] Tomimura T, Matsuda Y, Zhang X and Fujii M 2000 *Jpn. Soc. Mech. Eng. Int. J.* B **43** 665
[11] Jpn. Soc. Mech. Eng. 1983 *JSME Data Book: Heat Transfer 3rd Edition* (Tokyo: The Japan Society of Mechanical Engineers) pp 176-177
[12] National Astronomical Observatory 2013 *Chronological Scientific Tables* (Tokyo: Maruzen Publishing Co., Ltd.) p 389

Criterion of detachment for grip-and-release devices with slanted multi-beam structure using Ti–Ni wire

Kazuhito Emura*, Pasomphone Hemthavy*, Shigeki Saito, Kunio Takahashi***

*International Development Engineering, Tokyo Institute of Technology I4-11. O-okayama 2-12-1, Meguro-ku, Tokyo, Japan

** Department of Mechanical and Aerospace Engineering, Tokyo Institute of Technology I1-47. O-okayama 2-12-1, Meguro-ku, Tokyo,

E-mail: emura.k.aa@m.titech.ac.jp

Abstract. A multi-beam structure using Ti–Ni wire to grip rough surfaces and control detachment was manufactured. The grip-and-release mechanism of the structure was investigated. The experimental results were analyzed with a model assuming beam theory and using a fracture criterion at the adhered interface. The fracture criterion of the adhesion interface is suggested to depend on the shear stress.

1. Introduction

Geckos can grip rough surface with adhesion force and move quickly with their foot hairs, which are called setae and spatulae [1]. These hairs generate an adhesion force, and the grip-and-release is controlled by a tangential force [2]. These hairs inspired the development of a multi-beam structure. The structure consists of many beams with adhesive applied to the tips of the beams, as shown in figure 1. The structure grips a rough surface with the adhesion force and controls detachment with the tangential force. The adhesion devices can be applied to non-magnetic materials and in a vacuum. Clarifying the detachment criterion is important for the design of the adhesion devices. Although Wang et al. investigated the detachment mechanism of the multi-beam structure [3], their beams were vertical and too hard to absorb the surface roughness. In the present study, soft and slanted beams were considered. Ti–Ni superelastic wires were used for the beams. The mechanism of the detachment was experimentally investigated and analyzed according to beam theory.

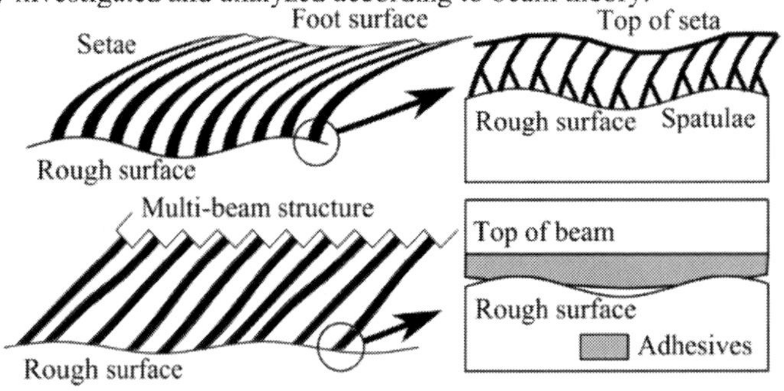

Figure 1. Model of multi-beam structure inspired from gecko foot hairs.

Content from this work may be used under the terms of the Creative Commons Attribution 3.0 licence. Any further distribution of this work must maintain attribution to the author(s) and the title of the work, journal citation and DOI.
Published under licence by IOP Publishing Ltd

2. Multi-beam structure

Figure 2 shows the construction of the multi-beam structure used in this study. The beams comprise Ti–Ni super-elastic wires with diameters of 0.10 mm. Ten wires are fixed to 0.2 mm thick copper sheets and 0.1 mm thick tin foil by brazing. The tips of the beams are polished at 45°.

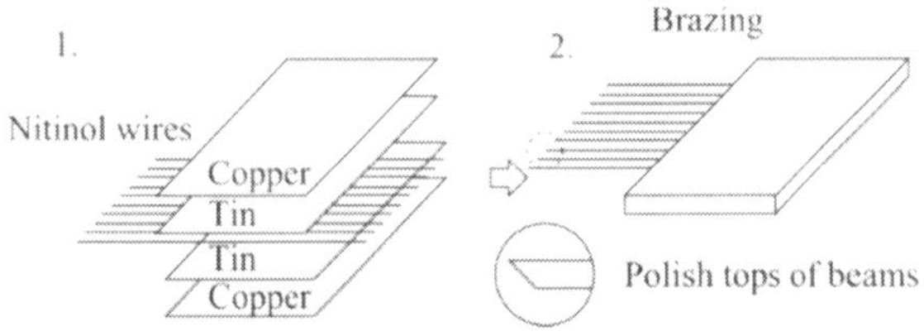

Figure 2. Construction of multi-beam structure.

3. Experiment to investigate detachment mechanism

3.1. Model of detachment mechanism

The structure adheres to a surface at the beam tips and is dragged, as shown in figure 3. The detachment mechanism of the structure was assumed to be a superposition of the mechanism for a single beam. The normal load F_n and tangential force F_t are applied to the tips of the structure side for each beam. F_n corresponds to the supported weight. The mechanism of the beam was considered using beam theory and the fracture criterion of the adhesion interface [4]. The maximum normal stress σ_{max} is taken from beam theory and expressed in equation (1); it is shown in figure 4 with various values of F_t, where S is the contact area, l is the length of the beams, θ is the angle of the beams, and Z is the section modulus. When σ_{max} is greater than the critical value σ_{cr}, the adhesion interface is broken, and the beam detaches from the surface. σ_{cr} was assumed to be constant and unique at the interface. By controlling F_t, σ_{max} can be controlled. In other words, the detachment from the surface can be controlled. The fracture criterion is defined as σ_{cr} and is expressed in equation (2). The detachment criterion of a single beam is defined as F_n and F_t when detachment occurs and is expressed in equation (3); it is shown in figure 5.

$$\sigma_{max} = \frac{F_n}{S} + \frac{|F_n l\cos\theta - F_t l\sin\theta|}{2Z} \quad (1)$$

$$\sigma_{max} > \sigma_{cr} \quad (2)$$

$$F_n = \frac{Sd\sigma_{cr} \pm 4l\sin^2\theta F_t}{d \pm 4l\sin\theta\cos\theta} \quad (3)$$

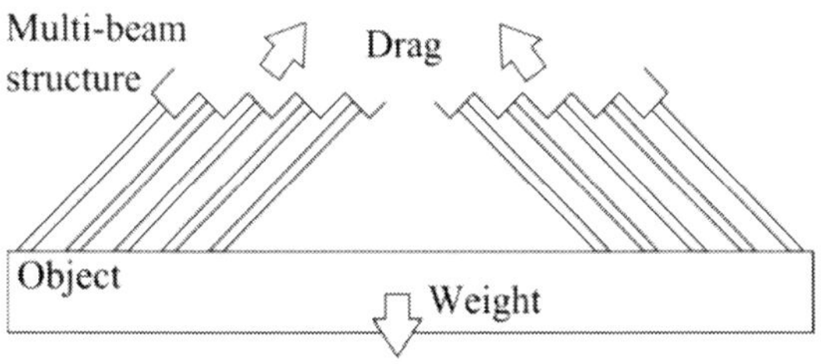

Figure 3. How multi-beam structure grips surface.

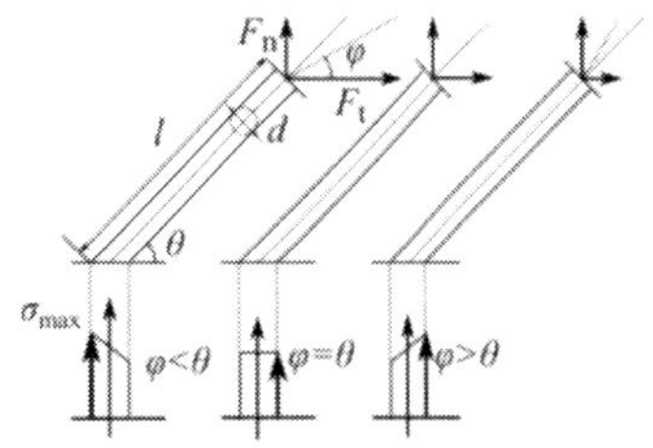

Figure 4. Normal stress distribution at adhesion interface.

IOP Conf. Series: Materials Science and Engineering **61** (2014) 012041 doi:10.1088/1757-899X/61/1/012041

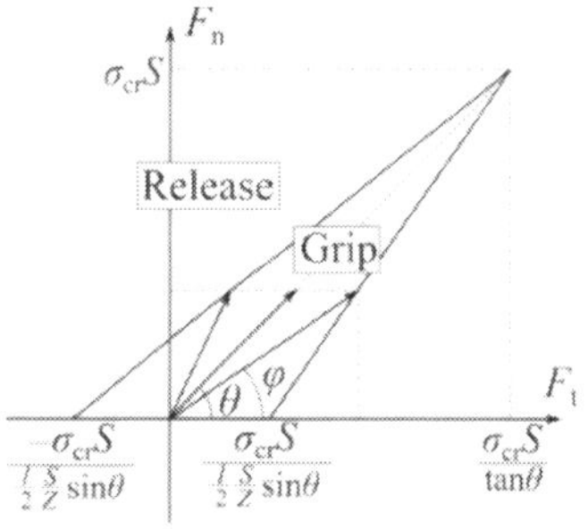

Figure 5. Detachment criterion of single beam.

3.2. Experimental system and process

To investigate the detachment criterion of the structure, the adhesion force between the beam tips and urethane sheet acting as the adhesive was measured. Figure 6 and table 1 show the experimental setup for the adhesion force measurements. A 0.5 mm thick urethane sheet is placed on the tilting surface as the adhesive. The structure was moved vertically by a motorized stage. The maximum adhesion force F was recorded by an electric balance. The measurement was carried out for various tilt angles φ. The tilt angle φ corresponds to the angle of resultant force of F_t and F_n. F_t and F_n were evaluated from F and are expressed in equation (4).

$$\begin{cases} F_t = F\cos\varphi \\ F_n = F\sin\varphi \end{cases} \tag{4}$$

The experiment was carried out according to the following process. (1) The structure was moved vertically at 1 μmsec^{-1} and adhered to the surface. (2) After contact, the structure was removed from the surface vertically at 1 μmsec^{-1}. The detachment criterion was evaluated using F and φ. The experiments were carried out from $\varphi = 25^\circ$ to $\varphi = 65^\circ$ at intervals of 5°.

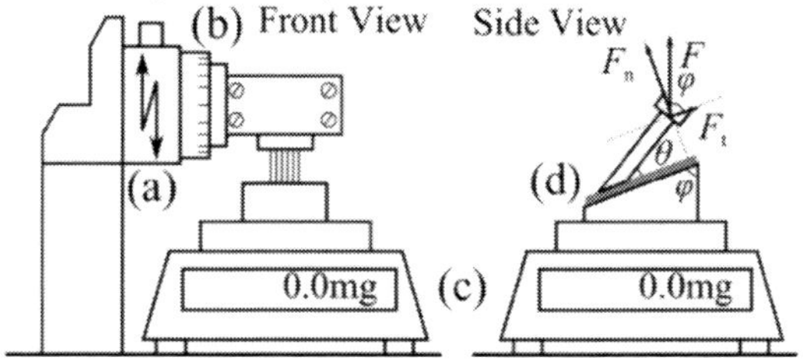

Figure 6. Measurement of adhesion force.

Table 1. Experimental setup.

Name	Explanation
(a) KZL060(Suruga seiki)	Motorized stage, resolution: 1μm
(b) B43-60N(Suruga seiki)	Manual rotation stage, resolution: 0.2°
(c) AY-220(SHIMADZU)	Electric balance, resolution: 0.1mg
(d) CT-05W-500(Exseal corp.)	Adhesives, urethane sheet 0.5mm thick

IOP Conf. Series: Materials Science and Engineering **61** (2014) 012041 doi:10.1088/1757-899X/61/1/012041

4. Results and discussion

Figure 7 shows the experimental results. The plotted points represent the maximum adhesion force at each φ. Figure 8 shows the detachment criterion. The plotted points represent the experimental detachment criterion evaluated from the experimental results and equation (4). The theoretical detachment criteria calculated from equations (1) and (2) are also plotted in figure 8; the solid and broken lines correspond to $\sigma_{cr} = 100(\text{kPa})$ and $\sigma_{cr} = 10(\text{kPa})$, respectively. The results from equations (1) and (2) showed poor agreement with the experimental results. Figure 9 shows σ_{max} calculated from the results using equation (2). The disagreement can be observed around $\varphi = 45^\circ = \theta$. In the model, because the normal stress distribution was uniform when $\varphi = \theta$, the structure hardly detached from the surface and could support heavier weights than when $\varphi > \theta$ and $\varphi < \theta$. Assuming that the shear stress distribution is uniform, the shear stress τ_{ave} at the interface can be expressed as

$$\tau_{ave} = \frac{F_t}{S} = \frac{F\cos\varphi}{S}. \tag{5}$$

Figure 10 shows the relationship between τ_{ave} and φ. Figures 9 and 10 indicate that σ_{max} decreases as τ_{ave} increases; in other words, σ_{cr} decreases when shear stress is applied at the adhesion interface. For the case $\varphi = \theta$, the more a beam becomes slanted, the more σ_{cr} decreases. Thus, the fracture criterion needs to be modified for design of the structure.

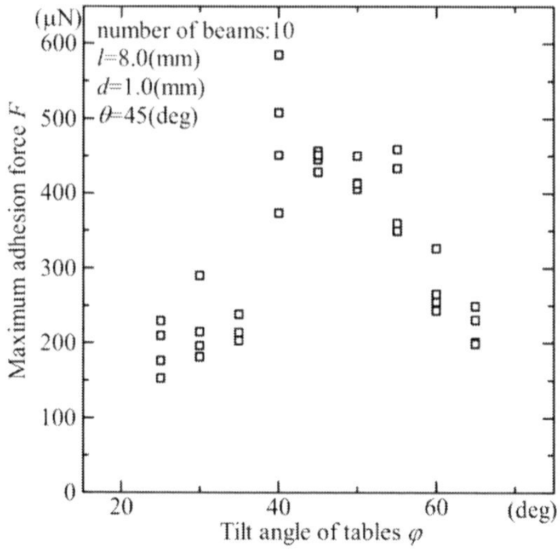

Figure 7. Maximum adhesion force at various tilt angles for tables.

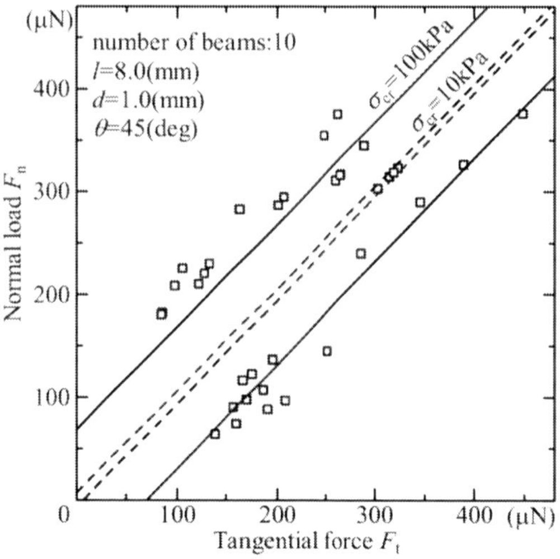

Figure 8. Experimental and theoretical detachment criteria.

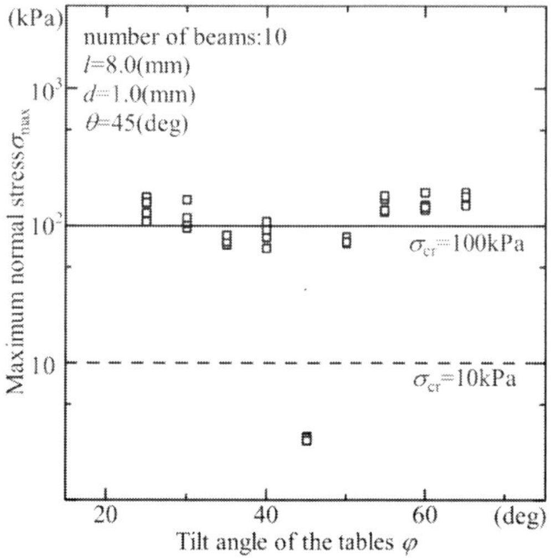

Figure 9. Maximum normal stress of each tilt angle for tables.

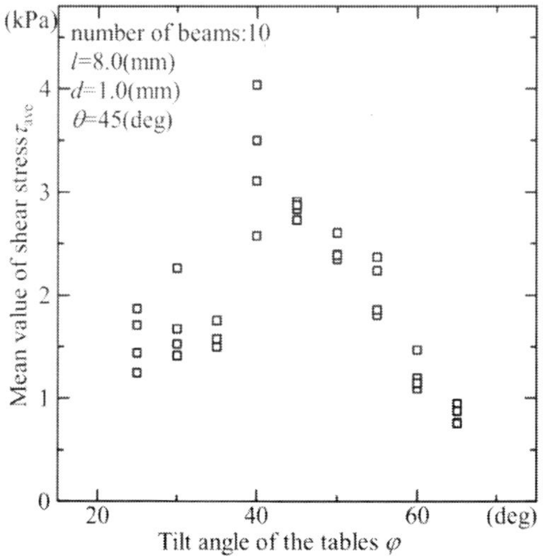

Figure 10. Mean shear stress of each tilt angle for tables.

5. Conclusion

A slanted multi-beam structure using Ti–Ni wire was fabricated. The structure grips rough surfaces with adhesion force and controls detachment from the surface. The maximum adhesion force between the structure and a urethane sheet acting as the adhesives was measured. The detachment mechanism of the structure was investigated and analyzed according to beam theory and the fracture criterion of the adhesion interface. The fracture criterion is suggested to be a function of the shear stress.

References

[1] Walker G 1993 *Jounal of adhesion & adhesive.* **13** 3-7.
[2] Takahashi K, 2006 *International Journal of adhesion & adhesive.* **26** pp. 639-64.
[3] Wang M, Sekiguchi Y, Hemthavy P, Saitou S, Takahashi K 2010 *16th symposium on Microjoining and Assembly Technology in Electronics.* **16** 435-438.
[4] Watanabe J, Hemthavy P, Saitou S, Takahashi K 2008 *14th symposium on Microjoining and Assembly Technology in Electronics.* **14** 369-372.

Analysis of adhesive elastic contact between a silica glass lens and silicone rubber using the JKR theory

Dooyoung Baek, Pasomphone Hemthavy and Kunio Takahashi

Department of International Development Engineering, Tokyo Institute of Technology, I4-210, O-okayama 2-12-1, Meguro-ku, Tokyo 152-8552, Japan

E-mail: baek.d.aa@m.titech.ac.jp

Abstract Contact between a silica glass lens and silicone rubber is experimentally investigated by simultaneously measuring displacement, force and contact radius. The relationship between these three parameters is derived using elastic theory. The discrepancy between the theoretical relationship and the experimental results is observed to increase as the deformation of the silicone rubber increases. Under smaller deformation conditions, the elastic theory shows good agreement with the experimental results, although infinite stress on the edge of the contact area is predicted in the theory, and time dependence and adhesion hysteresis are observed in all experiments. It is suggested that time dependence and adhesion hysteresis in contact are not induced by the deformation of the bulk of the silicone rubber, but are induced by surface effects. The result suggests that the applicability limit of the elastic theory must be carefully considered in the JKR analysis of point contact for polymers.

1. Introduction

Adhesion phenomena play a significant role in the engineering fields of contact mechanics and tribology, and are essential for general production methods that rely on adhesive forces. Understanding the adhesive elastic contact that takes place in adhesion phenomena has important applications in device design, e.g. for grip-and-release devices [1]. To describe adhesive elastic contact, Johnson, Kendall and Roberts proposed the JKR model [2]. In the JKR model, an elastic parabolic punch is compressed against a flat rigid surface, or conversely, a hard parabolic punch is compressed against a flat elastic substrate (i.e. the point contact model). This model is the simplest model for adhesive elastic contact, and assumes elastic deformation while taking into account the work of adhesion between materials. Since the JKR model assumes the linear elasticity of a material (the static state) and the equilibrium state of total energy, it is difficult to guarantee a good approximation for a practical contact system of polymeric materials which does not insure perfectly elasticity. However, the JKR model is sometimes used when considering the contact between polymeric materials [3]. Barthel [4] has reviewed many experimental studies and discussed the effects of the non-elasticity on the contact, including non-linearity, time dependence, etc. However, this has been dealt with based on the equilibrium state of total energy in the contact model. Although the polymer has non-elasticity, it must exhibit linear elasticity in a smaller deformation condition. In the present study, the theoretical consideration of the non-equilibrium state is discussed using the JKR model. The contact between a silica glass lens and silicone rubber is investigated experimentally, where the displacement, the force, and the contact radius of the contact model are measured

Content from this work may be used under the terms of the Creative Commons Attribution 3.0 licence. Any further distribution of this work must maintain attribution to the author(s) and the title of the work, journal citation and DOI.
Published under licence by IOP Publishing Ltd

simultaneously. By analyzing our experimental results while varying the amount of deformation in the silicone rubber, the applicability limit of the elastic theory is investigated.

2. Theoretical considerations

Figure 1 shows a schematic illustration of the point contact model describing contact between a rigid parabolic punch and a semi-infinite elastic body. This model is axisymmetric, and we assume linear elasticity, small deformation, and no friction between the two bodies.

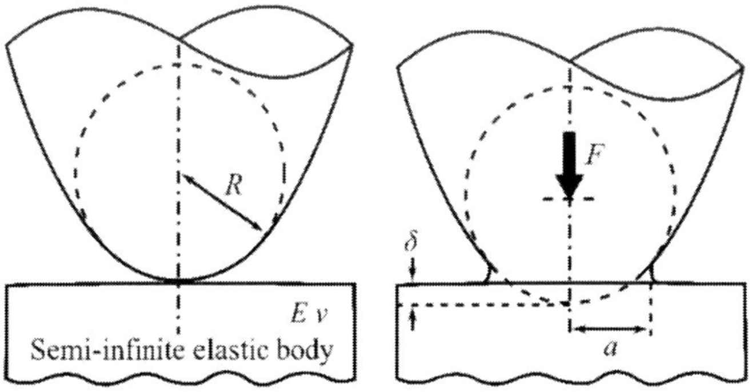

Figure 1. Schematic illustration of the point contact model, where R is the radius of curvature of the rigid parabolic punch, E and ν are the Young's modulus and Poisson's ratio of the semi-infinite elastic body, respectively, F is the force between the two bodies, δ is the vertically deformed displacement of the semi-infinite elastic body, and a is radius of the contact area raised by contact.

The theory of point contact can be expressed using the JKR theory. However, it is difficult to realize the equilibrium state of total energy in experiments, as attaining a perfectly static state during an experimental process is nearly impossible. Therefore, we must extend the JKR theory to the non-equilibrium state. In Figure 1, the force F between the rigid parabolic punch and the semi-infinite elastic body is balanced using the relation $F = P_H - P_B$ when the displacement δ is given. Here P_H is the compressive force due to the Hertz stress distribution and P_B is the adhesive force due to the Boussinesq stress distribution. Note that infinite stress is predicted on the edge of the contact area by the JKR stress distribution [5]. Then, the force can be expressed as

$$F = \frac{2Ea}{1-\nu^2}\left(\delta - \frac{a^2}{3R}\right). \tag{1}$$

The above expression shows the relationship between the force F, the contact radius a and the displacement δ. Using equation (1), the Young's modulus can be experimentally evaluated by measuring these three parameters, and the linear elasticity can be discussed based on experimentally obtained values of the Young's modulus.

3. Experimental procedure

In the experiments, the force F, the contact radius a, and the displacement δ must be measured, and these values are used in equation (1). To accomplish this we constructed an experimental system as schematically illustrated in Figure 2. In the experiment, the displacement δ was controlled by a vertical motorized stage, while the force F and the contact radius a were measured simultaneously by digital balance and microscope, respectively. The experiment was set up on the clean bench of a vibration isolation table, and the room temperature was set to 20 °C with 35% humidity. A silicone rubber sheet with thickness 10 mm was used as the semi-infinite elastic body, and a silica glass lens with radius of curvature 207.6 mm and diameter 40 mm was used as the rigid parabolic punch. The experiment was performed by controlling the displacement of the lens against the silicone rubber using the motorized stage. The displacement was controlled in steps of 1 μm, and the waiting duration between steps was set to 60 seconds to account for the relaxation time of the silicone rubber. First, we set the displacement to 0 μm when the two bodies are in contact. Second, for the loading process, we varied the motorized stage from 0 μm to the maximum loading displacement. Third, for the unloading process, we varied the motorized stage until the lens detached from the silicone rubber. The combined loading and unloading processes were considered as one cycle. The maximum loading displacements were chosen to be 30, 60 and 120 μm on each cycle for confirming the elasticity. The force and the contact radius were measured just before the displacement of the lens changed.

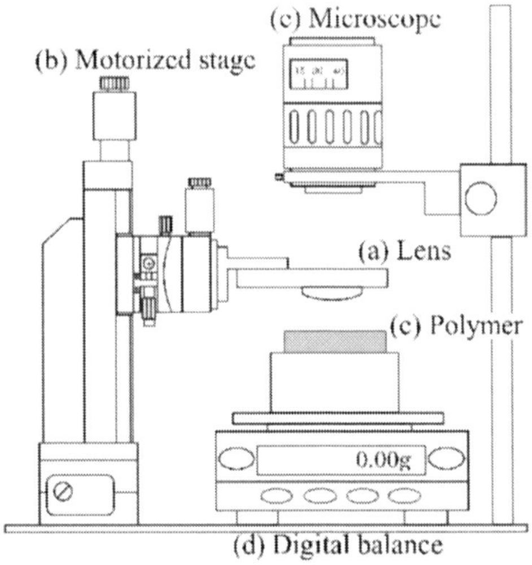

Figure 2. Schematic illustration of the experimental system. (a) Lens: Silica glass lens with radius of curvature R=0.2076 m and diameter 40 mm, (b) Motorized stage: Motorized vertical movement with resolution 1 μm, (c) Polymer: Silicone rubber sheet with thickness 10 mm, (d) Digital balance: Strain gauge type with resolution 0.01 g, (e) Microscope: Resolution 1600×1200 pixel.

4. Results and discussion

Figures 3 and 4 show the measured force and contact radius for each cycle as functions of the displacement, for maximum displacements of 30, 60, and 120 μm. The loading and unloading processes in Figures 3 and 4 refer to increasing and decreasing of the displacement, respectively (i.e. compressing and decompressing the lens to the rubber). Regions for which the contact radius is constant are observed, and are indicated in Figure 4 as i, ii, and iii. These regions correspond to regions i, ii, and iii in Figure 3 of each cycle. In the constant-radius regions, the deformed equation from equation (1) can be used,

$$F = \frac{2Ea_{\text{const.}}}{1-\nu^2}\left(\delta - \frac{a_{\text{const.}}^2}{3R}\right), \tag{2}$$

where a_{const} is the constant value of contact radius in regions i, ii, or iii. Equation (2) indicates that the force F is a linear function of the displacement δ. Furthermore, the slopes of regions i, ii, and iii in Figure 3 can be approximated linearly. Using equation (2) and the approximated slope, the Young's modulus E is evaluated as

$$E = \frac{(1-\nu^2)\alpha}{2a_{const}}, \quad (3)$$

where α is the approximated slope of regions i, ii, and iii in Figure 3. The approximated slope α, the constant value of contact radius a_{const}, and the evaluated Young's modulus E in each cycle are calculated in Table 1.

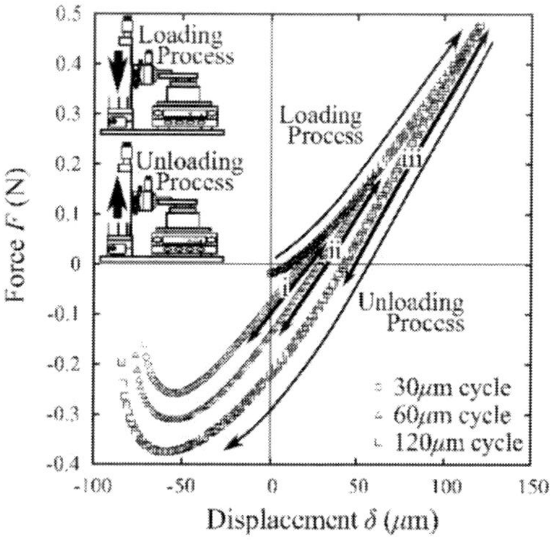

Figure 3. In the loading process, an increase of compressive force is observed. In the unloading process, the compressive force is decreased until the maximum adhesive forces, and then the lens is detached from the silicone rubber. The slopes of regions i, ii, and iii can be approximated linearly.

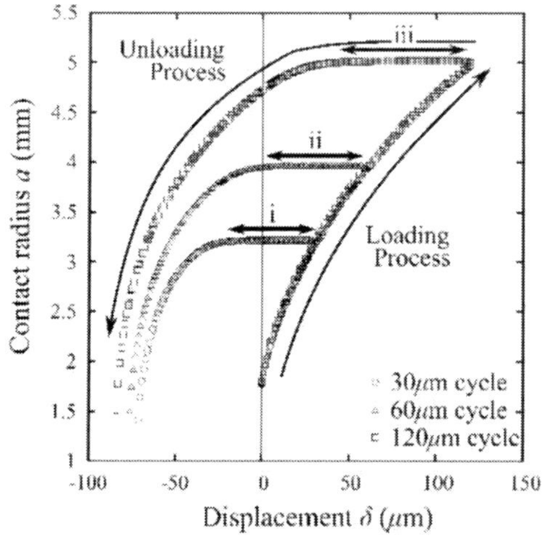

Figure 4. In the loading process, an increase in the contact radius is observed. In the unloading process, the contact radius is constant in regions i, ii, and iii of each cycle at first, and then progressive decreasing of the contact radius is observed.

Table 1. Evaluation results.

	α (N/m)	a_{const} (mm)	E (Pa)
i	4580.3	3.18	0.539
ii	5432.1	3.93	0.518
iii	6281.8	4.99	0.471

From Table 1, a decrease in the evaluated Young's modulus is observed as the maximum loading displacement is increased. This tendency exhibits typical strain-stress characteristics of polymers in large deformation, which means that the linear elasticity is not satisfied. Therefore, there is the limit of the linear elasticity in the experimental contact system.

The displacement was controlled in 1 μm step-movements with a waiting duration of 60 seconds between steps. The maximum strain of vertical deformation was chosen under 1.2% and the contact radius was observed to be smaller than 2.4% of the lens radius R=0.2076 m. Although the experimental conditions appear to satisfy the assumptions of the theory, a decrease in the Young's modulus is observed as shown in Table 1. It is considered that this tendency is attributed to the increase of the effects from the infinite stress distribution on the edge of the contact area by the Boussinesq. However, it is clear that the Young's modulus should have a constant value in limit of small deformations. Hence, it can be concluded that the Young's modulus of the silicone rubber is 0.539 MPa in the linear elasticity regime.

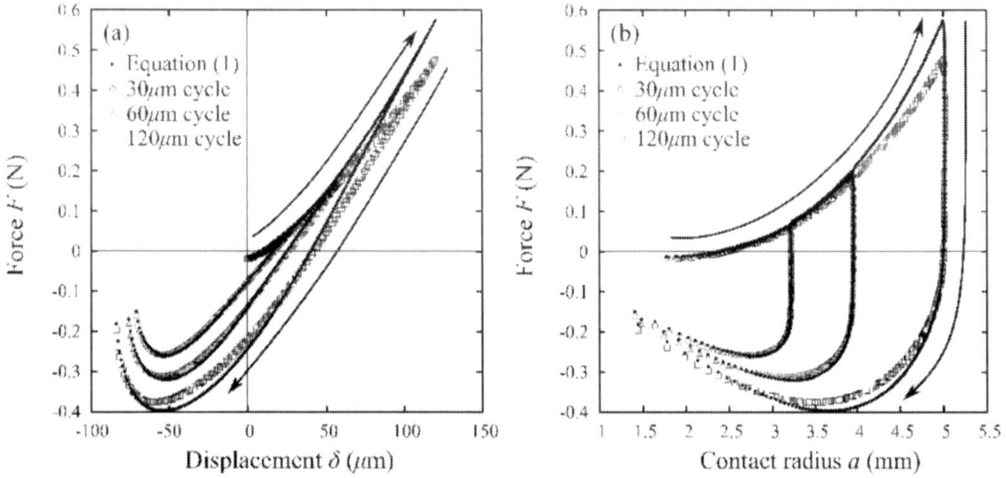

Figure 5. The calculated results of the expected force using the estimated E=0.539 MPa with equation (1) as a function of the displacement (a) and the contact radius (b). The experimental results from each cycle are also plotted.

5. Conclusions

Using the experimentally determined Young's modulus E=0.539 MPa, the expected force of the linear elastic state can be calculated using equation (1) with the experimentally measured values of displacement δ and contact radius a. The calculated results of the expected force and the experimental results for all cycles are plotted in Figure 5 as a function of the displacement (a) and the contact radius (b). As shown in Figure 5 (a) and (b), the cycle of maximum displacement 120 μm shows a larger discrepancy than the cycle of maximum displacement 30 μm. This discrepancy seems to increase as the displacement increases. However, the expected force for the cycle with maximum displacement 30 μm shows good agreement with the experimental results. This result suggests that the applicability limit of the elastic theory must be carefully considered in the JKR approximation of point contact using polymeric materials. Furthermore, in all experiments, time dependence is observed in the contact radius and the force, for a given displacement. Adhesion hysteresis [6], which is the difference between the loading path and the unloading path, is observed in Figure 5 (a). This means that the relationships among parameters show the elasticity, even when the time dependence and adhesion hysteresis are observed in case of extremely small deformation contacting processes. This suggests that the time dependence and the adhesion hysteresis in the contact is not induced by the deformation of the bulk of the silicone rubber, but would be induced by other effects such as a surface effect.

References

[1] Johnson K L, Kendall K and Roberts A D 1971 *Proc. R. Soc. London* A **324** 301
[2] Sekiguchi Y, Hemthavy P, Saito S and Takahashi K 2012 *J. Adhesion Sci. Technol.* **26** 2615
[3] Deruelle M, Hervet H, Jandeau G and Leger L 1998 *J. Adhesion Sci. Technol.* **12** 225
[4] Barthel E 2008 *J. Phys. D: Appl. Phys.* **41** 20
[5] Takahashi K, Mizuno R and Onzawa T 1995 *J. Adhesion Sci. Technol.* **9** 1451
[6] Silberzan P, Perutz S, Kramer E J and Chaudhury M K 1994 *Langmuir* **10** 2466

Effect of shape of elastic beam hair on its adhesion with wavy surfaces

Pasomphone Hemthavy[1], Takehiko Yazaki[1], Boqing Wang[1], Yu Sekiguchi[2] and Kunio Takahashi[1]

1 Department of International Development Engineering, Tokyo Institute of Technology, 2-12-1 O-okayama, Meguro-ku, Tokyo 152-8552, JAPAN

2 Precision and Intelligence Laboratory, Tokyo Institute of Technology, 4259-R2-31 Nagatsuta, Midori-ku, Yokohama 226-8503, JAPAN

E-mail: pasomphone@ide.titech.ac.jp

Abstract. An analysis on a tapered elastic beam whose side surface partially adhered to a rigid surface was carried out to study the effect of the beam shape on the gripping force. Considering the total energy of the system, the relation between the gripping force and the displacement was obtained analytically in closed form. The analytical result is significant because it provides an intuitive picture of the gripping force. Although, an individually tapered beam can generate less gripping force for flat or slightly wavy surfaces, compared to a rectangular beam, the analysis result suggests that the tapered beam has more ability to absorb surface waviness. This result can be applied to a multi-beam structure.

1. Introduction

Geckos have a myriad of hierarchical hair structures on their foot fingers, consisting of setae and spatulae. The mechanism of adhesion of the gecko's foot hair was observed to be mainly due to intermolecular surfaces forces [1, 2]. Nano-scaled spatulae at the top of the setae generate an adhesion force so that geckos can hang from roof surfaces as well as climb vertical walls [3]. Though a single foot hair demonstrates a negligible adhesion force, the myriad of slant hair structures enables adhesion to smooth surfaces with high interfacial shear strength [2, 4]. The gecko's foot hair was observed to have a tapered shape towards its tip with a patch on the tip of each spatula [1]. Since these discoveries, many attempts have been made to replicate the functional structures of gecko's foot hair by utilizing polymers of a precise size [5-10]. Multi-beam structures with mushroom tip shapes or tapered beams have also been fabricated and studied experimentally [5, 10]. However, theoretical study on the functional structure of the gecko's foot hair has not been as extensive. Takahashi et al. have researched a release mechanism that allows geckos to take quick steps and their ability to grip rough surfaces from the macroscopic point of view by assuming the seta as a curved slant beam, which can be controlled by normal and tangent forces [3]. Sekiguchi et al. have investigated the adhesive force between a side surface of a rectangular elastic beam and a flat surface of a rigid body [11]. However, the effects of the beam shape have not been clearly studied. In this study, we assume a tapered elastic beam, i.e., one with bending stiffness tapering along its longitudinal axis, to study the effect of the beam shape on the gripping force. By considering the minimum energy condition of the total energy of the system, the effect of the adhesion phenomenon is included in the process for obtaining the relation

Content from this work may be used under the terms of the Creative Commons Attribution 3.0 licence. Any further distribution of this work must maintain attribution to the author(s) and the title of the work, journal citation and DOI.
Published under licence by IOP Publishing Ltd

between the gripping force and the displacement of the beam. The ability of a multi-tapered beam structure to absorb surface waviness is also discussed.

2. Adhesion model between elastic tapered beam and flat rigid surface

Figure 1 shows a schematic illustration of the elastic tapered beam used in this paper for modeling the function of a single gecko foot hair. The shape of this tapered beam can be expressed by assuming the varying bending stiffness along the longitudinal axis of the beam as

$$EI = E_0 I_0 \left(1 - \frac{x}{L}\right)^n \qquad (0 \le n < 1), \tag{1}$$

where E is the Young's modulus and I is the moment of inertia of the area of the elastic beam, $E_0 I_0$ is the bending stiffness at the base of the beam, L is the length, and x represents the distance along the longitudinal axis of the beam. The width W of the beam is supposed to be constant in order to study the one-directional function of the beam. Figure 2 shows the distribution of the bending stiffness along the longitudinal axis of the beam, plotted by equation (1).

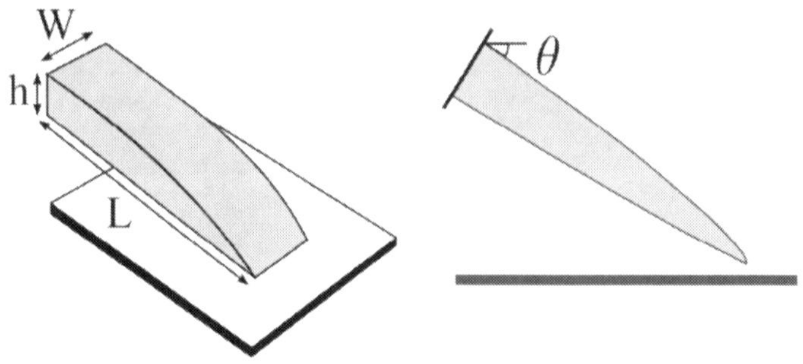

Figure 1. Schematic illustration of a tapered beam model used. The figure on the left-hand side shows the entire view of the beam with its geometries. The figure on the right-hand side is the side view of the beam.

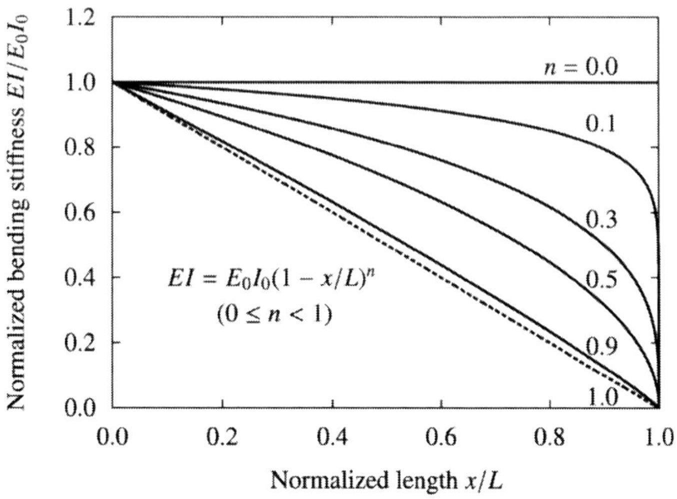

Figure 2. Distribution of the bending stiffness over the distance on the longitudinal axis of the beam.

As shown in figure 2, when $n=0$ the bending stiffness has a contant value (horizontal straight line), which is equivalent to a rectangular beam, for example. When $n=1$, the bending stiffness varies linearly along the longitudinal axis toward one end of the beam (dashed line). However, for $n=1$, there are singularities in the analytic result (described in section 3) which complicate the calculation of the gripping force. Consequently, without losing the essence of the study, we take the range of n to be $0 \leq n < 1$.

Figure 3 illustrates the loading and unloading process of the cantilever tapered beam to enable it to grip and release a flat rigid surface. As shown in the figure, the beam is fixed at the base with an inclined angle θ to the horizontal line. The loading process is performed by applying the displacement to the base of the beam, which brings it into contact with the rigid surface, as shown in figure 3(a). When the beam reaches the rigid surface, it starts contacting the surface from its tip edge, which is here defined as "line contact," as shown in figure 3(b). When further displacement is continually applied, the rigid surface presses the edge of the beam with a force f, giving rise to its deformation. As a consequence, the beam adheres to the rigid surface on its bottom side, which is defined as "area contact," as shown in figure 3(c). In static contact, the adhered area of this area contact will settle at an equilibrium that depends on the magnitude of the given displacement.

In the unloading process, starting from area contact, the beam is gradually moved up by applying a displacement in the opposite direction to that of the loading process; this decreases the adhered area, as shown in figure 3(d). When the adhered area finally decreases to line contact, as shown in figure 3(e), the beam finally detaches from the rigid surface, as shown in figure 3(f). This is considered the separation condition of the beam.

In this paper, we adopt the following assumptions for the adhesion model so that the linear beam theory can be applied. The beam is assumed to have small lateral deformation without longitudinal deformation. In addition, the weight and rotation of the beam and the frictional slip between the beam and rigid surface are assumed to be negligibly small.

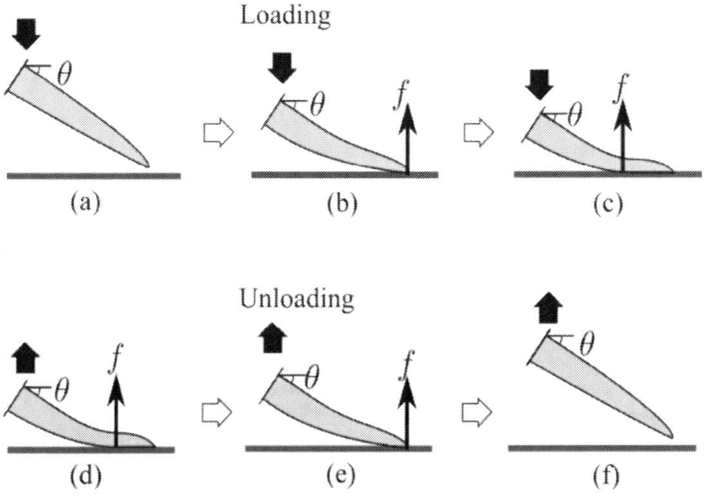

Figure 3. Schematic illustration of the loading process (a)-(c) and unloading process (d)-(f) of the beam. The beam approaches the rigid surface by the applied displacement (a); it starts contacting the surface at its tip edge (line contact) (b); a part of its bottom side adheres to the surface (area contact) (c). Unloading the beam decreases the contact area (d); the contact changes from area contact to line contact (e); the beam finally detaches from the surface (f).

3. Force between the elastic beam and the rigid surface

Our main task here is to obtain the relation between the gripping force of the beam to the rigid surface and the applied displacement. A mathematical model of the beam for line contact and for area contact are assumed, as shown in figures 4 and 5, respectively. A system of coordinate axes (x,y) with the origin located on the longitudinal axis at the fixed end of the beam is used. The displacement d applied to the beam is measured from the rigid surface to the base of the beam with a negative upward direction. When the beam is in contact with the rigid surface, the force f between the rigid surface and the beam acts perpendicularly to the rigid surface.

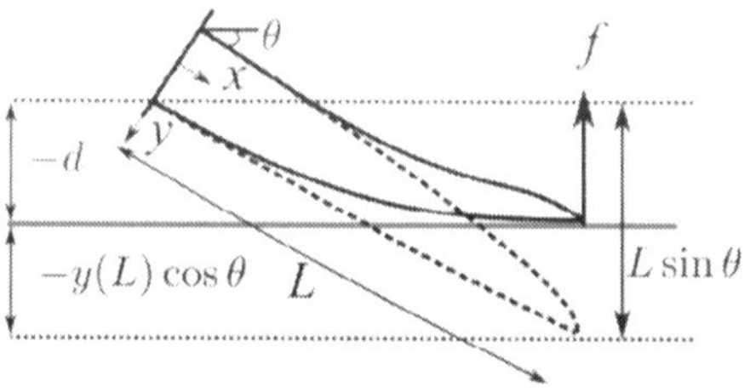

Figure 4. Schematic illustration of line contact of the beam with the rigid surface.

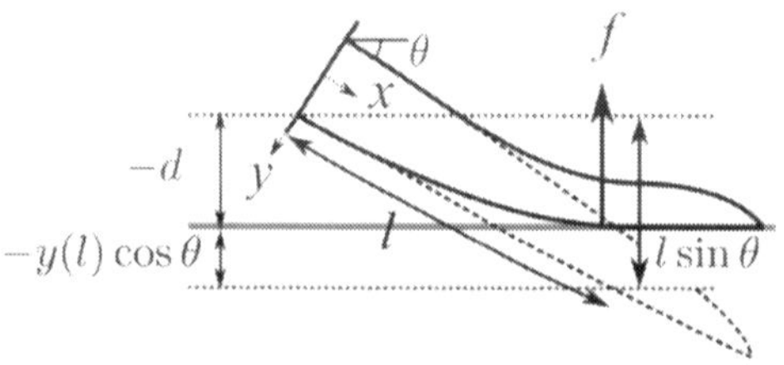

Figure 5. Schematic illustration of area contact of the beam with the rigid surface.

3.1. Force during line contact

During line contact, as shown in figure 4, the force f acting on the tip of the beam causes the beam to deflect slightly. The corresponding deflection of the beam, i.e., $y(x)$, is required to obtain the relation between the force f and the applied displacement d. On the other hand, according to the assumption that the base of the beam is fixed, the boundary conditions $y(0)=0$ and $dy(x)/dx=0$ at $x=0$ should be satisfied. From figure 4, the shearing force at any section of the beam, which is given as

$$S = -f\cos\theta \qquad (0 \le x \le L), \tag{2}$$

gives rise to the internal bending moment $M(x)$ of the beam, which is expressed as

$$M(x) = -f\cos\theta(L-x) \qquad (0 \le x \le L). \tag{3}$$

The basic equation of the deflection curve of the elastic beam is given as

$$\frac{d^2y}{dx^2} = -\frac{M(x)}{EI}. \tag{4}$$

From the bending stiffness of the beam, the internal bending moment, and the basic equation of the deflection curve (equations (1), (3), and (4), respectively), the deflection $y(x)$ of the beam, which satisfies the above-mentioned conditions, is obtained as

$$y(x) = -\frac{fL^3\cos\theta}{(n-2)(n-3)E_0I_0}\left\{\left(1-\frac{x}{L}\right)^{3-n} - (n-3)\frac{x}{L} - 1\right\}. \tag{5}$$

From the geometry of the beam shown in figure 4, and the deflection of the beam's tip, i.e., $y(L)$, it yields the relation between the displacement and the deflection of the beam as

$$L\sin\theta = -d - y(L)\cos\theta. \tag{6}$$

Finally, by substituting equation (5) into equation (6), we can obtain the relation between the force f and the applied displacement d for line contact as

$$\tilde{f}\frac{\cos\theta}{\tan\theta} = \frac{(3-n)}{12}\left(\frac{\tilde{d}}{\sin\theta}+1\right), \tag{7}$$

where $\tilde{f}$ *and* $\tilde{d}$ are dimensionless quantities, normalized as $\tilde{f} = f/(12E_0I_0/L^2)$ and $\tilde{d} = d/L$, respectively.

If further displacement is applied to the base of the beam, the beam's tip deflects. Therefore, when the beam deflects so that the slope of the tip equals that of the rigid surface, satisfying the following condition

$$\left.\frac{dy}{dx}\right|_{x=L} = -\tan\theta, \tag{8}$$

the bottom side of the beam begins to adhere to the rigid surface. This is area contact, as shown in figure 5. From equations (5) and (8), the shifting condition from line contact to area contact is obtained as

$$\frac{\tilde{d}}{\sin\theta} = \frac{1}{n-3}. \tag{9}$$

3.2. Force during area contact

The relation between the force f and the applied displacement d for area contact can be obtained in the same manner as described in the previous subsection, except for the following considerations. During area contact, the adhered portion of the beam conforms to the rigid surface, so that it can be considered the fixed end. Therefore, the force f acting on the edge of the adhered portion causes the rest of the beam to deflect, as shown in figure 5. The area of the adhered portion is defined as $(L-l)\times W$, where l is the length of the portion of the beam that is not adhered to the rigid surface.

Both ends of the beam are fixed, in contrast to line contact, so the bending moment of the beam is given as

$$M(x) = -f\cos\theta\, x + M_l \qquad (0 \le x \le l), \tag{10}$$

where M_l is determined by the boundary condition of the slope of the beam at $x=l$. The shearing force and the boundary conditions at $x=0$ are the same as those for line contact. The relation between the force and the displacement of the beam for area contact is obtained as

$$\tilde{f}\frac{\cos\theta}{\tan\theta} = \frac{(n-1)(n-2)^2(n-3)}{12} \times \frac{-[(1-\tilde{l})^{1-n}-1]\left\{\dfrac{\tilde{d}}{\sin\theta} + \dfrac{(1-\tilde{l})^{1-n}[(n-1)\tilde{l}-1]+1}{(n-2)[(1-\tilde{l})^{1-n}-1]}\right\}}{n(n-2)(n-3)\tilde{l}^2(1-\tilde{l})^{1-n} + (n-3)[(1-\tilde{l})^{2-n}-1]^2 - 2(n-2)[(1-\tilde{l})^{1-n}-1][(1-\tilde{l})^{3-n}-1]}, \tag{11}$$

where $\tilde{l}$ is the normalized non-adhesion length of the beam, defined as $\tilde{l} = l/L$.

However, $\tilde{l}$ is still unknown. In order to determine $\tilde{l}$, it is necessary to take into account the minimum energy condition of the total energy of the system. We define the total energy of the system U_{total} as the sum of the elastic energy U_{elastic} and the surface energy U_{surface}, denoted as

$$U_{\text{total}} = U_{\text{elastic}} + U_{\text{surface}}. \tag{12}$$

The elastic energy due to the deflection of the beam is given as

$$U_{\text{elastic}} = \int_0^l \frac{M^2}{2EI}dx, \tag{13}$$

and the surface energy of the adhered area is

$$U_{\text{surface}} = -\Delta\gamma \times (L-l)\times W, \tag{14}$$

where $\Delta\gamma$ is the work of adhesion required to separate a unit area of adhered surfaces.

The total energy of the system is obtained as

$$\tilde{U}_{\text{total}} = 12\tan^2\theta\left(\tilde{f}\frac{\cos\theta}{\tan\theta}\right)^2\left\{\frac{[(1-\tilde{l})^{2-n}-1]^2-(n-2)^2\tilde{l}^2(1-\tilde{l})^{1-n}}{(n-2)^2(n-3)[(1-\tilde{l})^{1-n}-1]}\right\}$$

$$+\frac{\tan^2\theta}{12}\frac{n-1}{(1-\tilde{l})^{1-n}-1}-\Delta\tilde{\gamma}(1-\tilde{l})\tilde{W}, \qquad (15)$$

where $\tilde{U}_{\text{total}} = U_{\text{total}}/(6E_0I_0/L)$ and $\Delta\tilde{\gamma} = 3\Delta\gamma W/(6E_0I_0/L^2)$. According to the total energy obtained above, $\tilde{l}$ can be determined by searching for the minimum energy, which occurs when

$$\frac{\partial\tilde{U}_{\text{total}}}{\partial\tilde{l}} = 0. \qquad (16)$$

4. Calculated results

For convenience in the calculation, we define here the adhesion parameter related to the work of adhesion and the width of the beam as

$$\frac{\Gamma}{\tan\theta} = \frac{\sqrt{3\Delta\tilde{\gamma}\tilde{W}}}{\tan\theta}.$$

Solving equation (16) gives us the value of $\tilde{l}$ for each adhesion parameter. Then, by substituting the value of $\tilde{l}$ into equation (11), we consequently obtain the relation between the normalized force and the normalized displacement, as shown in figures 6 and 7.

As shown in figure 6, when the beam approaches the rigid surface, the tip of the beam starts touching the rigid surface at P_0, i.e., when $d/(L\sin\theta) > -1$, and the force for line contact expressed by equation (7) will have a positive value. When further displacement is applied between P_0 and P_1, the

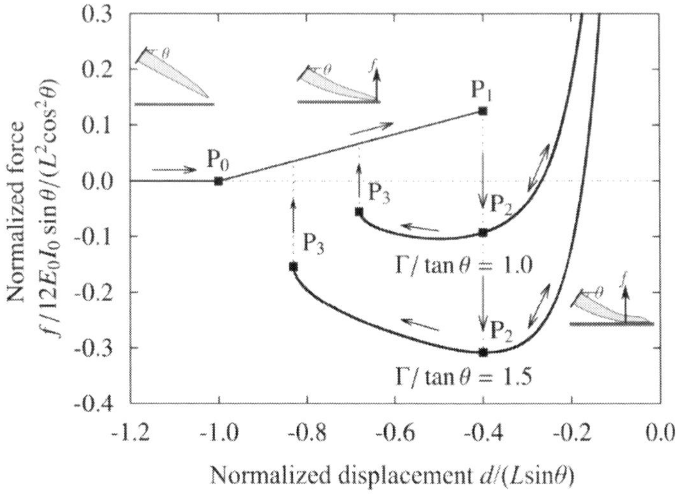

Figure 6. Relation between the normalized force and the normalized displacement, calculated for the adhesion parameter $\Gamma/\tan\theta = 1.0$ and 1.5, and n=0.5.

beam remains in line contact, and the force, which satisfies equation (7), varies along the straight line. The force during line contact is repulsive. After the displacement reaches P_1, which is the shifting condition from line contact to area contact expressed by equation (9), the contact changes to area contact and the force no longer satisfies equation (7). In consequence, the force jumps from P_1 to P_2 to follow equation (11) instead. During area contact, the force curve has several paths according to the value of the adhesion parameter, and the force is attractive. The force varies along each curve, responding to the applied displacement. If an appropriate displacement is applied for the adhesion parameter, the minimum energy exists, and the beam will adhere to the rigid surface at equilibrium. When the beam is unloaded, the force varies along the curve to the left of P_2. When the given displacement reaches the condition $\tilde{l} = 1$ at P_3, the beam finally detaches from the rigid surface. The maximum gripping force of the beam can be considered as the measurement of the minimum of each force curve.

Figure 7 shows the relation between the normalized force and the normalized displacement, which is plotted for different values of n. When n=0, i.e., the beam is uniformly rectangular, the analysis result of the present study exactly coincides with the result of the previous work of Sekiguchi [11].

According to the shifting condition from line contact to area contact, expressed by equation (9), the displacement is in inverse proportion to n. This implies that as the value of n increases, the displacement necessary for shifting to area contact decreases. In other words, the tapered beam could have more ability to absorb the surface roughness compared to a uniformly rectangular beam. However, the maximum gripping force generated by an individual tapered beam decreases as the value of n increases.

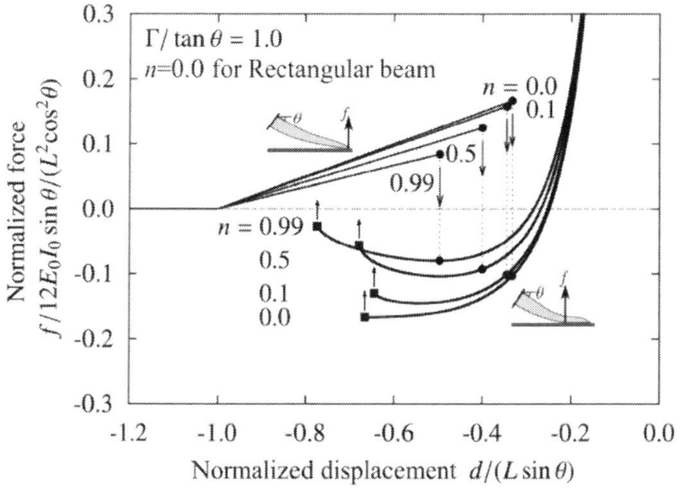

Figure 7. Relation between the normalized force and the normalized displacement for different values of n.

5. Effect of the shape on the gripping force of a multi-beam structure adhering to a wavy surface

To discuss the effect of the beam shape on the gripping force of a multi-beam structure adhering to a wavy surface, we assume a one-dimensional multi-beam structure consisting of a number of individual beams (100 beams were used in this study) fixed to a substrate with a uniform interval and declined angle θ, as shown in figure 8. The waviness of a rigid surface is assumed to be sinusoidal of single wavelength λ and amplitude A. In order to apply the adhesion model of this study, it is assumed that each area on the sinusoidal surface to which the beam's tip adheres is considerably in parallel to the substrate of the multi-beam structure, as shown in figure 9. The multi-beam structure is loaded until its substrate touches the convex part of the surface, before unloading to measure the gripping force. The total gripping force of the multi-beam structure is considered as the superposition of each individual beam.

Figure 10 shows the analysis result of the normalized gripping force against the normalized amplitude of the surface waviness. The white rectangular mark denotes the results of the multi-rectangular beam structure, and the black triangle mark denotes the results of the tapered beam structure. As shown in the figure, the multi-rectangular beam structure can generate a larger gripping force compared to the multi-tapered beam structure when the rigid surface is flat or slightly wavy.

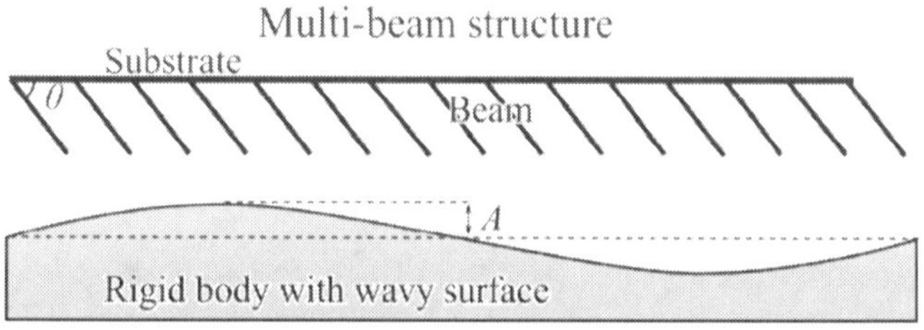

Figure 8. Schematic illustration of a one-dimensional multi-beam structure contacting to a sinusoidal wavy surface.

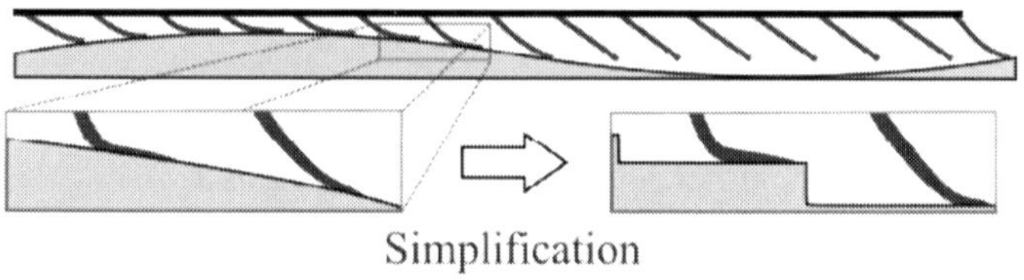

Figure 9. Simplification of the sinusoidal surface as steps parallel to the substrate to measure the gripping force.

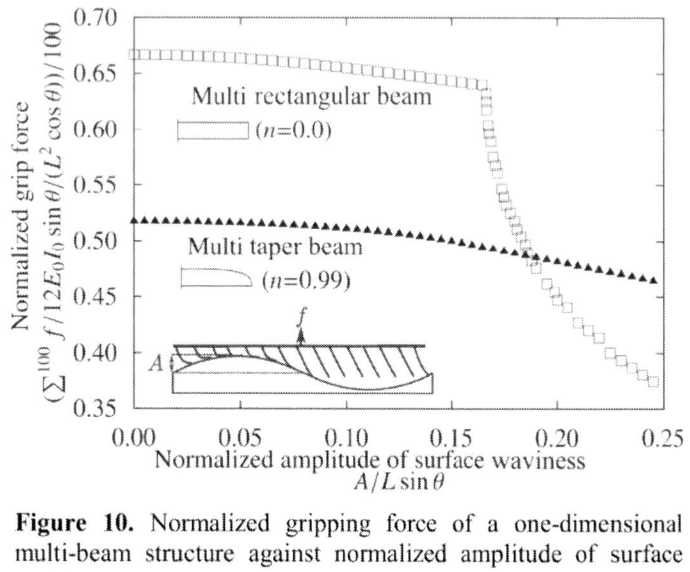

Figure 10. Normalized gripping force of a one-dimensional multi-beam structure against normalized amplitude of surface waviness.

However, when the normalized amplitude $A/(L\sin\theta)$ of the surface waviness exceeds a certain value, e.g., 0.16 for this analysis, its gripping force rapidly decreases and becomes lower than that of the multi-tapered beam structure. The analysis result suggests that the multi-tapered beam structure could have more ability to absorb the surface waviness compared to the multi-rectangular beam structure.

6. Conclusion

A tapered beam model with a varying bending stiffness along its longitudinal axis was assumed to study the relation between the gripping force and the applied displacement analytically. The analysis results were obtained in closed form, which provides an intuitive picture of the gripping force. Although each tapered beam can generate less gripping force than a rectangular beam, increasing the value of n can enable the tapered beam to absorb surface waviness. When the beams are compiled to form a multi-beam structure, rectangular beams create a greater gripping force for flat or slightly wavy surfaces, whereas tapered beams create a greater gripping force for more wavier surfaces. The result of this study is considered significant for designing a grip-and-release device using tapered beams.

References

[1] Autumn k, Liang Y A, Hsieh S T, Zesch W, et al. 2000 *Nature* **405** 681
[2] Autumn K, Sitti M, Liang Y A, Peattie A M, Hasen W R, Sponberg S, Kenny T W, Fearing R, Israelachvili J N and Full R J 2002 *PNAS* **99**(19) 12252
[3] Takahashi K, Berengueres J O L, Obata K J and Saito S 2006 *International Journal of Adhesion and Adhesives* **26** 639
[4] Murphy M P, Aksak B and Sitti M 2009 *Small* **5**(2) 170
[5] Sitti M and Fearing R S 2002 *IEEE-NANO* 137
[6] Aksak B, Murphy M P and Sitti M 2007 *Langmuir* **23** 3322
[7] Murphy M P, Kim S and Sitti M 2009 *Applied Materials & Interfaces* **1**(4) 849
[8] Kim S, Sitti M, Xie T and Xiao X 2009 *Soft Matter* **4** 3689
[9] Yu J, Chary S, Das S, Tamelier J, Turner K L and Israelachivili J N 2012 *Langnuir* **28** 11527
[10] Asbeck A, Dastoor S, Parness A, Fullerton L, Esparza N, Soto D, Heyneman B and Cutkosky M 2009 *IEEE International Conference on Robotics and Automation* 2675-80
[11] Sekiguchi Y, Hemthavy P, Saito S and Takahashi K 2012 J. Adhesion Sci. Technol. 26 23 2615

AUTHOR INDEX

AUTHOR INDEX

Institute of Physics
Dirac House, Temple Back
Bristol BS1 6BE UK

proceedings.com